AIR CONDITIONING SYSTEMS

AIR CONDITIONING SYSTEMS
DESIGN, COMMISSIONING AND MAINTENANCE

Roger Colby Legg

B T Batsford Ltd · London

Typeset by Latimer Trend & Co Ltd, Plymouth
Printed in Great Britain by
Dotesios, Trowbridge, Wilts

Published by B T Batsford Ltd
4 Fitzhardinge Street, London W1H 0AH

A CIP catalogue record for this book is available from the British Library

ISBN 0 7134 5644 2

Contents

The general antiphlogistic remedies are . . . free admission of pure cool air.

John Aitkin, 'Elements of Surgery', 1779

. . . the dreadful consequences which have been experienced from breathing air in situations either altogether confined or ill ventilated . . . if others are in the same apartment the breath from each person passes from one to another and it is frequently in this way that diseases are communicated.

The Marquis de Chabames, 1818

The very first rule of nursing . . . is this: To keep the air he breathes as pure as the external air, without chilling him.

Florence Nightingale, 1863

Foreword

Traditionally, air conditioning has been looked upon as a luxury but it is increasingly becoming commonplace even in the more temperate parts of the world, particularly in commercial and public buildings. In addition many new buildings, such as hospitals, computer suites and clean rooms, have special needs in terms of their environment which can only be provided by specialized environmental engineering systems.

The ability to control the internal air conditions effectively and efficiently has had a dramatic influence on modern architecture, with taller and larger plan-form buildings dominating city centres. Indeed, the proportion of the costs of engineering services in new buildings can often be as much as 60 per cent of the total cost, making it imperative to integrate those services closely with the overall building design.

The rapidly increasing demand for air conditioning, together with the increasing interest in Quality Assurance in buildings and services, has led to a demand for more qualified air conditioning design engineers. There is also a great need for other professional people involved in the design, construction and operation of modern buildings to understand the underlying principles of the topics covered in this volume.

The book is intended to provide a technical foundation for air conditioning design and covers a great deal of the syllabus requirements of most academic courses. The style is one of illustrating theory with relevant worked examples. By using data from both the *CIBSE Guide* and the *ASHRAE Handbooks*, the latter now being published in SI units, it is hoped that this text will appeal to students, practising engineers and others involved in air conditioning, in Europe and North America.

The author is well qualified to understand the needs of the student, having taught the subject for more than twenty years at the Institute of Environmental Engineering (formerly the National College for Heating, Ventilation, Refrigeration and Fan Engineering), South Bank Polytechnic, London.

Michael J Farrell

London 1991 (formerly Chair, Institute of Environmental Engineering)

9

Acknowledgment

I am indebted to my colleagues at the Institute of Environmental Engineering for much practical help, encouragement and advice in the writing of this book. In particular I am most grateful to Ron James who contributed the chapter on refrigeration and heat pump systems, to John Missenden who provided the text for control valves (pages 345–352) and to Stan Marchant for the text on cooling towers (pages 189–192). Tim Dwyer read the original manuscript and made many useful suggestions as well as providing some of the line drawings. Other figures were drawn by Bill Davis and my son Mark; photographs for Plates 1.5, 1.12, 9.7, 12.24, 12.25, 15.8, 15.12 and 15.14 were taken by Vernon Parker. Lastly, my thanks are due to my wife Iris for her loyal support and encouragement.

Bromley 1991   RCL

The author and publishers thank the following for permission to use certain material from books and articles, and to use illustrations as a basis for figures in this volume:

Tables 1.4, 7.2 and 13.2, and Figures 1.18, 4.5, 7.2, 13.1 from *CIBSE Guide* by permission of the Chartered Institute of Building Services Engineers.
Figure 3.2 (redrawn) by permission of McGraw-Hill Book Co.
Table 4.1 by permission of the Controller of Her Majesty's Stationery Office.
Table 4.5 from *ASHRAE Handbook* by permission from the American Society of Heating, Refrigeration and Air-Conditioning Engineers, Inc.
Figure 7.7 based on illustrations, courtesy of Trox Brothers Ltd.
Figure 8.4 courtesy of ICI Chemicals and Polymers Ltd.
Figure 8.4 courtesy of York International.
Figure 9.13 (redrawn) by permission of the Institute of Refrigeration.
Plates 10.3 and 10.8 supplied by Thermal Technology Ltd.
Figures 11.2 (b) and 11.4 supplied by Vokes Ltd.
Figures 12.7, 13.7, 13.11 and 13.13 (based on figures in *Internal Flow Systems* (2nd Edition) 1990, BHRA, Cranfield, UK) by permission of D. S. Miller.
Figure 14.15 (based on figures from *Fan Application Guide* 1981) by permission of the Fan Manufacturers' Association.
Figure 15.9 (a) courtesy of Holmes Valves Ltd.
Figure 15.9 (b), 15.11 and 15.15 courtesy of Crane Ltd.
Plates 15.18 and 15.19 courtesy Perflow Instruments Ltd.
Figure 18.1 and 18.2 by permission of the Building Services Research and Information Association.

1 Properties of Humid Air

Air is the working fluid for air conditioning systems. It is therefore important for the engineer to have a thorough understanding of the properties of air, before going on to consider the processes that occur when air passes through the various plant items which make up systems. The word *psychrometry* is often used for the science which investigates the properties of humid air, and the chart which shows these properties graphically is known as the *psychrometric* chart.

In this chapter the various air properties are defined and the appropriate equations given. In deriving the equations, it is usual to consider the air as consisting of two gases, dry air and water vapour. Even though one of these is strictly a vapour, both are considered to obey the ideal gas laws. Lastly, the tables and chart, from which numerical values of the air properties are obtained for practical calculations, are described and illustrated.

ATMOSPHERIC PRESSURE

At any point in the earth's atmosphere there exists a pressure due to the mass of air above that point – the atmospheric pressure. Standard atmospheric pressure at sea level is 1013.25 mbar (often approximated to 1013 mbar in this book) but due to changes in weather conditions there are variations from this standard pressure. For example, the minimum and maximum values recorded in London are 948.7 mbar (in 1821) and 1048.1 mbar (in 1825) respectively; those recorded for North America are 892 mbar (Long Key, Florida in 1935) and 1074 mbar (Yukon Territory, Canada in 1989).[1] The variations of pressure recorded at the time of the great storm which occurred over south-east England in October 1987 are shown on the barograph chart in Figure 1.1.

Atmospheric pressure varies with height above sea level, and for altitudes at which mankind lives the rate of decrease (lapse rate) for a standard atmosphere may be taken as a *reduction* of 0.13 mbar per metre of height *above* sea level and an *increase* of 0.13 mbar per metre of depth *below* sea level.

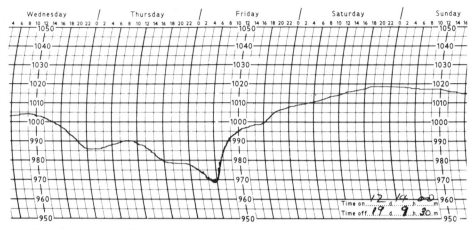

1.1 Barograph chart for 13 to 17 October 1987, London, UK

Example 1.1   Determine the standard atmospheric pressure for Nairobi which is at an altitude of 1820 m above sea level.

Solution

standard sea level atmospheric pressure	1013
lapse rate = -1820×0.13	-237
standard atmospheric pressure for Nairobi	776 mbar

Atmospheric pressure may be measured by using a number of instruments. In the laboratory it is usual to use a Fortin barometer, while for site work an aneroid barometer is the most usual instrument. For continuous recording a barograph is used.

DRY AIR AND WATER VAPOUR

Dry air consists of a number of gases, but mainly of oxygen and nitrogen. It is necessary to know the molecular mass of the dry air and this is calculated from the proportion each individual gas makes in the mixture. Table 1.1 gives this data, together with the calculation.

The sum of the molecular mass fractions is 28.97 and this is the value taken as the mean molecular mass of dry air.

Water vapour is said to be *associated* with the dry air. Its molecular mass is obtained from the masses of its chemical composition, H_2O, ie:

$$M_{h_2o} = (2 \times 1.01) + (1 \times 16)$$

$$= 18.02.$$

gas	proportion by volume	molecular mass	molecular mass fraction
	(1)	(2)	$\dfrac{(1) \times (2)}{100}$
	%	M	
nitrogen, N_2	78.03	28.02	21.86
oxygen, O_2	20.99	32.00	6.72
carbon dioxide, CO_2	0.03	44.00	0.01
hydrogen, H_2	0.01	2.02	0.00
argon, Ar	0.94	39.91	0.38

molecular mass fraction total   28.97

Table 1.1 Determination of molecular mass of dry air

VAPOUR PRESSURE

Saturated vapour pressure

Consider the vessel shown in Figure 1.2. The contents are at temperature t°C and the atmosphere above the water contains water vapour which exerts a pressure known as *saturated vapour pressure* (SVP). The SVP is the maximum vapour pressure that can occur at a given temperature. When

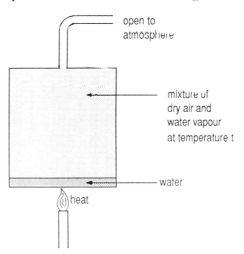

1.2 Vessel with saturated vapour

heat is applied to the vessel, more water evaporates and as the temperature rises, the SVP increases. Eventually, with heat still being supplied, the water will boil and this happens when the SVP is equal to atmospheric pressure. The variation of saturated vapour pressure against temperature is shown in Figure 1.3.

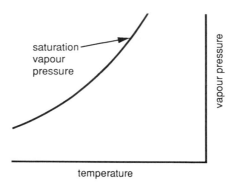

1.3 Saturation vapour pressure versus temperature

The values of SVP have been determined by experiment and published in the form of steam tables, selected values of which are given in Table 1.2.

dry-bulb temperature	SVP	dry-bulb temperature	SVP
°C	mbar	°C	mbar
0	6.108	20	23.37
2	7.055	25	31.66
4	8.129	30	42.42
6	9.346	35	56.22
8	10.72	40	73.75
10	12.27	50	123.4
12	14.02	60	199.2
14	15.98	70	311.6
16	18.17	80	473.6
18	20.63	100	1013.2

Table 1.2 Saturation vapour pressures

There is no simple relationship between temperature and SVP. The following equations are the relevant curve fits published by the National Engineering Laboratory[2]:

for water above 0°C:

$$\log_{10} p_{ssw} = 28.59 - 8.2 \log_{10} T + 0.00248\, T - 3142/T$$

where p_{ssw} is the SVP in bar over water at absolute temperature TK

for ice below 0°C:

$$\log_{10} p_{ssi} = 10.538 - 2664/T$$

where p_{ssi} is the SVP in bar over ice at absolute temperature TK.

These equations are suitable for use in computer programs in which air property values are required; they are not used in this text.

Superheated vapour

If all the water in the vessel shown in Figure 1.2 evaporates before boiling point has been reached and heat continues to be applied, the water vapour becomes superheated with the vapour pressure remaining constant. Therefore, on Figure 1.3, the superheated vapour is in the region to the right hand side of the SVP curve. Air conditioning engineers will normally be interested only in the variations in vapour pressure in the temperature range −20°C to 60°C.

RELATIVE HUMIDITY

Definition: Relative humidity is the percentage ratio of the vapour pressure of water vapour in the air to the saturated vapour pressure at the same temperature.

From the definition, relative humidity of air at condition A in Figure 1.4 is therefore given by:

$$\varphi = \frac{p_s}{p_{ss}} \times 100 \qquad (1.1)$$

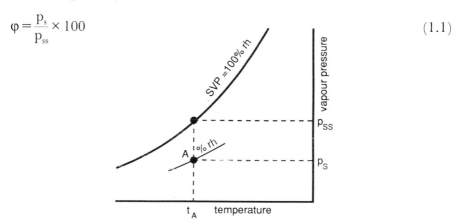

1.4 Definition of relative humidity

100%rh corresponds to the saturated vapour pressure, p_{ss}.

Example 1.2 Air at 20°C has a vapour pressure of 13 mbar. Determine the relative humidity.

Solution

From Table 1.2, the saturation vapour pressure at 20°C is 23.37 mbar.

Using Equation 1.1:

$$\varphi = \frac{p_s}{p_{ss}} \times 100 = \frac{13}{23.37} \times 100 = 55.6\%.$$

From the discussion on vapour pressure it should be noted that atmospheric pressure only determines the boiling point of water; it has no effect on saturated vapour pressure, or vapour pressure, and therefore atmospheric pressure has no effect on relative humidity.

Relative humidity can be measured directly by a number of instruments, in particular with a thermohydrograph as illustrated in Figure 1.5. However, for more accurate measurements, as well as for calibrating other humidity measuring devices, it is more usual to measure it indirectly by using dry and wet-bulb temperature measurements. These can then be referred to tables of humid air properties, or to a psychrometric chart, to determine the relative humidity.

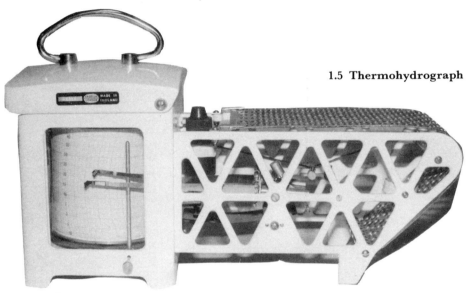

1.5 Thermohydrograph

IDEAL GAS LAWS

The ideal gas laws are used to derive a number of humid air properties. The errors in the numerical values of the air properties due to departures from the ideal laws are very small. For a discussion on this point, see JONES.[3]

DALTON's law of partial pressures

DALTON's law of partial pressures states that the pressure of a mixture of gases is equal to the sum of the partial pressure which each individual gas would exert by itself at the same volume and temperature.

$$p_t = p_1 + p_2 + p_3 \ldots \tag{1.2}$$

where p_t = total pressure of the mixture of gases

$p_1, p_2 \ldots$ = partial pressures of the individual gases.

DALTON's law is illustrated in Figure 1.6; in this two gases, A and B, at pressures p_a and p_b, which individually occupy the same volume, are combined in one of the vessels to give the total pressure p_t.

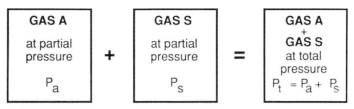

1.6 DALTON's law of partial pressures

Example 1.3 If the atmospheric pressure is 1013 mbar and the water vapour pressure is 40 mbar, determine the partial pressure of the dry air.

Solution

total air pressure (atmospheric)	1013
partial pressure water vapour	− 40
partial pressure of dry air	973 mbar.

General gas law

BOYLE'S law states that, at constant temperature, the product of the pressure, p, and volume, V, of a gas remains constant, ie:

$pV = $ constant.

CHARLES's law states that the volume of a gas, V, is proportional to its absolute temperature, T, the pressure remaining constant, ie:

$V/T = $ constant.

BOYLE's and CHARLES's laws combine to give the general gas law:

$pV = mRT.$ (1.3)

Note that the absolute temperature $T = (273 + t_a)$, the air dry-bulb temperature, t_a, being in degrees Celsius.

Individual gas constants are calculated from the universal gas constant, R_o, and the molecular mass, M, of the gas, ie:

$R = R_o/M.$ (1.4)

The value of R_o is 8314.66 J/kmolK and the molecular mass is expressed in kg/kmol.

Example 1.4   Determine the gas constants for dry air and water vapour.

Solution

As determined previously, the molecular masses for dry air and water vapour are respectively 28.97 and 18.02. The gas constants are therefore calculated as follows:

Using Equation 1.4:

> for dry air:
> $R_a = 8314.66/28.97 = 287\,J/kgK$

> for water vapour:
> $R_s = 8314.66/18.02 = 461\,J/kgK.$

DENSITY OF AIR

The density of air can be calculated using the general gas equation and this is illustrated by the following example.

Example 1.5   Determine the density of air with a temperature 20°C (normal temperature) and at an atmospheric pressure of 1013 mbar (standard pressure).

Solution

The air density is given by:

$$\rho = \frac{m_a}{V} = \frac{P_{at}}{R_a T} = \frac{1013 \times 100}{287 \times (273 + 20)} = 1.205\,kg/m^3.$$

The standard value of air density is usually given as $1.20\,kg/m^3$.

Again, consideration of the general gas law shows how the standard air density of $1.20\,kg/m^3$ can be corrected for atmospheric pressure and temperatures which differ from those on which the standard is based, ie:

$$\rho = 1.2 \frac{P_{at}(273 + 20)}{1013(273 + t_a)}$$

$$\rho = 0.347 \frac{P_{at}}{(273 + t_a)}. \tag{1.5}$$

Example 1.6 Determine the density of air which has a temperature of 30°C and an atmospheric pressure of 980 mbar.

Solution

Using Equation 1.5, the air density is calculated as:

$$\rho = 0.347 \frac{P_{at}}{(273 + t_a)} = 0.347 \frac{980}{(273 + 30)} = 1.122 \text{ kg/m}^3.$$

MOISTURE CONTENT

Definition: The moisture content of humid air is the mass of water vapour present in 1 kg of dry air.

This air property is variously referred to as either *humidity ratio, specific humidity* or *absolute humidity*.

It is important to recognize at this point in the discussion on humid air that some of its properties are based on 1 kg of *dry air*, unlike the properties of most other fluid mixtures which are based on 1 kg of the mixture. The derivation is as follows.

Using the general gas law, Equation 1.3:

for dry air:

$$p_a V_a = m_a R_a T_a$$

for water vapour:

$$p_s V_s = m_s R_s T_s.$$

For the mixture $V_a = V_s$ and $T_a = T_s$. Therefore from the definition of moisture content given above:

$$g = \frac{m_s}{m_a} = \frac{R_a p_s}{R_s p_a} = \frac{287}{461} \frac{p_s}{p_a}.$$

From DALTON's law of partial pressures, Equation 1.2:

$$p_a = p_{at} - p_s$$

$$\therefore g = 0.622 \frac{p_s}{(p_{at} - p_s)}. \tag{1.6}$$

Example 1.7 Determine the moisture content for air at a temperature of 20°C and a vapour pressure of 13 mbar when the atmospheric pressure is 1013 mbar.

Solution

Using Equation 1.6:

$$g = 0.622 \frac{p_s}{(p_{at} - p_s)} = 0.622 \frac{13}{1013 - 13}$$

$$= 0.00809 \text{ kg/kg}_{da}.$$

SATURATION MOISTURE CONTENT

If the vapour pressure, p_s, in Equation 1.6 is at saturation vapour pressure, p_{ss}, then the moisture content becomes the saturation moisture content. In the same way that saturated vapour pressure varied with temperature, saturation moisture content also varies with temperature. This is illustrated graphically in Figure 1.7, the resulting curve being a prominent feature of the psychrometric chart. Some typical values of saturation moisture contents are given in Table 1.3.

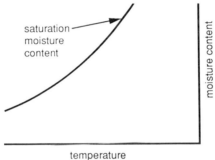

1.7 Saturation moisture content versus temperature

dry-bulb temperature	moisture content	dry-bulb temperature	moisture content
°C	kg/kg$_{da}$	°C	kg/kg$_{da}$
0	0.00379	20	0.0147
2	0.00438	25	0.0202
4	0.00505	30	0.0273
6	0.00582	35	0.0367
8	0.00668	40	0.0491
10	0.00766	45	0.0653
12	0.00876	50	0.0868
14	0.01001	60	0.153
16	0.0114		
18	0.0130		

Table 1.3 Saturation moisture contents

PERCENTAGE SATURATION

Definition: Percentage saturation is the percentage ratio of the moisture content in the air to the moisture content at saturation at the same temperature.

The percentage saturation of air at condition **A** in Figure 1.7 is therefore given by:

$$\mu = \frac{g_s}{g_{ss}} \times 100 \qquad\qquad (1.7)$$

1.8 Definition of percentage saturation

Example 1.8 Calculate the percentage saturation for the air condition described in Example 1.7.

Solution

From the solution of Example 1.7, for the air condition specified, the moisture content = 0.00809 kg/kg$_{da}$.
From Table 1.3 the saturation moisture content at 20°C is 0.0147 kg/kg$_{da}$.

Using Equation 1.7:

$$\mu = \frac{g_s}{g_{ss}} \times 100 = \frac{0.00809}{0.0147} \times 100$$

$$= 55\%$$

Lines of constant percentage saturation appear on a psychrometric chart as shown in Figure 1.9. For practical purposes, values of percentage saturation are interchangeable with those of relative umidity. Percentage saturation is slightly dependent upon atmospheric pressure and this accounts for the small numerical differences that do exist between these two air properties and which will be noted in the tables of air properties.

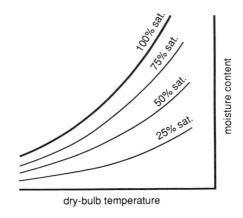

1.9 **Lines of constant percentage saturation**

SPECIFIC VOLUME

Definition: Specific volume is the volume of air containing 1 kg of dry air plus the associated moisture content.

The derivation is as follows:

Using the general gas law for dry air, Equation 1.3:

$$p_a V_a = m_a R_a T_a.$$

From the definition given above:

$$v = \frac{V_a}{m_a} = \frac{R_a T_a}{P_a}.$$

From DALTON's law of partial pressures:

$$P_a = P_{at} - P_s$$

$$\therefore v = \frac{R_a(273 + t_a)}{(P_{at} - P_s)}$$

$$v = \frac{287(273 + t_a)}{(P_{at} - P_s)}. \qquad (1.8)$$

Example 1.9 Determine the specific volume for air at 20°C, a vapour pressure of 14 mbar and an atmospheric pressure of 1013 mbar.

Solution

Using Equation 1.8:

$$v = \frac{287(273 + t_a)}{(P_{at} - P_s)} = \frac{287(273 + 20)}{(1013 - 14)\ 100}$$

$$= \frac{287 \times 293}{99900}$$

$$= 0.842 \ \text{m}^3/\text{kg}_{da}.$$

Lines of constant specific volume are drawn on the psychrometric chart as shown in Figure 1.10.

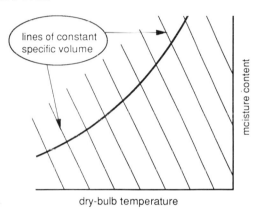

1.10 **Lines of constant specific volume**

Relationship between air density and specific volume

Air density is defined as the *mass of air per unit volume*, whereas the specific volume is defined in terms of *unit mass of dry air*. Therefore the relationship between the two is:

$$\rho = (1 + g)/v. \tag{1.9}$$

Though the difference between density and the reciprocal of specific volume is relatively small, the engineer should be aware of the difference compared with the true relationship, as given in Equation 1.9, when making calculations in different areas of work. So it is usual to use density when measuring air flow rates through pressure drop devices such as orifice plates and to use specific volume in air conditioning load calculations.

DRY-BULB AND WET-BULB TEMPERATURES

Dry-bulb and wet-bulb temperatures, measured together, are among the most popular methods for determining the air condition, and from these measurements other air properties may be derived. Dry and wet-bulb temperatures can be measured using a variety of instruments, eg mercury-in-glass, thermocouple and resistance thermometers.

Dry-bulb temperature

Definition: The dry-bulb temperature of air is the temperature obtained with a thermometer which is freely exposed to the air but which is shielded from radiation and free from moisture.

The word *dry* is used to make a distinction from the *wet*-bulb.

Wet-bulb temperature

Definition: The wet-bulb temperature of air is the temperature obtained with a thermometer whose bulb is covered by a muslin sleeve which is kept moist with distilled/clean water, freely exposed to the air and free from radiation.

The reading obtained is affected by air movement over the instrument. For this reason there are two wet-bulb temperatures – *sling* and *screen*:

(i) The *sling wet-bulb* is obtained in a moving air stream, preferably above 2 m/s. This is usually measured with either a sling hygrometer (Figure 1.11) or an Assman hygrometer, (Figure 1.12). However, a sling

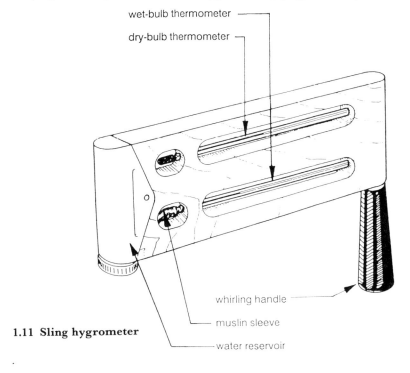

wet-bulb thermometer

dry-bulb thermometer

whirling handle

muslin sleeve

water reservoir

1.11 Sling hygrometer

reading may also be obtained if a wet-bulb thermometer is installed in a duct through which air is flowing at a reasonable velocity. The sling wet-bulb is considered to be more accurate than the screen wet-bulb temperature and for this reason is preferred by air conditioning engineers.

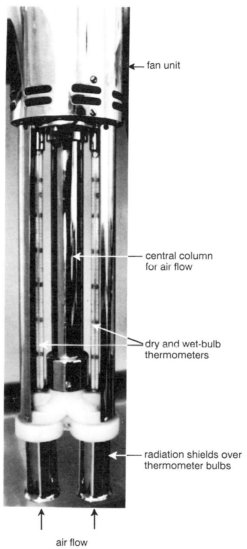

fan unit

central column for air flow

dry and wet-bulb thermometers

radiation shields over thermometer bulbs

air flow

1.12 Assman hygrometer

(ii) The *screen wet-bulb* is assumed to be in still air, usually installed in a Stevenson *screen* (from which this type of wet-bulb obtained its name), as used by meteorologists. A standard Stevenson screen is illustrated in Figure 1.13.

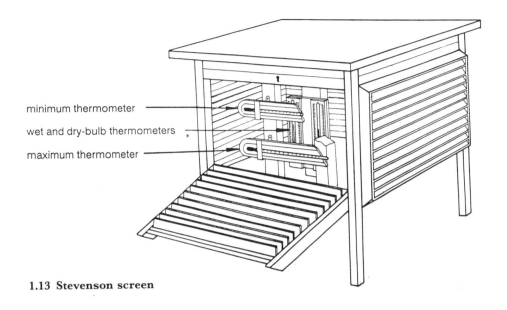

minimum thermometer

wet and dry-bulb thermometers

maximum thermometer

1.13 Stevenson screen

The psychrometric equation

The psychrometric equation relates the dry-bulb and wet-bulb temperatures with their corresponding vapour pressures and with the atmospheric pressure. To understand this relationship consider the diagram of the wet-bulb thermometer in Figure 1.13.

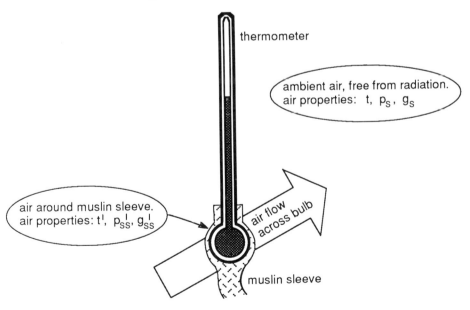

thermometer

ambient air, free from radiation. air properties: t, p_s, g_s

air around muslin sleeve. air properties: t^l, p_{ss}^l, g_{ss}^l

air flow across bulb

muslin sleeve

1.14 Diagram of a wet-bulb thermometer

Moisture is being evaporated from the surface of the muslin sleeve into the surrounding air. For evaporation to take place heat must be supplied and this can only come from the ambient air in the form of sensible heat, with the temperature of the bulb lower than that of the surrounding air. At equilibrium the latent heat loss due to moisture evaporation will equal the sensible heat gained. The air film at the surface of the muslin sleeve is considered to be at saturation moisture content, g_{ss}'. (Note the ' to indicate that the moisture content is at the wet-bulb temperature.) The latent heat loss is proportional to the moisture content difference between this air film and the ambient air, ie $(g_{ss}' - g)$. The sensible heat gained is proportional to the temperature difference between the bulb and the ambient air $(t - t')$, ie:

$$B(g_{ss}' - g) = C(t - t') \tag{1.10}$$

where B and C are constants related to parameters of heat and mass transfer, eg surface area and latent heat of evaporation. But from Equation 1.6:

$$g = 0.622 \frac{p_s}{p_{at} - p_s}$$

and

$$g_{ss}' = 0.622 \frac{p_{ss}'}{p_{at} - p_{ss}'}.$$

Since p_s and p_{ss}' are very small compared with p_{at} these equations may be written as:

$$g = 0.622 \ p_s/p_{at} \text{ and } g_{ss}' = 0.622 \ p_{ss}'/p_{at}.$$

Substituting these expressions of moisture content in Equation 1.10:

$$\frac{0.622 \ B}{p_{at}} (p_{ss}' = p_s) = C(t - t').$$

By rearranging the terms and grouping the constants the psychrometric equation is obtained:

$$p_s = p_{ss}' - p_{at}A(t - t') \tag{1.11}$$

where A is known as the *psychrometric constant*.

The numerical difference between the dry-bulb and the wet-bulb temperature is known as the *wet-bulb depression*.

Since the rate of moisture evaporation depends on the speed of the air over the wet-bulb, the wet-bulb temperature will also depend on the air speed. However, the wet-bulb becomes independent of the air velocity above 2 m/s. The two wet-bulb temperatures described above – sling and screen – cater for this with different values for the constant A.

Wet-bulb temperatures are also affected by air's being either above or below freezing point and again different values of A are necessary to deal with these conditions.

The psychrometric constants for a 4.8 mm bulb diameter are:

sling:
$A = 6.66 \times 10^{-4} \, K^{-1}$ when $t' > 0°C$
$A = 5.94 \times 10^{-4} \, K^{-1}$ when $t' < 0°C$

screen:
$A = 7.99 \times 10^{-4} \, K^{-1}$ when $t' > 0°C$
$A = 7.20 \times 10^{-4} \, K^{-1}$ when $t' < 0°C$.

When working with the psychrometric equation it is important to remember that the saturated vapour pressure, p_{ss}', is taken at the wet-bulb temperature.

Example 1.10 Calculate the vapour pressure for air with the following conditions:

dry-bulb temperature –	22°C
wet-bulb temperature (sling) –	14°C
atmospheric pressure –	1013 mbar

Solution

From Table 1.2 the saturation vapour pressure at $14°C = 15.98$ mbar. Since the air is above 0°C and the wet-bulb is a sling reading the psychrometric constant, A, is $6.66 \times 10^{-4} K^{-1}$.

Using Equation 1.11:

$$p_s = p_{ss}' - p_{at}A(t - t')$$
$$p_s - 15.98 - 1013 \times (6.66 \times 10^{-4}) \times (22 - 14)$$
$$= 15.98 - 4.05 = 11.93 \text{ mbar.}$$

Lines of constant wet-bulb temperature are drawn on the psychrometric chart, as illustrated in Figure 1.15.

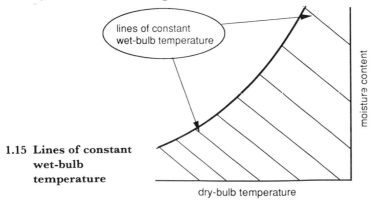

1.15 Lines of constant wet-bulb temperature

lines of constant wet-bulb temperature

moisture content

dry-bulb temperature

DEW-POINT TEMPERATURE

Definition: The dew-point temperature is the temperature of saturated air which has the same vapour pressure as the air condition under consideration.

Referring to Figure 1.16, when air at the original condition A is cooled at constant vapour pressure, ie at constant moisture content, the temperature of the air will eventually reach the saturation line and at this point water vapour will begin to condense. This temperature, a unique condition on the saturation line, is called the *dew-point temperature*, t_{dp}.

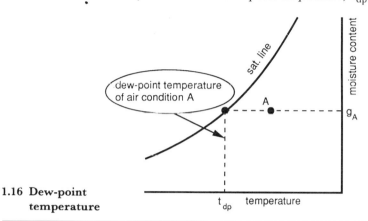

1.16 Dew-point temperature

Example 1.11 Air at a dry-bulb temperature of 40°C and having a moisture content of 0.202 kg/kg$_{da}$, is cooled at constant vapour pressure. At what temperature will dew begin to form?

Solution

Cooling air at constant vapour pressure is the same as cooling it at constant moisture content.
Referring to Table 1.3, the saturation moisture content of 0.0202 kg/kg$_{da}$ occurs at 25°C. Therefore the dew-point temperature of the given air condition is 25°C.

There are commercially available instruments which measure dew-point temperature directly. However, it is more usual to obtain its value by referring to tables of properties of humid air, or to a psychrometric chart, using measurements of other air properties.

SPECIFIC ENTHALPY

Definition: The specific enthalpy of humid air is a calculated property combining the sensible and latent heat of 1 kg of dry air plus its associated water vapour, relative to a datum at 0°C.

The equation for specific enthalpy is formulated as follows:

Consider 1 kg of dry air and the associated moisture content, g, at dry bulb temperature, t. The sensible heat, h_1, of 1 kg of dry air, relative to the datum 0°C, is given by:

$$h_1 = 1 \times c_{pa}\,(t-0)$$

where c_{pa} = specific heat of dry air
= 1.005 kJ/kgK

$$h_1 = 1.005\,t. \tag{1.12}$$

Similarly the sensible heat of the moisture content, h_2, is given by:

$$h_2 = g\,C_{ps})\,(t-0)$$

where C_{ps} = specific heat of water vapour
= 1.89 kJ/kgK

$$h_2 = 1.89\,g\,t. \tag{1.13}$$

The water vapour is considered to have evaporated at 0°C and therefore the latent heat of the moisture content, h_3, is given by:

$$h_3 = g\,h_{fg}$$

where h_{fg} = the latent heat of evaporation at 0°C
= 2501 kJ/kg

$$h_3 = 2501\,g. \tag{1.14}$$

The specific enthalpy of humid air, h, is obtained from the sum of Equations 1.12, 1.13 and 1.14:

$$h = h_1 + h_2 + h_3$$
$$h = 1.005t + g(1.89t + 2501). \tag{1.15}$$

Example 1.12 Determine the specific enthalpy of air at a dry-bulb temperature of 20°C and moisture content of 0.008 kg/kg$_{da}$.

Solution

Using Equation 1.15:

$$\begin{aligned} h &= 1.005t + g(1.89t + 2501)\\ &= 1.005 \times 20 + 0.008(1.89 \times 20 + 2501)\\ &= 40.4 \text{ kJ/kg}_{da}. \end{aligned}$$

Lines of constant enthalpy do not always appear on a psychrometric chart. For example, on the CIBSE chart (shown in Figure 1.18), to obtain the specific enthalpy of an air condition, a straight edge is used to join the corresponding enthalpy marks above the 100%sat line with those on either the bottom or the right hand side of the chart.

Humid specific heat

Equation 1.15 for specific enthalpy can be rearranged as follows:

$$h = (1.005 + 1.89\ g)\ t + 2501\ g$$
$$= c_{pas}\ t + 2501\ g$$

where $c_{pas} = (1.005 + 1.89\ g)$. $\qquad\qquad$ (1.16)

The term c_{pas} is known as the *humid specific heat*.

Example 1.13   Calculate the humid specific heat for the air condition in Example 1.12.

Solution

Using Equation 1.16, the humid specific is given by:

$$c_{pas} = 1.005 + 1.89\ g = 1.005 + 1.89 \times 0.008$$
$$= 1.02\ kJ/kgK.$$

The variations in the moisture content found in atmospheric air are such that for practical purposes an average value of $1.025\ kg/kg_{da}K$ may be used for c_{pas} in sensible heat loads calculations.

ADIABATIC SATURATION TEMPERATURE

An adiabatic process is one in which no external heat enters or leaves the system under consideration. In Figure 1.17 air is flowing through a duct in the bottom of which is an open water tank.

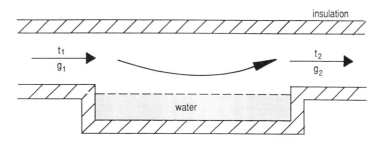

1.17 Adiabatic humidification process

The plant casing is considered to be perfectly insulated so that no heat flows into the duct from the surroundings or vice versa. Air enters the duct at dry-bulb temperature t_1 and moisture content g_1 and as it passes down the duct moisture will be evaporated so that at the end of the duct the air will have a moisture content g_2. For water to evaporate, heat must be supplied and, since this is an adiabatic process, this can come only from the air itself. Therefore the latent heat gained by the air must equal the sensible heat loss by the air. In other words, there must be a drop in air dry-bulb temperature to compensate for the increase in moisture content. If the air leaves at dry-bulb temperature t_2, then for each kilogram of dry air:

latent heat gained = sensible heat loss

$$(g_2 - g_1) \, h_{fg} = c_{pas}(t_1 - t_2). \tag{1.17}$$

If the tank is infinitely long, then the air at the end of the process will be at 100% sauration and the temperature at which this occurs is known as *adiabatic saturation temperature*, t^*, and the corresponding moisture content is g_{ss}^*. By substituting t^* for t_2 and g_{ss}^* for g_2 in Equation 1.17 and rearranging:

$$t^* = t - (g_{ss}^* - g)h_{fg}/c_{pas}. \tag{1.18}$$

To determine values of the adiabatic saturation temperature, t^*, an iterative solution of Equation 1.18 is required.

TABLES OF PROPERTIES OF HUMID AIR

The properties of humid air at the standard atmospheric pressure of 1013.25 mbar are given in tables published by CIBSE.[4] As an example of these tables the properties for a dry-bulb temperature of 20°C are given in Table 1.4. The special condition which is 100%sat with the relative humidity also at 100% should be noted. At this point the dry-bulb, wet-bulb, dew-point and adiabatic saturation temperatures are equal and the vapour pressure is the saturated vapour pressure.

When specifying an air condition it is usual to give the dry-bulb temperature and one other property. From these two values the other properties can be obtained. If the value of any property is not uniquely specified in the table, linear interpolations between adjacent conditions are justified.

THE PSYCHROMETRIC CHART

The psychrometric chart is a most useful design tool for air conditioning engineers. The chart published by CIBSE is shown in Figure 1.18. The properties plotted on the chart are:

– dry-bulb temperature;
– sling wet-bulb temperature;

% Saturation μ	% Relative Humidity φ	Moisture Content kg (g)	Specific Enthalpy kJ (h)	Specific Volume m³ (v)	Vapour pressure kPa (ps)	Dew Point Temperature °C (td)	Adiabatic Saturation Temperature °C (t*)	Wet Bulb Screen °C (t'sc)	Wet Bulb Sling °C (t')
100	100·00	0·014 75	57·55	0·8497	2·337	20·0	20·0	20·0	20·0
96	96·09	0·014 16	56·05	0·8489	2·246	19·4	19·6	19·6	19·6
92	92·17	0·013 57	54·56	0·8481	2·154	18·7	19·1	19·2	19·1
88	88·25	0·012 98	53·06	0·8473	2·062	18·0	18·7	18·8	18·7
84	84·31	0·012 39	51·56	0·8466	1·970	17·3	18·2	18·3	18·2
80	80·37	0·011 80	50·06	0·8458	1·878	16·5	17·7	17·9	17·7
76	76·43	0·011 21	48·57	0·8450	1·786	15·7	17·2	17·5	17·3
72	72·47	0·010 62	47·07	0·8442	1·694	14·9	16·7	17·0	16·8
70	70·49	0·010 33	46·32	0·8438	1·647	14·5	16·5	16·8	16·5
68	68·51	0·010 03	45·57	0·8434	1·601	14·0	16·2	16·5	16·3
66	66·53	0·009 736	44·82	0·8431	1·555	13·6	16·0	16·3	16·0
64	64·54	0·009 441	44·08	0·8427	1·508	13·1	15·7	16·1	15·8
62	62·55	0·009 146	43·33	0·8423	1·462	12·6	15·5	15·8	15·5
60	60·56	0·008 851	42·58	0·8419	1·415	12·1	15·2	15·6	15·3
58	58·57	0·008 556	41·83	0·8415	1·369	11·6	14·9	15·4	15·0
56	56·58	0·008 260	41·08	0·8411	1·322	11·1	14·7	15·1	14·7
54	54·59	0·007 966	40·33	0·8407	1·276	10·6	14·4	14·9	14·5
52	52·59	0·007 670	39·58	0·8403	1·229	10·0	14·1	14·6	14·2
50	50·59	0·007 376	38·84	0·8399	1·182	9·4	13·9	14·4	13·9
48	48·59	0·007 080	38·09	0·8395	1·136	8·8	13·6	14·1	13·7
46	46·59	0·006 785	37·34	0·8391	1·089	8·2	13·3	13·9	13·4
44	44·58	0·006 490	36·59	0·8388	1·042	7·6	13·0	13·6	13·1
42	42·58	0·006 195	35·84	0·8384	0·9945	6·9	12·7	13·4	12·8
40	40·57	0·005 900	35·09	0·8380	0·9480	6·2	12·4	13·1	12·5
38	38·56	0·005 605	34·34	0·8376	0·9011	5·5	12·1	12·8	12·2
36	36·55	0·005 310	33·60	0·8372	0·8541	4·7	11·8	12·6	12·0
34	34·53	0·005 015	32·85	0·8368	0·8070	3·9	11·5	12·3	11·7
32	32·52	0·004 720	32·10	0·8364	0·7600	3·0	11·2	12·0	11·4
30	30·50	0·004 425	31·35	0·8360	0·7127	2·1	10·9	11·8	11·1
28	28·48	0·004 130	30·60	0·8356	0·6656	1·2	10·6	11·5	10·7
24	24·43	0·003 540	29·10	0·8348	0·5710	−0·8	10·0	10·9	10·1
20	20·38	0·002 950	27·61	0·8341	0·4763	−3·0	9·3	10·4	9·5
16	16·32	0·002 360	26·11	0·8333	0·3814	−5·6	8·6	9·8	8·8
12	12·25	0·001 770	24·61	0·8325	0·2863	−8·9	8·0	9·2	8·2
8	8·17	0·001 180	23·11	0·8317	0·1910	−13·4	7·3	8·6	7·5
4	4·09	0·000 590	21·62	0·8309	0·0956	−20·8	6·5	8·0	6·8
0	0·00	0·000 000	20·11	0·8301	0·0000	—	5·8	7·3	6·1

Table 1.4 Properties of humid air at 20°C dry-bulb temperature
(Reproduced from Section C1 of the *CIBSE Guide*, by the permission of the Chartered Institute of Building Services Engineers)

− moisture content;
− specific enthalpy;
− specific volume;
− percentage saturation.

When working with the chart the following points should be noted:

(i) Sling wet-bulb is used in preference to screen wet-bulb temperature as it is considered to be the more consistent of the two measurements.

(ii) Vapour pressure is not given. This property is rarely required by the air conditioning engineer.

(iii) Dew-point temperature can be obtained for a given air condition as previously described.

(iv) The lines of constant dry-bulb temperature are not at right angles to the lines of constant moisture content. This is because the chart is based on enthalpy and moisture content and dry-bulb temperature is added subsequently, determined from the enthalpy equation. (See Example 2.1 in the following chapter.)

SYMBOLS

A	psychrometric constant
B,C	constants
c	specific heat
c_{pas}	specific heat of *humid* air
g	moisture content
g_{ss}	saturation moisture content
h	specific enthalpy of *humid* air
h_{fg}	latent heat of evaporation of water
M	molecular mass
m	mass
p	pressure
p_{at}	atmospheric pressure
p_{ss}	saturation vapour pressure
p_t	total pressure
R	particular gas constant
R_o	universal gas constant
T	absolute temperature
t	dry-bulb temperature
t′	wet-bulb temperature, sling
t'_{sc}	wet-bulb temperature, screen
t*	adiabatic saturation temperature
T_{dp}	dew-point temperature
V	volume
v	specific volume of humid air
μ	percentage saturation
ρ	air density
φ	relative humidity, percentage

Subscripts

a	air, dry air
s	water vapour, steam

Abbreviations

SVP	saturation vapour pressure
%rh	percentage relative humidity
%sat	percentage saturation

1.18 Psychrometric chart
(Reproduced from Section C1 of the *CIBSE Guide*, by permission of the Chartered ▶ Institute of Building Services Engineers)

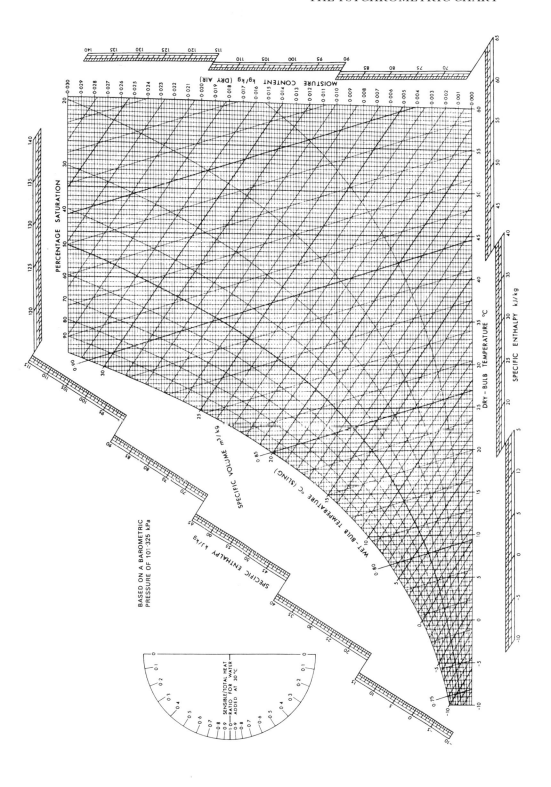

2 Air Conditioning Processes

Air conditioning plant items can be thought of as the *building blocks* from which systems are designed and constructed. It is necessary to understand the psychrometric processes that can be achieved with each *block* before dealing with the complete *system*. These processes can be shown most easily on a psychrometric chart, or as a psychrometric sketch as in the diagrams which follow. The air conditions are given as letters and these correspond with those on a diagram of the equipment itself. The process line on the chart is usually shown as a straight line, even though the actual conditions of the air as it passes *through* the plant item might, to some extent, deviate from that line. Generally the air conditioning systems engineer is interested only in the state of the air as it *enters* and *leaves* the item of plant.

MIXING OF TWO AIR STREAMS

Air streams at different conditions are often mixed within an air condition-ing system, the most usual case being that of air from outdoors (via the fresh air intake grille) mixing with air returned from the air conditioned space.

Figure 2.1 shows two air streams, **A** and **B**, mixing to produce condition **M**. It is assumed that the mixing process is adiabatic, ie there is no leakage of air into or out of the ductwork and no miscellaneous heat gains or losses. Because of the way the psychrometric chart has been constructed and from the laws of conservation of mass and energy, the mixing process can be drawn as the straight line **AMB** on the chart. If x is the proportion of air stream **A** in the total air *mass* flow rate leaving the system, then the air properties of the mixed air condition are determined as follows:

specific enthalpy:

$$h_M = xh_A + (1-x)h_B \tag{2.1}$$

moisture content:

$$g_M = xg_A + (1-x)g_B \tag{2.2}$$

dry-bulb temperature (approximately):

$$t_M = xt_A + (1-x)t_B. \tag{2.3}$$

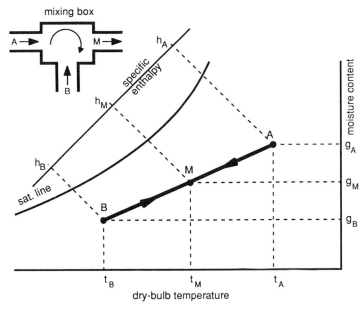

2.1 Mixing of two air streams

Example 2.1   In an air conditioning plant, 0.5 kg/s of outdoor air mixes with 1.5 kg/s or recirculated air. Determine the specific enthalpy, moisture content and dry-bulb temperature for the mixed air stream for the following air conditions:

air stream		dry-bulb temperature	moisture content
		(°C)	(kg/kg$_{da}$)
recirculated air	**A**	22	0.010
outdoor air	**B**	4	0.002

Solution
Air mass flow rate of the mixture $= 0.5 + 1.5 = 2.0$ kg/s
Proportion of outdoor air to total air, $x = 0.5/2.0 = 0.25$
Refer to Figure 2.2.

The specific enthalpies of air streams **A** and **B** are obtained from tables of air properties (or less precisely from the psychrometric chart):

$$h_A = 47.54 \text{ kJ/kg}_{da}$$
$$h_B = 9.04 \text{ kJ/kg}_{da}.$$

The mixed air enthalpy is obtained from Equation 2.1:

$$h_M = xh_A + (1-x)h_B$$
$$= 0.25 \times 9.04 + (1 - 0.25)47.54 = 37.92 \text{ kJ/kg}_{da}.$$

The mixed air moisture content is obtained from Equation 2.2:

$$g_M = xg_A + (1-x)g_B$$
$$= 0.25 \times 0.002 + (1-0.25)0.01 = 0.008 \text{ kg/kg}_{da}.$$

The mixed air temperature is obtained from Equation 2.3, approximately:

$$t_M = xt_A + (1-x)t_B$$
$$= 0.25 \times 4 + (1-0.25)22 = 17.5°\text{C}.$$

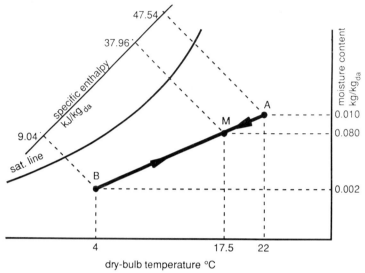

2.2 Air mixing: Example 2.1

The value of the dry-bulb temperature determined from Equation 2.3 is sufficiently accurate for practical air conditioning calculations. If a precise value is required the temperature of the mixed air stream, t_m, should be calculated using the enthalpy Equation 1.15, with values of specific enthalpy and moisture content determined from Equations 2.1 and 2.2 respectively.

Example 2.2 Using the data from Example 2.1, calculate a precise value for the mixed air temperature.

Solution
Using Equation 1.15:

$$h_M = (1.005 + 1.89\,g_M)t_M + 2501\,g_M$$
$$37.92 = (1.005 + 1.89 \times 0.008)t_M + 2501 \times 0.008$$
$$t_M = (37.96 - 20.01)/1.02 = 17.6°\text{C}.$$

This value of the mixed air temperature agrees closely with the approximate value determined in Example 2.1.

SENSIBLE HEATING COILS

A sensible heating process, occurring at constant moisture content, is one in which the dry-bulb temperature of the air is increased when the air passes over a hot, dry surface. The heater might be a pipe coil using hot water or steam, or electrical resistance elements or one of a number of the alternative air-to-air heat recovery units described in Chapter 10. Heaters are required in air conditioning systems for frost protection, as pre-heaters for humidifiers and as after-heaters for space temperature control.

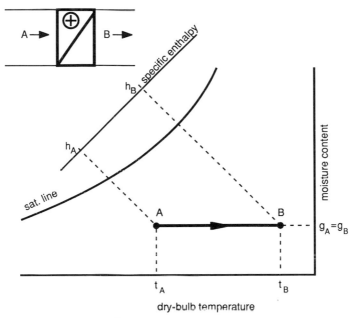

2.3 Sensible heating

In Figure 2.3 air passes through such a heater, the air dry-bulb temperature rising from condition **A** to condition **B**, the moisture content remaining constant. The specific enthalpy of the air will also increase. The load on the heater battery is then given by one of the following equations:

$$Q_h = \dot{m}_a \, c_{pas} \, (t_B - t_A) \qquad (2.4)$$

$$Q_h = \dot{m}_a \, (h_B - h_A). \qquad (2.5)$$

The choice of equation will depend on the function of the heater within the system. Thus Equation 2.4 is usually appropriate for a heater which deals with room or zone sensible loads which are related to the dry-bulb temperature, whereas Equation 2.5 is preferred for a preheater used in conjunction with an adiabatic humidifier when the heating load is related to both sensible and latent heat exchange in the plant. However, both equations will give the same load.

Example 2.3 An air flow rate of 1.5 kg/s passes through a heater battery, the dry-bulb temperature rising from 10°C to 24°C. Calculate the load on the heater battery.

Solution
Refer to Figure 2.4. Using air temperatures and Equation 2.4:

$$Q_h = \dot{m}_a \, c_{pas} \, (t_B - t_A)$$
$$= 1.5 \times 1.02 \, (24 - 10) = 21.4 \, \text{kW}.$$

Using Equation 2.5. Assuming a moisture content of 0.006 kg/kg$_{da}$ the enthalpies are obtained from either tables or chart.

$$Q_h = \dot{m}_a \, (h_B - h_A)$$
$$= 1.5 \, (36.9 - 22.7) = 21.3 \, \text{kW}.$$

The small difference between the two loads is due to rounding errors; the difference is not significant for practical calculations.

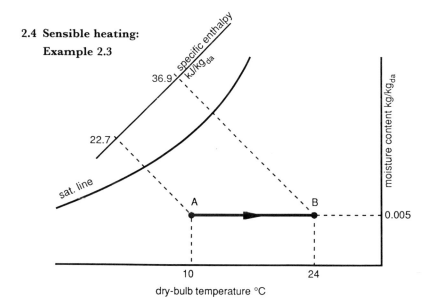

2.4 Sensible heating: Example 2.3

COOLING COILS

Coolers are required in air conditioning systems to reduce the supply air temperature and moisture content, particularly in summer when there are heat gains to the rooms, to provide for both room temperature and humidity control. An air cooling coil uses either a heat transfer medium such as chilled water or the direct expansion of refrigerant into a pipe coil. The cooling of air may be sensible only or it may be accompanied by dehumidification giving latent as well as a sensible cooling load for the coil.

Sensible cooling at constant moisture content

With sensible cooling, the air temperature is reduced and for this to occur at constant moisture content all parts of the coil (air-side) surface temperature must be above the dew-point temperature of the entering air stream.

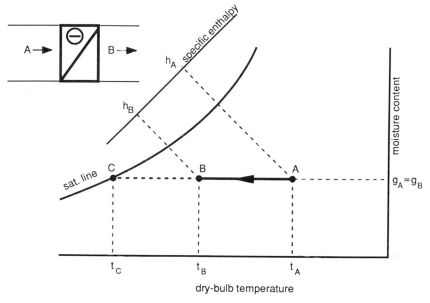

2.5 Sensible cooling at constant moisture content

In Figure 2.5 air passes through a cooler, the air dry-bulb temperature falling from condition **A** to condition **B**, the moisture content remaining constant. The load on the cooling coil is then given by either of the following equations:

$$Q_c = \dot{m}_a \, c_{pas} \, (t_A - t_B) \qquad (2.6)$$

$$Q_c = \dot{m}_a \, (h_A - h_B). \qquad (2.7)$$

Equation 2.6 would usually be used to determine the load, since no latent heat exchange is involved. But both equations will produce the same answer.

Sensible cooling with dehumidification

In Figure 2.6 the air passes through a cooler, both the air dry-bulb temperature and the moisture content falling from condition **A** to condition **B**. The heat in the water which has been condensed from the air stream will normally be very small relative to the total cooling load and therefore can be ignored in the calculations. The load on the cooling coil is then given by Equation 2.7. For the coil to dehumidify, the coil (air-side) surface temperature must be below the dew-point temperature of the entering air stream. The average coil temperature at air condition **C**, on the saturation line, is known as the *apparatus dew-point temperature* (ADP) and the line **ABC** is drawn as a straight line on the chart.

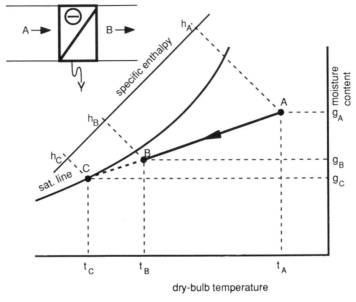

2.6 Sensible cooling with dehumidification

Example 2.4   An air flow rate of 2.4 kg/s passes through a cooling coil, the air having the following conditions:

air stream		dry-bulb temperature	moisture content
		($^{\circ}$C)	(kg/kg$_{da}$)
on-coil	A	24	0.0100
off-coil	B	12	0.0075

Determine the load on the coil.

Solution

Refer to Figure 2.7. From the psychrometric chart the specific enthalpies are:

$$\text{on-coil} \qquad h_A = 49.5 \text{ kJ/kg}_{da}$$
$$\text{off-coil} \qquad h_B = 31.0 \text{ kJ/kg}_{da}.$$

Using Equation 2.6:

$$Q_c = \dot{m}_a \, (h_A - h_B)$$
$$= 2.4 \, (49.5 - 31) = 44.4 \text{ kW}.$$

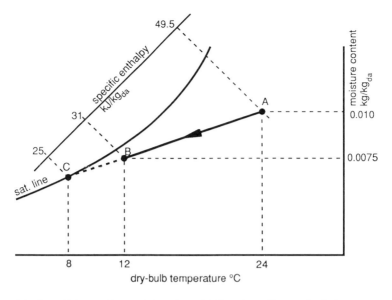

2.7 Sensible cooling with dehumidification: Example 2.4

Cooling coil contact factor

An important property of a coil is its ability to dehumidify. Referring to Figure 2.6, the minimum moisture content for a coil which is 100% efficient at dehumidifying the air would be the saturation moisture content at the apparatus dew-point, g_C. The (decimal) efficiency for dehumidification, known as the *contact factor* of the coil, is therefore defined by any one of the following equations:

by moisture content differences:

$$\beta = \frac{g_A - g_B}{g_A - g_C} \qquad (2.8)$$

by specific enthalpy differences:

$$\beta = \frac{h_A - h_B}{h_A - h_C} \qquad (2.9)$$

by dry-bulb temperature differences (approximately):

$$\beta = \frac{t_A - t_B}{t_A - t_C}. \qquad (2.10)$$

Example 2.5 Using the data in Example 2.4, determine the contact factor of the coil.

Solution
The process line has to be drawn on a psychrometric chart, as in Figure 2.7. The continuation of the line **AB** cuts the saturation line at **C**. For the conditions given, $t_C = 8°C$ at which condition $h_C = 25 \, kJ/kg_{da}$.
   Using Equation 2.9:

$$\beta = \frac{h_A - h_B}{h_A - h_C}$$

$$= \frac{49.5 - 31}{49.5 - 25} = 0.76.$$

The actual value of the contact factor will depend on the design of the coil and the air flow rate. The most important parameters are:

- number of rows of pipe coils;
- design of the heat transfer surface;
- air velocity across the face of the coil;
- drainage of condensate.

The plant diagram in Figure 2.6 shows a drain, trap and tundish to deal with the water which has been condensed from the air stream. This is a practical requirement, the trap preventing air from being blown out of, or drawn into, the air stream. The break between the trap and the tundish ensures that the air conditioning system is not contaminated through the drain; it also provides the plant operator with a convenient means to observe whether or not the coil is dehumidifying.

HUMIDIFIERS

Humidifiers are used to increase the moisture content of the air within air conditioning systems and these may be classified in three groups:

- adiabatic with recirculation of the spray water;
- adiabatic with no recirculation of the spray water;
- isothermal.

The psychrometric operating characteristics of these humidifiers are given here. For a general description of each, see Chapter 9.

Adiabatic humidifiers: recirculation of spray water
This group of humidifiers include the spray washer, the capillary washer and the sprayed cooling coil. Figure 2.8 shows a diagram of a typical unit. Water is pumped from a tank mounted at the bottom of the humidifying

chamber and the water which is not evaporated into the air stream falls back into the tank to be recirculated by the pump through the pipework to the nozzles. The theoretical process line follows the adiabatic saturation temperature, though in practice this is usually drawn following the wet-bulb temperature of the entering air condition. However, a close approximation, which assists the solution of a number of air conditioning calculations, is to consider the process occurring at constant specific enthalpy.

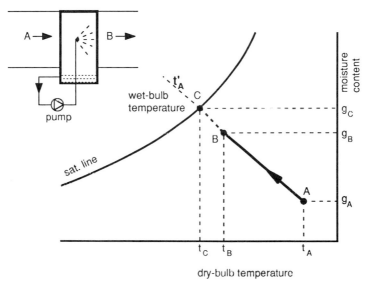

2.8 Adiabatic humidification with an air washer

The make-up water to be supplied to the humidifier tank is given by:

$$\dot{m}_w = \dot{m}_a (g_B - g_A). \tag{2.11}$$

The humidifying efficiency is usually expressed as a contact factor, β, which is defined by either of the following equations:

by moisture content differences:

$$\beta = \frac{g_B - g_A}{g_C - g_A} \tag{2.12}$$

by dry-bulb temperature differences (approximately):

$$\beta = \frac{t_B - t_A}{t_C - t_A}. \tag{2.13}$$

The contact factor depends on the nozzle design, the number and arrangement of spray nozzles and the water pressure at the nozzles. In the capillary type, water is sprayed on to cells packed with a suitable material which allows close contact between air and water. In the sprayed coil type, the wetted air-side surface of the cooling coil provides the majority of the surface from which water is evaporated into the air.

Example 2.6 Air at 25°C dry-bulb temperature and 15°C wet-bulb temperature enters a spray water humidifier with a contact factor of 0.7. Determine the moisture content of the air leaving the humidifier.

Solution
Refer to Figure 2.9. From the psychrometric chart the moisture content of the air entering the humidifier is obtained:

$$g_A = 0.0065 \text{ kg/kg}_{da}.$$

Assuming the humidifying process occurs at constant wet-bulb temperature the ADP is also at 15°C at which condition:

$$g_C = 0.0107 \text{ kg/kg}_{da}.$$

Using Equation 2.12:

$$\beta = \frac{g_B - g_A}{g_C - g_A}$$

$$0.7 = \frac{g_B - 0.0065}{0.0107 - 0.0065}$$

$$\therefore g_B = 0.0094 \text{ kg/kg}_{da}.$$

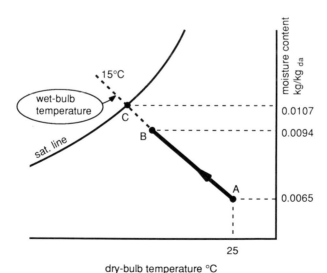

2.9 **Adiabatic humidification: Example 2.6**

Spray humidifiers as heat exchangers – non adiabatic processes

Spray-type humidifiers with water recirculation can be designed with heating and cooling equipment to achieve heating, cooling and dehumidification processes, in addition to the adiabatic humidification process previously described.

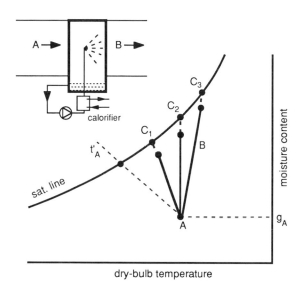

2.10 Air washer with heating of spray water

Figure 2.10 shows the pumped water heated by a calorifier. As the heat supply to the calorifier increases, the temperature of the spray water rises producing a corresponding rise in the ADP, above the adiabatic saturation temperature (point **C**) of the air entering the process. The process line will be one of a set, typically shown by AB_1, AB_2 and AB_3, depending on the amount of heat supplied to the spray water. With the AB_1 process, the increase in moisture content is accompanied by some sensible cooling. The AB_2 process is the special case of humidification occurring at constant dry-bulb temperature, ie isothermally. In the AB_3 process, humidification is occurring with sensible heating. In each case the heating load on the calorifier is given by:

$$Q_h = m_a(h_B - h_A). \tag{2.14}$$

The humidifying contact factor is given by Equation 2.12. It may also be defined in terms of specific enthalpy differences, ie:

$$\beta = \frac{h_B - h_A}{h_A - h_C}. \tag{2.15}$$

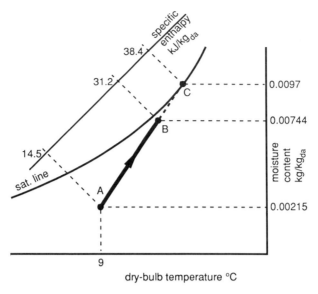

2.11 Air washer with heating of spray water: Example 2.7

Example 2.7   An air flow rate of 2.4 kg/s at a condition of 9°C dry-bulb temperature and 3°C wet-bulb temperature enters a spray water humidifier. If 40 kW of heat is provided by a calorifier to the spray water, and given that the contact factor is 0.8, determine the moisture content of the air leaving the humidifier.

Solution
Refer to Figure 2.11. From the chart, enthalpy and moisture content of air entering process:

$$h_A = 14.5 \text{ kJ/kg/kg}_{da}$$

$$g_A = 0.00215 \text{ kg/kg}_{da}.$$

Using Equation 2.14:

$$Q_h = m_a(h_A - h_B)$$

$$40 = 2.4 \ (h_B - 14.5)$$

$$\therefore h_B = 31.2 \text{ kJ/kg}_{da}.$$

Using Equation 2.15:

$$\beta = \frac{h_B - h_A}{h_C - h_A}$$

$$0.7 = \frac{31.2 - 14.5}{h_C - 14.5}$$

$$h_C = 38.4 \text{ kJ/kg}_{da}.$$

The process line is now drawn on the chart, joining condition **A** to condition **C** with an enthalpy of 38.4 kJ/kg$_{da}$ at which condition the moisture content $g_C = 0.0097$ kg/kg$_{da}$. Using Equation 2.10:

$$\beta = \frac{g_B - g_A}{g_C - g_A}$$

$$0.7 = \frac{g_B - 0.00215}{0.0097 - 0.00215}$$

$$\therefore g_B = 0.00744 \text{ kg/kg}_{da}.$$

Sprayed cooling coil

Consider the case of the sprayed cooling coil shown in Figure 2.12. With the coil receiving no coolant, the humidifying process will be adiabatic, with the ADP at condition **C**. With the coil receiving coolant, the ADP will be depressed below **C** giving a set of process lines, typically **AB$_4$**, **AB$_5$** and **AB$_6$**. With the **AB$_4$** process, humidification is occurring with sensible cooling, the **AB$_5$** process is the special case of sensible cooling at constant moisture content and the **AB$_6$** process is one of sensible cooling with dehumidification, the same process previously described for the dehumidifying cooling coil. The cooling load will be given by Equation 2.7 and the contact factor for cooling by Equations 2.8, 2.9 and 2.10.

At first sight it may seem strange that the processes **AB$_5$** and **AB$_6$** occur when water is being sprayed into the air stream. They do occur because the coil air-side surface temperature is at, or below, the dew-point temperature of the entering air.

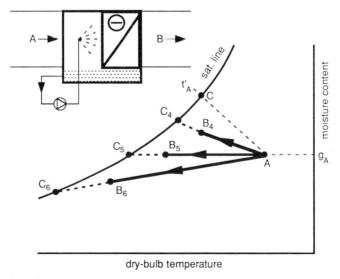

dry-bulb temperature

2.12 Sprayed cooling coil processes

The processes described for the sprayed cooling coil can also be obtained with the air washer by introducing chilling to the spray water. However, this once popular method of including a whole range of psychrometric processes within one piece of equipment is rarely used today because of the probability of scaling of heat transfer surfaces and problems associated with the hydraulic circuits, eg maintaining water levels in a water chilling refrigeration plant above the level of the humidifier tank.

Adiabatic humidifiers: no recirculation of spray water

These humidifiers include the spinning disc and direct water injection through a nozzle, the psychrometric process being shown in Figure 2.13.

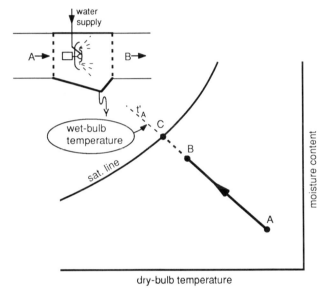

2.13 Direct water injection humidification

The amount of water supplied is for humidification purposes only, none is recirculated and any water not evaporated in the process is drained from the bottom of the unit. Efficiency of humidification, expressed as a contact factor and calculated using Equations 2.12 and 2.13, depends on the fineness of the water droplets produced, either by the operation of the spinning disc or by the pressurized water in the nozzle. Only a small amount of water is supplied relative to the mass of air passing through the unit. Therefore, heating or chilling the water supply will not bring about a significant departure from an adiabatic process and the practical process may be taken as following the wet-bulb temperature of the entering air condition.

Not all the water supplied to the humidifier is evaporated into the air stream and the water supplied to the unit must include for the overall humidifying efficiency which is defined by:

$$\eta_h = \frac{\dot{m}_w}{\dot{m}_a(g_B - g_A)} \times 100. \qquad (2.16)$$

Example 2.8   An air mass flow rate of 1.4 kg/s of air enters a spinning disc humidifier and leaves with a moisture content of 0.009 kg/kg$_{da}$. If the overall humidifying efficiency is 60%, determine the water supply rate to the humidifier. The moisture content of the entering air is 0.005 kg/kg$_{da}$.

Solution
Using Equation 1.34:

$$\eta_h = \frac{\dot{m}_w}{\dot{m}_a(g_B - g_A)} \times 100$$

$$60 = \frac{\dot{m}_w}{1.4(0.009 - 0.005)} \times 100$$

$$\dot{m}_w = 0.00336 \text{ kg/s} = 3.36 \text{ l/s}.$$

Steam humidifiers

At the present time steam humidifiers are the most popular method of increasing the moisture content of the air supply in an air conditioning system. There are two basic types:

- direct steam injection;
- pan.

Direct steam injection

Steam which is injected directly into the air stream is supplied either from a central boiler or through a local steam generator unit installed as part of the air conditioning system. With this type of humidifier all the latent heat for evaporation is added *outside* the air stream and the water vapour supplied to the air increases the enthalpy of the air. Since it occurs at near constant temperature, the process is usually referred to as an isothermal process.

Referring to Figure 2.14, the load on the humidifier is given by:

$$Q_s = \dot{m}_a(h_B - h_A). \qquad (2.17)$$

The steam supplied by the humidifier is given by:

$$\dot{m}_s = \dot{m}_a(g_B - g_A). \qquad (2.18)$$

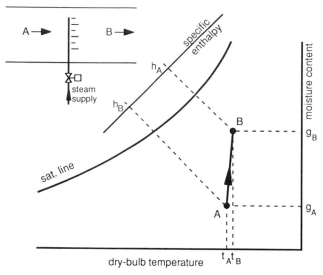

2.14 Steam humidification

The small rise in dry-bulb temperature from a steam humidifier is due to the sensible heating effect of the steam. This temperature rise is calculated by equating the sensible heat gained by the air to the sensible heat loss by the steam, ie:

$$\dot{m}_a \, c_{pas}(t_B - t_A) = \dot{m}_a(g_B - g_A)c_{ps}(t_S - t_B).$$

Example 2.9 An air supply rate of 1.4 kg/s of air at 15°C and with a moisture content of 0.004 kg/kg$_{da}$ enters a steam humidifier. Calculate the load on the humidifier and the steam supplied, if the air leaves with a moisture content of 0.008 kg/kg$_{da}$.

Solution
Referring to Figure 2.15 and assuming the process occurs at constant temperature. The enthalpies are obtained from the psychrometric chart:

$$h_A = 25 \text{ kJ/kg}_{da} \qquad {\scriptstyle 29}$$
$$h_B = 35.5. \qquad {\scriptstyle 39}$$

Using Equation 2.14:

$$Q_s = \dot{m}_a(h_B - h_A)$$
$$= 1.4(35.5 - 25) = 14.7 \text{ kW}.$$

The quantity of steam supplied is obtained using Equation 2.18:

$$\dot{m}_s = \dot{m}_a(g_B - g_A)$$
$$= 1.4(0.008 - 0.004) = 0.0056 \text{ kg/s}.$$

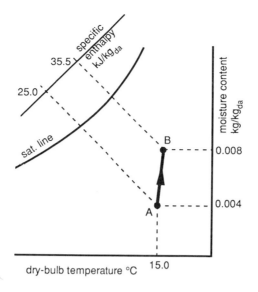

2.15 Steam humidification: Example 2.9

The specific heats are $c_{pas} = 1.025$, $c_{ps} = 1.89$ and the steam temperature is 100°C. Since $(t_S - t_B)$ will not be significantly different from $(t_S - t_A)$, the temperature rise in the air is given by:

$$(t_B - t_A) = 1.84(g_B - g_A)(100 - t_A). \qquad (2.19)$$

Example 2.10   Determine the temperature rise of the air for the steam humidifier in Example 2.9.

Solution
Using Equation 2.19:

$$\begin{aligned}(t_B - t_A) &= 1.84(g_B - g_A)(100 - t_A)\\ &= 1.84(0.008 - 0.004)(100 - 15) = 0.63 \text{ K}.\end{aligned}$$

In practice the process line for a steam humidifier may be drawn at constant temperature, as assumed in Example 2.9.

Pan steam humidifier

A pan type steam humidifier, shown in Figure 2.16, consists of a water tank mounted at the bottom of the air duct, together with a heating element. When heat is supplied, vapour is evaporated from the surface of the water to produce the humidifying effect. Some evaporative cooling also occurs, caused by the air flowing over the water surface and because of this there will be a corresponding drop in air dry-bulb temperature. This will be equivalent to a sensible to total heat ratio of approximately 0.2. (Sensible to total heat ratios on the psychrometric chart are described in the following section, relative to room process lines.)

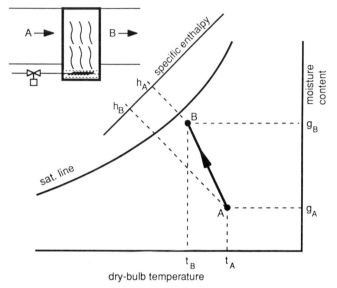

2.16 Steam pan humidification

The heat load on the humidifier and the steam supply are obtained from Equations 2.17 and 2.18 respectively.

AIR CONDITIONED SPACE: ROOM PROCESS LINES

To gain an understanding of the psychrometric processes that occur in an air conditioned space, it is useful to consider the sensible loads to be produced either by a cooler or a heater, and the latent loads by either a humidifier or a dehumidifier.

Taking the case of a room receiving both a sensible and latent heat gain in Figure 2.17. The room is maintained at air condition **R** and is supplied with air at condition **S**. The heat balances are as follows:

$$q_s = \dot{m}_a c_{pas}(t_R - t_s) \tag{2.20}$$

$$q_l = \dot{m}_a h_{fg}(g_R - g_s). \tag{2.21}$$

The ratio of the room sensible to latent heat gains is obtained by dividing Equation 2.20 by 2.21:

$$\frac{q_s}{q_l} = \frac{c_{pas}(t_R - t_s)}{h_{fg}(g_R - g_s)}.$$

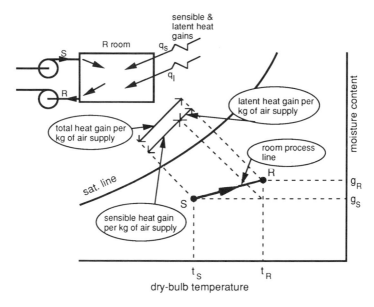

2.17 Room process line

With average values of $c_{pas} = 1.025 \, \mathrm{kJ/kgK}$ and $h_{fg} = 2450 \, \mathrm{kJ/kg}$ the following relationship is obtained:

$$\frac{q_s}{q_l} = 0.000418 \frac{(t_R - t_S)}{(g_R - g_S)} . \tag{2.22}$$

Equation 2.22 shows that, for a given room sensible to latent heat ratio, the temperature to moisture content ratio of the supply air is independent of the supply air mass flow rate.

Room ratio lines

The room process line is usually known as the *room ratio line* (RRL) and any supply condition on that line will satisfy the room condition at appropriate mass supply rates. When designing, it will be usual to draw the room ratio line on the psychrometric chart before the supply condition is determined and the most usual method of doing this is to make use of the pair of quadrants printed on the chart. Referring to Figure 2.18, the procedure is as follows:

(i) Calculate the sensible to total heat gain ratio, **y**.

$$y = \frac{q_s}{q_s + q_l} .$$

(ii) Enter the value of **y** on the appropriate quadrant, in this case the bottom quadrant. Measure the angle of **y** from the horizontal line (1.0) and this will give the slope of the RRL. Draw a line at this angle on the chart, passing through the required room condition, **R**.

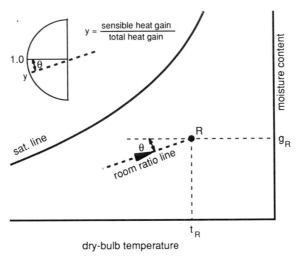

2.18 Plotting the room ratio line

In calculating the value of **y** all room loads are taken as positive values, irrespective of whether the load is a gain or a loss. The differences between loads are dealt with by the two quadrants forming a semi-circle and the choice of which one to use will depend on the orientation of the RRL on the chart. Depending on whether the room loads are gains or losses there are four possible positions of the RRL, ie:

sensible heat gain, latent heat gain – bottom quadrant
sensible heat loss, latent heat gain – top quadrant
sensible heat gain, latent heat loss – bottom quadrant
sensible heat loss, latent heat loss – top quadrant.

SYMBOLS

c_{ps} specific heat of steam
c_{pas} specific heat of humid air
g moisture content
h specific enthalpy of humid air
h_{fg} latent heat of evaporation of water
$\dot{m}_a$ mass flow rate of dry air
$\dot{m}_s$ mass flow rate of steam
$\dot{m}_w$ mass flow rate of water
Q_c cooling load
Q_h heating load
Q_s humidifier load
q_s sensible heat gain to air conditioned space
q_l latent heat gain to air conditioned space
t dry-bulb temperature
x fraction of outdoor air
y ratio of room sensible to total heat gain
β contact factor
η_h humidifying efficiency

Subscripts

A, B relate to specific air conditions
C air condition on saturation line
M mixed air condition
R room air condition

Abbreviation

ADP apparatus dew-point
RRL room ratio line

3 Indoor Design Conditions

When a building is to be air conditioned it will be necessary for the designer to decide the internal space conditions that should be maintained throughout the year when the building is occupied. Many systems are required to provide conditions which meet the thermal comfort conditions for the occupants; other systems provide conditions suitable for the efficient operation of machines and processes, the storage of food and artifacts. However, it will rarely be the case that air conditions have to be maintained at a constant level – variations are usually permitted about an optimum level. In this chapter, various aspects of these topics are examined, leading to a choice of appropriate indoor design conditions.

THERMAL COMFORT

In normal health a man or a woman has an internal body temperature of about 37°C and this temperature has to be supported for healthy living. Departures of a few degrees from normal body temperature are usually a sign of ill health, and even a danger to life itself; heat stroke and hypothermia are well known, if relatively rare, examples of high and low body temperatures. When these conditions occur, precautions must be taken against a further deterioration in that temperature.

The body generates a certain amount of heat due to the oxidation of food and this has to be dissipated if the body temperature is not to rise. Conversely, if too much heat is lost to the surroundings the body temperature will fall. The amount of heat produced will depend on the amount of physical activity being undertaken, or rate of work. At rest the body produces about 100 watts of heat and during hard work about 500 watts. It is the body's physiological mechanisms which regulate the rate of heat production together with the rate of heat loss, arranging the balance which maintains constant body temperature. When the body temperature falls, more heat is generated by imperceptible tensing of the muscles, a further fall by the onset of shivering. A rise in body temperature is countered by increased perspiration.

The nude body can only cope with a small range of external conditions in maintaining its body temperature and clothing is therefore used as insulation. The amount of clothing will affect the rate of heat loss and this in turn will affect the feeling of warmth.

The feeling of *warmth* depends on a balance between the rate of heat production and the rate of heat loss, which in turn depends on the environmental conditions. Generally the body loses heat through *convection*, *radiation* and *evaporation* but it may also *gain* heat by convection and radiation when the surrounding air and surface temperatures are higher than the body's surface temperature. Evaporation heat loss consists of insensible perspiration from the skin together with the water vapour expired from the lungs. When rates of physical activity are low in a normal indoor environment, the proportion of heat loss through these modes of heat transfer are of the order of 45%, 30% and 25% respectively. Relative to the surface conditions of the body:

- convection heat loss (or gain) depends on air dry-bulb temperature and air velocity;
- radiation heat loss/gain depends on the temperatures of the surrounding room surfaces, including the surfaces of heat-producing equipment within the room. The average temperature of all these surfaces is usually expressed as the *mean radiant temperature*;
- evaporation heat loss through insensible perspiration depends on the air vapour pressure and the air velocity.

The four variables of the physical environment that affect the heat loss from the body are therefore:

- air dry-bulb temperature;
- air vapour pressure (or relatively humidity);
- air velocity;
- mean radiant temperature.

For any individual, the sensation of thermal comfort is a complex subjective reaction to an environment, depending on a number of *personal factors*, such as age, sex and state of health. However, for a group of people there are only two personal factors which have a significant correlation with comfort, these being:

- amount of physical activity (rate of work);
- amount of clothing.

INDICES OF THERMAL COMFORT

The air dry-bulb temperature in itself is often a satisfactory index by which thermal comfort can be judged. But where the other variables have a significant effect on the rate of heat exchange they should be taken into account. A number of indices employing two or more of these variables have been devised to express, as a single numerical value, the sensation of thermal comfort. These indices include globe thermometer temperature, equivalent temperature, resultant temperature and effective temperature. The scale of each of these indices is a temperature against which the subjective assessment of thermal comfort may be measured. (See Bur-berry[1] for a description of the various comfort indices.)

COMFORT REQUIREMENTS

The literature on thermal comfort is now considerable. The first major work on the subject in the United Kingdom was carried out by Bedford[2] in 1936. In these original studies Bedford established optimum conditions and comfort zones* for workers engaged in light manual or sedentary work. The comfort conditions were based on the comfort votes of a large group of people. At the optimum condition there was a small percentage of those who felt some thermal discomfort, eg there might be a feeling of being slightly cool or slightly warm. As the comfort temperature deviates from the optimum condition the level of dissatisfaction increases. This is illustrated in Figure 3.1.

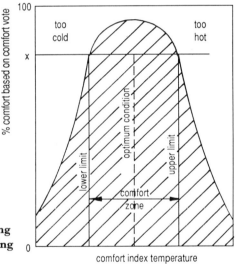

3.1 **Diagram showing basis for selecting a comfort zone**

Twenty years or so after Bedford's study, Hickish[3] determined the requirements for comfort for a similar group of workers in summer. The work by Hickish indicated that the zones established by Bedford were still appropriate, but with the optimum summer condition one to two degrees Kelvin higher compared with winter. This difference, similar to the findings of researchers in the USA, was mainly attributable to the lighter clothing worn in summer compared with winter. Though it has been suggested that the standards of thermal comfort have risen since these original studies, it is probable that the comfort zones still apply today.

*A comfort zone is an arbitrary range of (comfort) temperatures in which there is a risk of a certain number of people experiencing discomfort.

The temperatures applicable to thermal comfort, obtained by BEDFORD and HICKISH, are given in Table 3.1. The comfort zones are based on a satisfaction level (comfort vote) of 80% of the occupants. A lower limit was not available from the survey of summer conditions. It was considered that the upper limit is determined by the onset of sweating.

thermal index	winter			summer	
	lower limit $^{\circ}$C	optimum condition $^{\circ}$C	upper limit $^{\circ}$C	optimum condition $^{\circ}$C	upper limit $^{\circ}$C
air temperature	15.5	18.5	22.0	19.5	24.0
globe temperature	15.5	18.5	23.5	20.6	24.0
equivalent temperature	14.5	16.7	21.0	19.0	23.0

Table 3.1 Comfort zones of factory workers doing light manual or sedentary work

THE COMFORT EQUATION

The most commonly accepted work on thermal comfort at the present time is that by FANGER[4] who derived the *comfort equation*. This equation is based on the heat balance for the human body when in thermal equilibrium with the environment, correlated with the two personal factors of activity level and clothing, the four environmental factors and subjective observations of the state of comfort.

The internal heat produced by the body during a particular activity is expressed in units known as the *met*, one *met* being equal to 58 watts per square metre of body surface area, (corresponding to the *activity* of *sitting*). The insulation value of clothing is measured in units known as the *clo*, one *clo* being equal to 0.155 m^{2}K/W. Values of these personal variables have been determined; those for typical activities and types of clothing are given in Tables 3.2 and 3.3 respectively.

activity	met
sleeping	0.8
sitting	1.0
typing	1.2
standing	1.4
ordinary standing work in shop, laboratory, kitchen	1.6-2
slow walking (3km/h)	2.0
normal walking (5km/h)	2.6
fast walking (7km/h)	4.0
ordinary carpentry and brick-laying work	3.0
running (10km/h)	8.0

Table 3.2 Heat production in the body at different typical activities

clothing	clo
naked	0
bikini	0.05
shorts	0.1
ordinary tropical clothing: shorts, open shirt with short sleeves, light underwear	0.3
light summer clothing: long light-weight trousers, open shirt with short sleeves, light underwear	0.5
light tropical suit	0.8
typical business suit	1.0
traditional heavy Northern European suit with waistcoat, long underwear, singlet with long sleeves	1.5
polar clothing	3-4

Table 3.3 Thermal resistance of different clothing ensembles

The comfort equation is complex and its solution requires multiple iterations. Therefore, for direct application, diagrams have been prepared using computer analysis of the comfort equation for all the relevant combinations of the variables. Some of these charts are reproduced in Figure 3.2, the curved lines giving the optimum comfort condition and indicating the combination of conditions which satisfy the comfort equation.

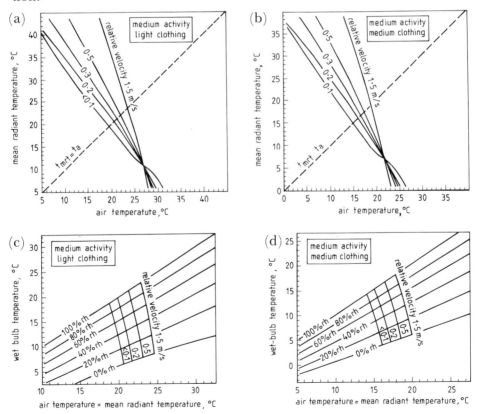

Example 3.1 In a small factory the occupants are doing light bench work (medium activity). For both cases determine the optimum air temperature, assuming that the MRT is close to the air temperature and that the air velocity is in the range 0.1 to 0.2 m/s.

Solution
It is assumed that in summer light clothing (clo = 0.5) is worn and in winter, medium clothing (clo = 1.0).
  From Figure 3.2(a), at an air velocity of 0.15 m/s:

$$\text{summer air temperature} = 20°C.$$

From Figure 3.2(b), at an air velocity of 0.15 m/s:

$$\text{winter air temperature} = 15.5°C.$$

At the optimum comfort condition there is likely to be a level of dissatisfaction of about 10% and as the conditions deviate from the optimum, the level of discomfort will increase, as indicated previously in Figure 3.1.
  The comfort equation may also be used to examine the range of conditions which will be suitable for a particular design.

Example 3.2 For the conditions given in Example 3.1, determine the effect on the optimum dry-bulb temperature of relative humidity in the range 30 to 70%.

Solution
Summer From Figure 3.2(c), at an air velocity of 0.15 m/s dry-bulb temperatures are read as:

$$\text{at } 30\%\text{rh} = 20.5°C$$
$$\text{at } 70\%\text{rh} = 19.8°C.$$

Winter From Figure 3.2(d), at an air velocity of 0.15 m/s dry-bulb temperatures are read as:

$$\text{at } 30\%\text{rh} = 16.0°C$$
$$\text{at } 70\%\text{rh} = 15.0°C.$$

PLEASANT INDOOR ENVIRONMENTS

Creating a comfortable environment within an air conditioned space depends not only on obtaining the optimum *average* room conditions but also attempting to produce within the room a *pleasant and refreshing* atmosphere. Conversely, local feelings of discomfort have to be avoided.

◄ **3.2 Typical comfort line diagrams derived from the comfort equation**

The term *freshness* is the most commonly used word to describe this aspect of comfort. The ventilation rate, usually the supply of outdoor air, required to reduce odours and other air-borne contaminants to acceptable levels, plays an important role in creating a *fresh* atmosphere. Recommended ventilation rates for this aspect of design are given in Chapter 5. Freshness has also been associated with ionization of the air and is discussed by CROOME-GALE and ROBERTS.[5]

CONSIDERATIONS OTHER THAN THERMAL COMFORT

The indoor air conditions that have to be maintained within an air conditioned space depend on a number of considerations other than thermal comfort. These include the following:

- operating efficiency of machines and processes;
- preservation of contents;
- prevention of electrostatic charges;
- prevention of condensation.

Sometimes a compromise has to be struck between the conditions required for comfort and these other requirements. To consider this it is convenient to discuss each of the environmental factors separately as they affect thermal comfort and freshness, together with a consideration of other space requirements.

Dry-bulb temperature

In most buildings the dry-bulb temperature is usually the most significant variable affecting comfort and therefore the most closely controlled. As important as the average, or controlled temperature, is the variation of air temperature in the space, especially the variations from floor to ceiling. An ideal environment is one in which the temperature at floor level is slightly greater than at head level, though in practice this is difficult to achieve with conventional air conditioning systems. Excessive vertical temperature gradients produce a feeling of stuffiness in the atmosphere. These temperature gradients are almost always associated with a lack of air movement in the occupied zone of the room, particularly at head height, this being brought about by inadequate air diffusion within the space. Almost certainly the failure of many early air heating systems to produce satisfactory conditions was due to this effect. BEDFORD[6] suggests that the maximum temperature difference from floor to head height should not be greater than 3 K and should preferably be less than this.

Relative humidity

Reference to FANGER's comfort diagrams, and illustrated by Example 3.2, shows that relative humidity has little effect on the overall sensation of thermal comfort. A change of relative humidity of 40% can be compensated by approximately one degree Kelvin in dry-bulb temperature and therefore a wide tolerance of the room humidity can be accepted where comfort is the main requirement. However, a significant correlation has been found between humidity and freshness. BEDFORD[7] advises that, to create a pleasant and invigorating environment, the relative humidity should not exceed 70% and preferably be well below this figure. It has been suggested by SEELEY[8] that drastic changes in moisture in the nasal cavity, which can occur with a sudden change in the air condition, should be avoided. This implies that the outdoor air conditions should be taken into account in the choice of the controlled indoor conditions, a warmer humid climate would require an indoor air condition which was relatively more humid than that of a warm, dry climate. Although an optimum level of humidity may also be the optimum for a feeling of freshness, achieving this level in summer when cooling and dehumidifying is required will be more expensive than maintaining relative humidity at a reduced level. Humidity is only one factor in the creation of a pleasant indoor atmosphere and if other criteria are met then humidity becomes less important.

The build-up of static electricity on machinery and carpets often causes problems. For example, with printing presses an electrostatic charge on the press will cause the paper to stick to the machinery; in this case the relative humidity should be at least 55%. For carpets to minimize the risk of electrostatic shocks the relative humidity should be maintained above 40% unless underfloor heating is involved, (causing very dry carpets), when it should be a minimum of 55%.

Process work will often require close control of humidity; some typical conditions quoted in the *ASHRAE Handbook*[9] are given in Table 3.4.

activity	dry-bulb temperature	relative humidity
	(°C)	(%rh)
cereal packaging	24 to 27	45 to 50
tea packaging	18	65
ceramics:		
clay storage	16 to 27	35 to 65
decorating room	24 to 27	48
electrical instruments manufacture	21	50 to 55
match manufacture	22 to 23	50
optical lens grinding	27	80
pharmaceuticals:		
milling room	24	35
packing and storage	24	35
animal rooms	24 to 27	50
photographic print room	21 to 22	45 to 55

Table 3.4 Typical process design conditions

To prevent condensation on internal surfaces and interstitial condensation within the building fabric it may be necessary to limit the room humidity. For example in winter, with low outdoor air temperatures, high room humidity will result in condensation on single glazed windows.

The humidity levels may also have an important effect on system energy consumption. It has been shown that for some systems the theoretical annual cost savings on refrigeration compressor energy consumption can be halved if the room relative humidity is allowed to rise to 60% compared with maintaining 50%.[10]

Air movement

It is generally well known that wind accompanied by low air temperatures produce a cooling effect on the body far greater than that caused by temperature alone. Daily weather forecasts in winter often quote wind-chill conditions, an index of cooling which combines temperature with wind speed.

Cold draughts in occupied spaces are also a well known phenomenon, though more difficult to predict than the overall cooling effect of a low temperature wind. A draught is a localized sensation, also caused by a combination of temperature and velocity of the air stream, felt mostly by the back of the neck and the ankles. Warm draughts may also occur but these do not present the designer of air conditioning systems with the same scale of problem as a cold draught.

Unless the air diffusion is carefully planned, it is often difficult to avoid producing cold draughts at times of large cooling loads. To avoid the sensation, it is usual to design for air velocities in the occupied zone of the room in the range 0.1 to 0.2 m/s.[11] An impression of freshness can be created by a variable air movement (small scale turbulence), promoted by means of the air diffusion system, in particular by means of the supply air outlets. The level of sophistication of the supply air outlets to achieve satisfactory results will be reflected in the capital cost of these items. With variable air volume systems it is possible that less than satisfactory air diffusion will be obtained when the outlet is operating at low flow rates.

Mean radiant temperature (MRT)

The temperatures of the surrounding surfaces have a significant effect on thermal comfort. Apart from the use of radiant heating/cooling systems, the MRT may not be directly controlled since it is a function of the internal fabric surfaces, radiation effects of the sun, and the warm surfaces of machines and lights. For new buildings with higher fabric insulation standards, the indoor surface temperatures in winter will be close to the air temperature, leading to the conclusion in this case that the control of air temperature will be satisfactory for comfort. However, in spring, summer and autumn the MRT will often be significantly higher than the air temperature, especially where heat absorbing glass and/or internal blinds are used. Where the air dry-bulb temperature differs appreciably from the

MRT, the air temperature may be adjusted through the control system to give an equivalent comfort sensation. For freshness, the surrounding wall surfaces should be uniformly at a somewhat higher temperature than the air temperature.[12]

INDOOR DESIGN CONDITION ENVELOPES

In choosing indoor design conditions it is also necessary to decide on the allowable deviation from an optimum condition, in order that the amount of zoning and level of sophistication of the control system may be determined. The aim of the design engineer should be to satisfy the large majority of the occupants within the comfort zone; care should be taken not to specify precise values for the indoor conditions when a range of air properties would be equally acceptable, since this may incur unnecessary expenditure. Once these ranges of indoor air conditions have been agreed they may be shown on a psychrometric chart.

Figure 3.3(a) shows an envelope for a room in which the control of air temperature and relative humidity are required. Figure 3.3(b) shows an envelope in which only the air temperature is controlled, with the absolute humidity (or dew-point temperature) set only at high and low limits. This latter envelope is similar to the ASHRAE comfort zone[13] which shows overlapping zones for summer and winter, but since it is not practical to delineate the boundaries of the seasons for an *indoor* climate there seems to be little virtue in making this distinction when dealing with a year-round air conditioning system. The limits of the ASHRAE comfort zone are:

dry-bulb temperature   20 to 26.7°C
moisture content   0.004 to 0.011 kg/kg_{da}.

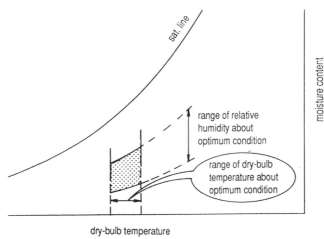

dry-bulb temperature

3.3 Indoor design condition envelopes

(a) with range of relative humidity

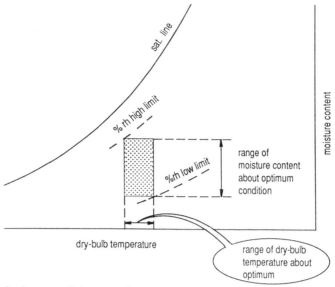

3.3 Indoor design condition envelopes

   (b) with range of moisture contents

The plant should be designed and controlled to achieve the space conditions throughout the year, to a given tolerance, within the envelope.

MEASUREMENT OF ROOM CONDITIONS

Comfort meters are available for evaluating the indoor environment. These instruments measure the four environmental variables, combining them according to the comfort equation to produce a single numerical value. Such instruments are possibly satisfactory for environments which are *already* comfortable, but when room conditions are less than satisfactory it will be necessary to measure the individual variables. Suitable dry-bulb temperature and humidity instruments are described in Chapter 1.

Air velocities may be measured with a variety of instruments such as vane anemometers and hot wires; for the low air speeds associated with comfort air conditioning a Kata thermometer is often used.

A globe thermometer is suitable for the measurement of the mean radiant temperature; the instrument consists of a thermometer whose bulb is at the centre of a blackened, hollow sphere.

The *globe* temperature, t_g, measured with a 150 mm sphere, is related to the air temperature, velocity and MRT and is given by:

$$t_g = \frac{t_r + 2.35 t_a \sqrt{v}}{1 + 2.35 \sqrt{v}}. \tag{3.1}$$

The *resultant* temperature, t_{res}, measured with a 100 mm sphere, is given by:

$$t_{res} = \frac{t_r + 10 t_a \sqrt{v}}{1 + 10\sqrt{v}}.$$ (3.2)

The resultant temperature is also used in heat loss and heat gain calculations and as an index of thermal comfort.

SYMBOLS

t_a air temperature
t_g globe thermometer
         temperature
t_r mean radiant temperature
t_{res} resultant thermometer
         temperature
v   air velocity

Abbreviations

clo   insulation value of clothing
met   heat produced by human
         body, metabolic rate
MRT mean radiant temperature

4 Outdoor Air Design Conditions

Suitable outdoor air conditions must be selected for the design of air conditioning systems so that the maximum plant loads can be determined. This applies to both summer and winter system load requirements. The designer, making an inappropriate choice of design conditions, may either

select conditions which are too extreme, leading to:

– oversized plant;
– inflated capital cost;
– poor plant efficiency;
– increased operating costs;
– poor plant control;

or

select conditions which are insufficiently stringent, leading to:

– undersized plant;
– indoor design conditions not being maintained.

METHODS OF SELECTING OUTDOOR DESIGN CONDITIONS

Recorded weather data is used to establish suitable outdoor design conditions; this data is generally available in two forms:

(i) Standard tables, published by the Meteorological Office,[1] for various cities and towns in the majority of the countries of the world;
(ii) Tables of frequency of occurrence of outdoor air conditions, published by, or available from, several authorities.

STANDARD METEOROLOGICAL OFFICE TABLES

The standard tables published by the Meteorological Office give daily, monthly and absolute maxima and minima dry-bulb temperatures for each month of the year, together with the daily relative humidity at two specific times of the day. The daily and monthly air conditions are the average values for the 30 year period 1931–60.* Temperatures and

*Revised tables are due to be published for the period 1961–90 early in this decade.

Period 1931–60 Bibliography 61, 62, 63, 64	Temperature						Relative humidity		Precipitation			Bright sunshine			
	Average daily		Average monthly		Absolute		Average of observations at		Average monthly fall	Maximum fall in 24 h	Average No. of days with 0·25 mm or more	Average monthly duration	Average per cent of possible	Maximum duration in one day	Average No. of days with no sun
	Max.	Min.	Max.	Min.	Max.	Min.	0900	1500							
			degrees Celsius				per cent		millimetres			hours		hours	
January	5·3	1·6	11·4	−4·6	13·3	−11·7	89	82	74	42	17	43	16	7·3	14
February	6·0	1·5	12·0	−4·0	15·6	−8·9	89	76	54	26	15	58	21	9·0	9
March	9·1	2·7	15·6	−2·4	20·6	−7·2	85	68	50	33	13	98	27	11·2	8
April	12·2	4·7	18·8	0·0	23·9	−1·7	75	58	53	38	13	139	33	13·1	3
May	15·6	7·3	23·1	2·2	29·4	−1·1	74	58	64	33	14	167	34	15·3	3
June	18·8	10·4	26·1	6·1	30·6	2·8	74	59	50	33	13	180	36	15·3	2
July	20·2	12·3	26·5	8·4	32·2	6·1	75	62	69	36	15	166	33	15·4	2
August	20·0	12·1	26·0	8·1	32·8	6·1	80	64	69	49	14	159	35	14·0	2
September	17·2	10·3	23·4	5·3	27·2	2·8	84	67	61	39	14	117	31	11·9	3
October	13·0	7·3	18·6	1·4	25·0	−2·2	88	73	69	41	15	86	26	9·9	6
November	8·8	4·7	13·9	−0·7	19·4	−4·4	90	80	84	38	17	48	18	7·7	12
December	6·4	2·9	11·9	−3·1	14·4	−6·1	90	84	67	47	18	38	16	7·0	14
Year	12·7	6·5	28·7*	−5·8*	32·8	−11·7	83	69	764	49	178	1299	29	15·4	78
No. of years	30	30	30	30	30	30	16	13	30	30	30	30	30	30	30

Table 4.1 Meteorological data for Birmingham, UK

humidities for Birmingham (UK), given in Table 4.1, are typical of this data.

Since these tables are published for the whole world (unlike the frequency distribution data), they provide a universal means of obtaining outdoor design conditions.

Summer design conditions

The procedure recommended in the *CIBSE Guide*[2] for selecting a summer design condition, using data of the type given in Table 4.1, is as follows:

(a) Select the month with the highest mean monthly maximum dry bulb temperature (DBT). That month becomes the design month and the mean monthly maximum DBT is taken as the design DBT.

(b) For the design month, use the daily maximum DBT with the relative humidity at 15.00 hrs to establish a moisture content, which is then associated with the design DBT to give the design wet-bulb temperature (WBT).

The rationale for this procedure is as follows: the maximum monthly DBT will be exceeded, *on average*, once every two years and this is considered a reasonable tolerance for most commercial applications. Daily maximum dry-bulb temperatures are assumed to occur at about the same time as the 15.00 hrs humidity reading.

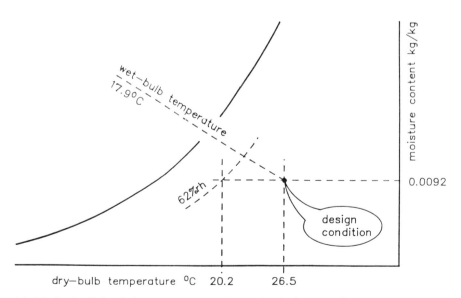

4.1 Method of obtaining summer outdoor air design conditions using standard meteorological data

Example 4.1 Using the data in Table 4.1, determine an outdoor air design condition for Birmingham.

Solution

Referring to the psychrometric sketch in Figure 4.1: The highest mean monthly maximum temperature occurs in July. July is therefore the design month and the temperature of 26.5°C is the design DBT.

The daily DBT of 20.2°C is associated with the relative humidity of 62%, giving a moisture content of 0.0092 kg/kg$_{da}$. Associating this with the design DBT, the design WBT of 17.9°C (sling) is obtained from the chart.

Modified method

Some adjustment of the design wet-bulb temperature may be necessary for a site close to a large area of water (sea, lake or river) where more evaporation of water vapour occurs on hotter days compared with average days. The engineer must judge this, using any local data that has been accumulated and which can be compared with the published tables. At present there is no relevant published information to guide the engineer in the level of adjustment to make. A suggested modified procedure is illustrated by the following example.

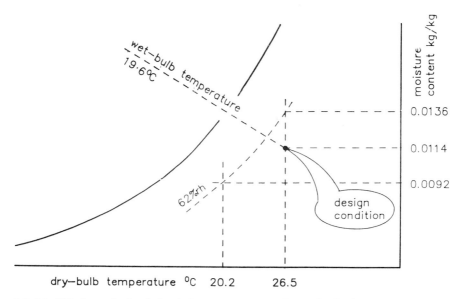

4.2 Modified method of obtaining summer outdoor air design conditions

> **Example 4.2** Using the data in Table 4.1, determine a summer outdoor design air condition for a building which is to be built close to a lake near Birmingham.
>
> *Solution*
> Referring to the psychrometric sketch in Figure 4.2. Initially the procedure is the same as that described in Example 4.1. A second moisture content of $0.0136 \text{ kg/kg}_{da}$ is obtained by projecting the 62% relative humidity to the design dry-bulb temperature of 26.5°C. The arithmetic mean between this new moisture content and the original of $0.0092 \text{ kg/kg}_{da}$ becomes the design moisture content. The resulting moisture content of $0.0114 \text{ kg/kg}_{da}$ is associated with the DBT of 26.5°C, giving a design WBT of 19.6°C (sling).

Winter design conditions

Winter outdoor design conditions may be obtained by a procedure suggested by FOWLER:[3]

- The month is selected with the lowest monthly minimum temperature and this minimum temperature becomes the winter design temperature.
- The 9.00 am relative humidity is associated with the daily minimum dry-bulb temperature to obtain a design moisture content which in turn is used with the design DBT.

The reasoning here is that, for an *average* day, daily minimum air temperature in winter occurs close to the time of the maximum, or morning, humidity reading.

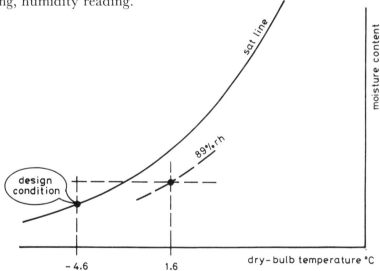

4.3 Method of obtaining winter outdoor air design conditions using standard meteorological data

Example 4.3 Using the data in Table 4.1, determine a winter outdoor design air condition for Birmingham.

Solution
Referring to the psychrometric sketch in Figure 4.3: The lowest mean monthly minimum temperature occurs in January. January is therefore the winter design month and the temperature of $-4.6°C$ is the design DBT.

The daily minimum DBT of $1.6°C$ is associated with the relative humidity of 89% giving a moisture content of 0.0038 kg/kg$_{da}$. Since the saturation moisture content at the design DBT is less than this, the design condition will be $-4.6°C$, 100%sat.

FREQUENCY OF OCCURRENCE OF OUTDOOR AIR CONDITIONS

The Meterological Office of the United Kingdom has published a set of reports[4] dealing with the combined frequency distribution of dry-bulb temperature and screen wet-bulb temperature for 28 stations throughout the UK. The original measurements of DBT and WBT (screen) are held on magnetic tape. A number of analyses[5] have been made of this data so that they should be presented in a form suitable for use by air conditioning engineers. Thus percentage frequency distributions were obtained for the following air properties:

(i)   dry-bulb temperature;
(ii) wet-bulb temperature, sling;
(iii) specific enthalpy;
(iv) dry-bulb temperature in association with moisture content.

These frequency distributions, for the whole of the annual 8760 hourly observations, were for the following daily periods:

(a) 24 hours
(b) 12 hours   07.00–18.00 GMT
     12 hours   19.00–06.00 GMT
(c)   8 hours   02.00–09.00 GMT
      8 hours   10.00–17.00 GMT
      8 hours   18.00–01.00 GMT.

Typical 24-hour frequency distributions for Heathrow for dry-bulb temperature, sling wet-bulb temperature and specific enthalpy are given in Tables 4.2, 4.3 and 4.4 respectively. In each of these tables, the frequency of occurrence of the measured air condition, for a range of the air property, is given as a percentage of the total hours in the year. The frequency of occurrence of the DBT in association with moisture content is given on a psychrometric chart in Figure 4.5; in this case each *box* contains the (percentage × 100) of the measured air conditions which were within the limits of 2 K dry-bulb temperature and 0.001 kg/kg$_{da}$ moisture content.

dry-bulb temperature 1K intervals (°C)	frequency f_1 (%)	dry-bulb temperature 1K intervals (°C)	frequency f_1 (%)	dry-bulb temperature 1K intervals (°C)	frequency f_1 (%)
		0.0 to 0.9	2.19	20.0 to 20.9	1.71
		1.0 to 1.9	2.78	21.0 to 21.9	1.27
		2.0 to 2.9	3.30	22.0 to 22.9	0.98
		3.0 to 3.9	3.85	23.0 to 23.9	0.63
		4.0 to 4.9	4.16	24.0 to 24.9	0.46
		5.0 to 5.9	5.12	25.0 to 25.9	0.31
<-13.1	–	6.0 to 6.9	5.46	26.0 to 26.9	0.20
-13.0 to-12.1	0.00	7.0 to 7.9	5.82	27.0 to 27.9	0.12
-12.0 to-11.1	0.00	8.0 to 8.9	6.21	28.0 to 28.9	0.07
-11.0 to-10.1	0.01	9.0 to 9.9	6.01	29.0 to 29.9	0.06
-10.0 to -9.1	0.01	10.0 to 10.9	6.08	30.0 to 30.9	0.04
-9.0 to -8.1	0.01	11.0 to 11.9	5.76	31.0 to 31.9	0.02
-8.0 to -7.1	0.02	12.0 to 12.9	5.63	32.0 to 32.9	0.02
-7.0 to -6.1	0.04	13.0 to 13.9	5.47	33.0 to 33.9	0.01
-6.0 to -5.1	0.08	14.0 to 14.9	5.08	34.0 to 34.9	0.00
-5.0 to -4.1	0.16	15.0 to 15.9	4.97	>35.0	–
-4.0 to -3.1	0.30	16.0 to 16.9	4.40		
-3.0 to -2.1	0.51	17.0 to 17.9	3.57		
-2.0 to -1.1	0.83	18.0 to 18.9	2.82		
-1.0 to -0.1	1.38	19.0 to 19.9	2.09		

Table 4.2 Percentage frequency distribution of hourly values of outdoor air dry-bulb temperature, annual 24 hour periods, Heathrow, UK
Note: temperatures measured at a frequency <0.005% are listed as 0.00

wet-bulb temperature 1K intervals (°C)	frequency f_2 (%)	wet-bulb temperature 1K intervals (°C)	frequency f_2 (%)	wet-bulb temperature 1K intervals (°C)	frequency f_2 (%)
		0.0 to 0.9	2.77	20.0 to 20.9	0.13
		1.0 to 1.9	3.46	21.0 to 21.9	0.04
		2.0 to 2.9	4.18	22.0 to 22.9	0.01
		3.0 to 3.9	4.76	>23.0	–
		4.0 to 4.9	5.63		
		5.0 to 5.9	5.89		
<-13.1	–	6.0 to 6.9	6.47		
-13.0 to-12.1	0.00	7.0 to 7.9	6.85		
-12.0 to-11.1	0.00	8.0 to 8.9	6.78		
-11.0 to-10.1	0.01	9.0 to 9.9	6.94		
-10.0 to -9.1	0.01	10.0 to 10.9	6.38		
-9.0 to -8.1	0.01	11.0 to 11.9	6.67		
-8.0 to -7.1	0.03	12.0 to 12.9	6.72		
-7.0 to -6.1	0.06	13.0 to 13.9	6.39		
-6.0 to -5.1	0.11	14.0 to 14.9	5.36		
-5.0 to -4.1	0.23	15.0 to 15.9	3.98		
-4.1 to -3.1	0.47	16.0 to 16.9	2.77		
-3.0 to -2.1	0.80	17.0 to 17.9	1.38		
-2.0 to -1.1	1.38	18.0 to 18.9	0.70		
-1.0 to -0.1	2.29	19.0 to 19.9	0.35		

specific enthalpy	frequ-ency	specific enthalpy	frequ-ency
2 kJ/kg intervals	f_3 (%)	2 kJ/kg intervals	f_3 (%)
-10.0 to -8.1	0.00	30.0 to 31.9	5.58
-8.0 to -6.1	0.01	32.0 to 33.9	5.54
-6.0 to -4.1	0.01	34.0 to 35.9	5.36
-4.0 to -2.1	0.03	36.0 to 37.9	5.06
-2.0 to -0.1	0.07	38.0 to 39.9	4.40
0.0 to 1.9	0.20	40.0 to 41.9	3.72
2.0 to 3.9	0.46	42.0 to 43.9	2.99
4.0 to 5.9	0.92	44.0 to 45.9	2.16
6.0 to 7.9	1.67	46.0 to 47.9	1.42
8.0 to 9.9	2.98	48.0 to 49.9	0.85
10.0 to 11.9	3.45	50.0 to 51.9	0.51
12.0 to 13.9	4.33	52.0 to 53.9	0.33
14.0 to 15.9	4.98	54.0 to 55.9	0.18
16.0 to 17.9	5.43	56.0 to 57.9	0.10
18.0 to 19.9	6.21	58.0 to 59.9	0.05
20.0 to 21.9	6.18	60.0 to 61.9	0.02
22.0 to 23.9	6.45	62.0 to 63.9	0.01
24.0 to 25.9	6.23	64.0 to 65.9	0.00
26.0 to 27.9	6.18	>66.0	-
28.0 to 29.9	6.01		

4.4 Percentage frequency distribution of hourly values of outdoor air conditions, annual 24 hour periods, Heathrow, UK (percentage × 100)

The values given as 0.00% (zero in Figure 4.4) are the extreme conditions recorded with a frequency of less than 0.05%.

This data may be used to determine outdoor air design conditions, system operating requirements at off-peak conditions and annual energy consumption (see Chapter 17).

Using the tables of frequency of occurrence of hourly values to determine the outdoor design conditions will give a more accurate solution than would the standard meteorological data. Calculations using the frequency tables do not have to be precise; the nearest 0.5 K is sufficient for temperatures and 1.0 kJ/kg$_{da}$ for enthalpies. Though the frequencies of dry-bulb temperature, wet-bulb temperature and specific enthalpy are not coincident with each other, they can be considered to be self-consistent since the air properties were derived from the same set of measurements.

◄ Table 4.3 Percentage frequency distribution of hourly values of outdoor air sling wet-bulb temperature, annual 24 hour periods, Heathrow, UK
Note: temperatures measured at a frequency <0.005% are listed as 0.00

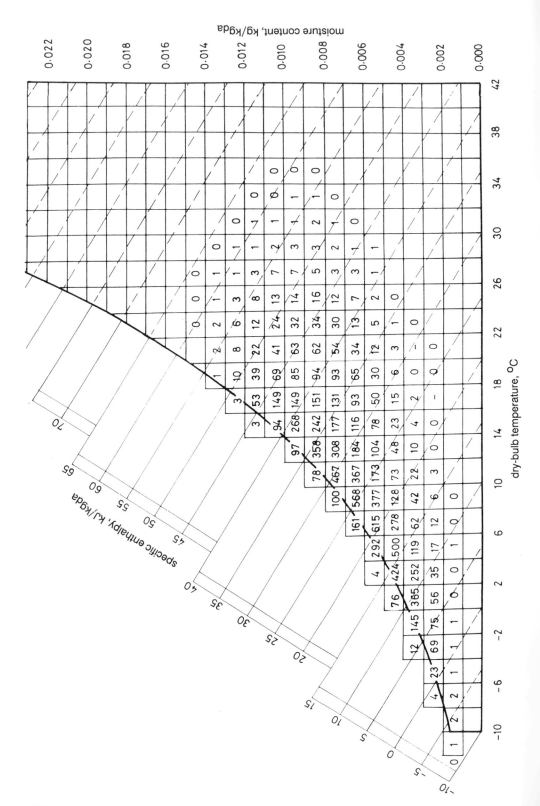

Summer design conditions

To determine a design condition for summer, a *target* of a total accumulated percentage (percentile) of annual hours is set, at which the outdoor air condition is reached or exceeded. This percentile is a considered judgement made by the design engineer. Individual frequencies of the appropriate air property are then summed from the extreme condition until the target is reached, within reasonable accuracy.

Example 4.4   Determine the outdoor design conditions for an air conditioning system for a building near Heathrow, London. A 99.65% percentile (equivalent to the design condition being exceeded for 30 hours per year) is considered an appropriate target.

Solution
To obtain the summer design conditions, the frequencies are summed commencing from the maximum property in the table until the target of (100 to 99.65)% is reached. All 0.00% values are ignored. The calculations are set out in the tables below.

Considering *dry-bulb temperature*:

Refer to Table 4.2 and commence at the extreme value of 34°C:

dry-bulb temperature	frequency
1 K intervals ($°C$)	f_1 ($\%$)
27.0 to 27.9	0.12
28.0 to 28.9	0.07
29.0 to 29.9	0.06
30.0 to 30.9	0.04
31.0 to 31.9	0.02
32.0 to 32.9	0.02
33.0 to 33.9	0.01
$\Sigma(f_1)$ = 0.34	

$$\Sigma(f_1) = 0.34$$

design condition $= 27°C$.

◄ Table 4.4 **Percentage frequency distribution of hourly values of outdoor air specific enthalpy, annual 24 hour periods, Heathrow, UK**

Wet-bulb temperature (sling)
Refer to Table 4.3 and commence at the extreme value of 23°C:

wet-bulb temperature 1 K intervals (°C)	frequency f_2 (%)
19.5 to 19.9	0.17
20.0 to 20.9	0.13
21.0 to 21.9	0.04
22.0 to 22.9	0.01
$\Sigma(f_2)$ = 0.35	

$$\Sigma(f_2) = 0.35$$

design condition = 19.5°C.

Specific enthalpy
Refer to Table 4.4 and commence at the extreme value of 64 kJ/kg$_{da}$:

specific enthalpy intervals (kJ/kg$_{da}$)	frequency f_3 (%)
54.0 to 55.9	0.18
56.0 to 57.9	0.10
58.0 to 59.9	0.05
60.0 to 61.9	0.02
62.0 to 63.9	0.01
$\Sigma(f_3)$ = 0.36	

$$\Sigma(f_3) = 0.36$$

design condition = 54 kJ/kg$_{da}$.

The summer design conditions for the air conditioning system(s) will therefore be:

dry-bulb temperature	27°C
wet-bulb temperature	19.5°C
specific enthalpy	54 kJ/kg$_{da}$.

Winter design conditions
Similar calculations can be made for winter design conditions. In this case a target percentile is set, in which the air property is at or below the design condition. The individual frequencies of the air property being considered are then summed from the extreme minimum condition until the percentile is reached.

Example 4.5 Determine the outdoor design conditions for an air conditioning system which is being specified for a building near Heathrow, London. A percentile of 0.35% (equivalent to 30 hours per year) is considered suitable.

Solution
To obtain the winter design conditions, the frequencies are summed commencing from the minimum property in the table until the target percentile of 0.35% is reached. All 0.00% values are ignored. The calculations are set out in the tables below.

Dry-bulb temperature
Refer to Table 4.2 and commence at the extreme value of $-11°C$:

dry-bulb temperature 1 K intervals ($°C$)	frequency f_1 (%)
-11.0 to -10.1	0.01
-10.0 to -9.1	0.01
-9.0 to -8.1	0.01
-8.0 to -7.1	0.02
-7.0 to -6.1	0.04
-6.0 to -5.1	0.08
-5.0 to -4.1	0.16
$\Sigma(f_1)$ =	0.33

$$\Sigma(f_1) = 0.33$$

design condition $= -4°C$.

Wet-bulb temperature
Refer to Table 4.3 and commence at the extreme value of $-11°C$:

wet-bulb temperature 1 K intervals ($°C$)	frequency f_2 (%)
-11.0 to -10.1	0.01
-10.0 to -9.1	0.01
-9.0 to -8.1	0.01
-8.0 to -7.1	0.03
-7.0 to -6.1	0.06
-6.0 to -5.1	0.11
-5.0 to -4.5	0.11
$\Sigma(f_2)$ =	0.34

$$\Sigma(f_2) = 0.34$$

design condition $= -4.5°C$.

Specific enthalpy
Refer to Table 4.4 and commence at the extreme value of -8 kJ/kg:

specific enthalpy	frequ-ency
intervals (kJ/kg$_{da}$)	f_3 (%)
-8.0 to -6.1	0.01
-6.0 to -4.1	0.01
-4.0 to -2.1	0.03
-2.0 to -0.1	0.07
0.0 to 1.9	0.20
$\Sigma(f_3)$ =	0.32

$$\Sigma(f_3) = 0.32$$

$$\text{design condition} = 2 \text{ kJ/kg}_{da}.$$

The winter design conditions for the air conditioning system(s) will therefore be:

dry-bulb temperature	$-4.0°C$
wet-bulb temperature	$-4.5°C$
specific enthalpy	2 kJ/kg$_{da}$.

The choice of which air properties to use as the design condition(s) will depend on the application. For example, dry-bulb temperature will be used for sensible heating processes, wet-bulb temperatures for cooling tower selection and enthalpy for dehumidifying cooling coils.

ISOTHERMS OF SUMMER DESIGN CONDITIONS

Another method for the United Kingdom, given in the *CIBSE Guide*,[6] makes use of isotherms of DBT and WBT (screen) drawn on two pairs of maps. The first pair, reproduced in Figure 4.5, give temperatures which are reached or exceeded for 1% of the total hours of the four months June to September (equivalent to 29 hours). The second pair of maps is based on a 2.5% criteria, equivalent to 73 hours per year. It is recommended that a lapse rate correction for altitude of (minus) 0.6 K for every 100 m of height above sea level should be applied to the temperatures obtained from these maps.

WORLD-WIDE DATA IN *ASHRAE Handbook*

The *ASHRAE Handbook*[7] tabulates outdoor air conditions for more than 2500 localities throughout the USA, together with data for other countries. Typical data from these tables is given in Table 4.5. The recommended

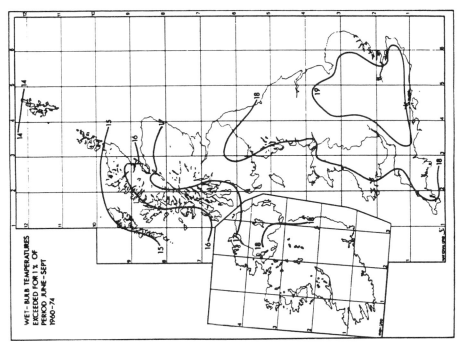

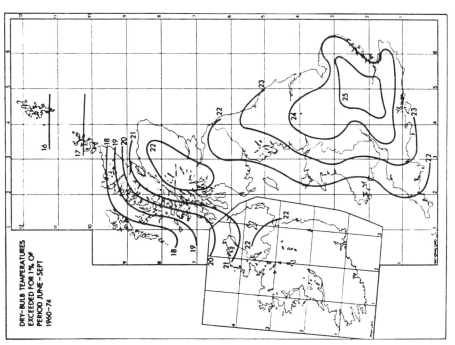

4.5 Design condition isotherms for the UK (1% criteria)
(Reproduced from Section A2 of the *CIBSE Guide*, by permission of the Chartered
Institute of Building Services Engineers)

summer design conditions given in Column 6 are based on 1%, 2.5% and 5% total hours in which the dry-bulb temperatures are reached or exceeded in the four summer months June to September (December to March in southern hemisphere). These percentages are equivalent to 29, 73 and 146 hours per annum, respectively. The coincident wet-bulb temperature listed with each dry-bulb temperature is the mean of all the wet-bulb temperatures occurring at the specific dry-bulb temperature. Additionally column 8 gives the wet-bulb temperatures at the 1%, 2.5% and 5% frequency levels, which are computed independently of, and are not coincident with, the dry-bulb temperatures of Column 6.

Recommended winter conditions of dry-bulb temperature in Column 5 are based on 99% and 97.5% criteria. Both frequency levels represent the total hours in the three winter months December, January and February (June, July and August for southern hemisphere), during which the temperature levels were reached or exceeded. That is to say, those temperatures which are at or below the conditions given, for an average of 22 and 54 hours respectively, in those three months.

Col. 1	Col. 2	Col. 3	Col. 4	Winter,[b] °C Col. 5 Design Dry-Bulb		Summer,[c] °C Col. 6 Design Dry-Bulb and Mean Coincident Wet-Bulb			Col. 7 Mean Daily	Col. 8 Design Wet-Bulb			Prevailing Wind Col. 9		Temp. °C Col. 10 Median of Annual Extr.	
State and Station[a]	Lat.	Long.	Elev.													
	° '	° '	m	99%	97.5%	1%	2.5%	5%	Range	1%	2.5%	5%	Winter	Summer	Max.	Min.
ALABAMA																
Alexander City	32 57	85 57	201	− 8	− 6	36/25	34/24	33/24	12	26	26	26				
Anniston AP	33 35	85 51	183	− 8	− 6	36/25	34/24	33/24	12	26	26	26	SW 3	SW	36.9	− 10.9
Auburn	32 36	85 30	199	− 8	− 6	36/25	34/24	33/24	12	26	26	26			37.7	− 9.7
Birmingham AP	33 34	86 45	189	− 8	− 6	36/23	34/24	33/23	12	26	25	24	NNW 4	WNW	36.9	− 10.6
Decatur	34 37	86 59	177	− 12	− 9	35/24	34/23	33/23	12	26	25	24				
Dothan AP	31 19	85 27	114	− 5	− 3	34/24	33/24	33/24	11	27	26	26				
Florence AP	34 48	87 40	177	− 8	− 6	36/23	34/23	33/23	12	26	25	24	NW 4	NW		
Gadsden	34 01	86 00	169	9	7	36/24	34/24	33/23	12	26	25	24	NNW 4	WNW		
Huntsville AP	34 42	86 35	185	− 12	− 9	35/24	34/23	33/23	13	26	25	24	N 5	SW		
Mobile AP	30 41	88 15	64	− 4	− 2	35/25	34/25	33/24	10	27	26	26	N 5	N		
Mobile Co	30 40	88 15	64	− 4	− 2	35/25	34/25	33/24	9	27	26	26			36.6	− 5.4
Montgomery AP	32 23	86 22	52	− 6	− 4	36/24	35/24	34/24	12	26	26	26	NW 4	W	37.2	− 7.7
Selma-Craig AFB	32 20	87 59	51	− 6	− 3	36/26	35/25	34/25	12	27	27	26	N 5	SW	37.8	− 8.0
Talladega	33 27	86 06	172	− 8	− 6	36/25	34/24	33/24	12	26	26	26			37.6	− 11.6
Tuscaloosa AP	33 13	87 37	52	− 7	− 5	37/24	36/24	34/24	12	26	26	25	N 3	WNW		
ALASKA																
Anchorage AP	61 10	150 01	35	− 31	− 28	22/15	20/14	19/13	8	16	15	14	SE 2	WNW		
Barrow (S)	71 18	156 47	9	− 43	− 41	14/12	12/10	9/8	7	12	10	8	SW 4	SE		
Fairbanks AP (S)	64 49	147 52	133	− 46	− 44	28/17	26/16	24/15	13	18	17	16	N 3	S		
Juneau AP	58 22	134 35	4	− 20	− 17	23/16	21/14	19/14	8	16	15	14	N 4	W		
Kodiak	57 45	152 29	22	− 12	− 11	21/14	18/13	17/13	6	16	14	13	WNW 7	NW		
Nome AP	64 30	165 26	4	− 35	− 33	19/14	17/13	15/12	6	14	13	13	N 2	W		
ARIZONA																
Douglas AP	31 27	109 36	1249	− 3	− 1	37/17	35/17	34/17	17	21	21	20			40.2	− 10.0
Flagstaff AP	35 08	111 40	2136	− 19	− 16	29/13	28/13	27/12	17	16	16	15	NE 3	SW	32.2	− 24.2
Fort Huachuca AP (S)	31 35	110 20	1422	− 4	− 2	35/17	33/17	32/17	15	21	20	19	SW 3	W		
Kingman AP	35 12	114 01	1079	− 8	− 4	39/18	38/18	36/18	17	21	21	21				
Nogales	31 21	110 55	1159	− 2	0	37/18	36/18	34/18	17	22	21	21	SW 3	W		
Phoenix AP (S)	33 26	112 01	339	− 1	1	43/22	42/22	41/22	15	24	24	24	E 2	W	44.9	− 2.9
Prescott AP	34 39	112 26	1528	16	− 13	36/16	34/16	33/16	17	19	18	18				
Tucson AP (S)	32 07	110 56	780	− 2	0	40/19	39/19	38/19	14	22	22	22	SE 3	WNW	42.7	− 7.1
Winslow AP	35 01	110 44	1492	− 15	− 12	36/16	35/16	34/16	18	19	18	18	SW 3	WSW	39.3	− 18.0
Yuma AP	32 39	114 37	65	2	4	44/22	43/22	42/22	15	26	26	25	NNE 3	WSW	46.0	− .7

OUTDOOR AIR CONDITION ENVELOPES

Air conditioning systems will usually be required to operate throughout the year against a whole range of outdoor air conditions. The extent of these conditions is most usefully shown on what may be termed the *outdoor air condition envelope*, based on the frequency distributions similar to that shown in Figure 4.6.

These envelopes provide a useful tool for summarizing plant operations, eg see Chapter 6. The design conditions of dry-bulb temperature and specific enthalpy, determined in the previous example, are shown on the skeleton envelopes in Figure 4.6: the shaded areas on each envelope represent the *total* accumulated frequency in which the outdoor air condition falls above, or below, the design condition.

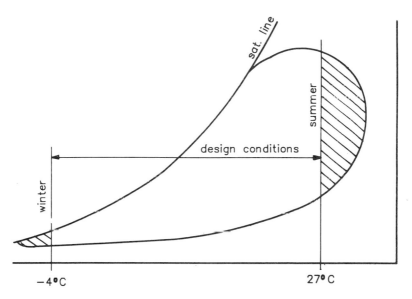

4.6 Outdoor air condition envelope and design conditions
    (a) dry bulb temperature

Continued

◀ Table 4.5 **Example outdoor air design conditions, USA**
        (Reprinted from the 1989 *ASHRAE Handbook – Fundamentals* with permission from the American Society of Heating, Refrigeration and Air-Conditioning Engineers Inc)

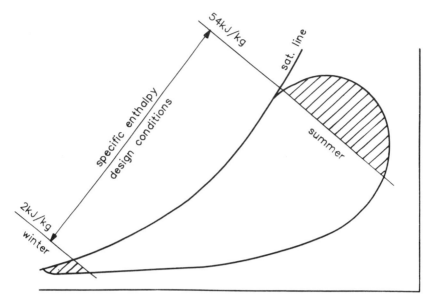

4.6 Outdoor air condition envelope and design conditions
   (b) **specific enthalpy**

Where suitable frequency data are not available a method of producing an approximate envelope has been suggested by JONES.[8] This method makes use of the type of data given in Table 4.1.

POINTS TO NOTE

- The conditions arrived at are only an average prediction based on historical data. It is most unlikely that, for any one year, these conditions would actually occur at the frequency predicted.
- Most of the published meteorological records were obtained on airfields or similarly exposed sites. Air conditions at a proposed site may differ significantly from those records and therefore the local meso and micro-climate must be considered if realistic conditions are to be obtained (eg cities act as *heat islands*). A similar comment applies to the interpolation of data for a location which is between data stations.
- Care should be taken in positioning the fresh air intake for the air conditioning system. Ambient conditions can be affected by solar-produced convection currents on some faces of the building, heat gains from the plant room and from *air discharges from exhaust systems.*
- The external design conditions will not be as critical if heat recovery from the exhaust air is employed in systems using 100% outdoor air or with systems employing a large percentage of recirculated air.

5 Room Heat Gains, Air Diffusion and Air Flow Rates

Having selected the indoor and outdoor design conditions, the design process then requires the calculation of the summer heat gains and winter heat losses. To maintain satisfactory indoor conditions, air supply and extract flow rates are determined based on a number of considerations, including the need to offset these gains and losses. This in turn requires an understanding of the way the air is supplied and extracted from the space through outlet grilles and diffusers so that a pleasant, draught-free environment is achieved. Air flow rates must also meet *ventilation* and *air movement control* requirements of the occupied building.

HEAT GAINS

The ideal structure, from the thermal point of view, would be one which modified the gains so that the temperatures are within the comfort zone. Since buildings are rarely built like this, either the excess gain has to be dealt with by mechanical cooling and/or ventilation or the temperature is allowed to rise, causing discomfort.

Detailed methods of calculating heat gains, and the provision of data, are given in other texts, particularly that by BURBERRY[1] and the *CIBSE Guide*[2] and *ASHRAE Handbook*.[3]

Sensible heat gains to the air conditioned space arise from the following sources:

- solar radiation through windows;
- solar radiation on the outside surface of the building structure (walls and roofs);
- heat transmission through the building fabric due to outdoor/indoor temperature difference;
- infiltration of outdoor air;
- occupants;
- lighting;
- machines;
- processes.

The total of the gains from these sources is the *design sensible heat gain*, q_s, which becomes the cooling load for the space. To obtain an accurate estimate of total gain, care must be taken when adding the individual sources together. Diversity factors and time lags should be applied where appropriate; if two or more of the gains vary through the day and are out of phase, then they must be added according to the time at which they occur.

Storage effect of the building structure

The effect of heat gains which are cyclic in nature, such as the gains through windows, is to reduce the *instantaneous* heat gain through the storage effect of the structure. This is illustrated in Figure 5.1. The net heat gain to the space equals the cooling load which has to be dealt with by the air conditioning system.

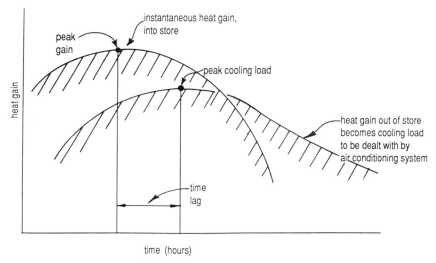

5.1 Instantaneous heat gain through glass and effect of structure in determining the air cooling load

Two classifications of structure are used for air conditioned buildings. These are:

- *heavyweight building* – a building with solid floors, ceilings and solid internal walls and ceilings;
- *lightweight building* – a building with lightweight demountable partitions and suspended ceilings. Floors are either solid with carpet or wood block finish or suspended type.

The cooling load arising from an instantaneous sensible heat gain will be reduced more by the former than by the latter.

If the indoor temperature is allowed to rise (swing) above the design value during peak periods, the plant size can be reduced by assessing the reduced cooling load for the space. Some information to allow a manual calculation to be made for this swing in summer indoor air conditions is tabulated in the *Carrier Engineering Handbook*[4] but the data are somewhat arbitrary. The *CIBSE Guide*[5] gives equations which allow the development of suitable computer programs to deal with both steady state and cyclic conditions, together with the modifying effect which the surfaces have, because of their ability to absorb and release heat.

REDUCTION OF HEAT GAINS

Some sources of heat gain can be excessive, imposing large loads on the air conditioning systems. Whilst details of the calculation methods are not given here, it is often important that architects and engineers should consider ways of reducing heat gains to the building. By limiting the heat gains *at source*, plant sizes can be reduced. The methods of reducing gains include the following:

Windows Solar heat gains through windows are often the main source of excessive cooling loads. However, gains through windows can be useful in offsetting heat loss in winter and thus reducing energy demands. A balance is required to optimize the window size/shading device to limit the maximum gain in summer as well as the maximum heat loss in winter, so that the overall plant sizes and operating costs can be reduced. Another important factor in the optimization of window size is the lighting requirement. Good daylighting can reduce the use of artificial lighting and hence limit heat gains and energy consumption. Solar heat gain through glazing may be reduced by one or more of the following:

- limiting window area;
- external solar shading devices;
- internal solar shading devices;
- blinds between panes of double glazing;
- various types of glass;
- orientation.

Glazing area Probably the most satisfactory way to reduce the solar heat gain through glazing is to limit the window area in relation to the outside wall. HARDY and MITCHELL[6] give some guidance to reasonable glazing areas as shown in Table 5.1.

building construction	maximum % of glazing area	
	air conditioning	mechanical ventilation
heavyweight	70	45
lightweight	50	20

Table 5.1 **Recommended maximum glazing area as a percentage of external wall area**

Shading devices on the external side of the window include shutters, awnings, canopies, blinds and projecting horizontal and vertical fins. Correctly designed, these sun-controls are the most effective of all for reducing solar radiation since the absorbed heat is dissipated externally. These devices which can be fixed, adjustable or retractable, have to be designed to prevent direct radiation falling on the window at appropriate times of the day and year. Fixed projections are most suitable in the tropics and sub-tropics where the sun's altitude is high, but less effective in temperate zones such as Britain. This is because they would have to project a distance further than the window height to give adequate protection at low sun angles. In so doing, daylight would be reduced as well as possibly impairing the visual environment. Balconies have been used to good effect in some buildings, a notable example being Guy's Hospital tower in London, a detail of which is shown in Figure 5.2 (the balcony also assists with window cleaning). Adjustable external louvres can have high maintenance costs; manual control is impractical if the building is more than two storeys high.

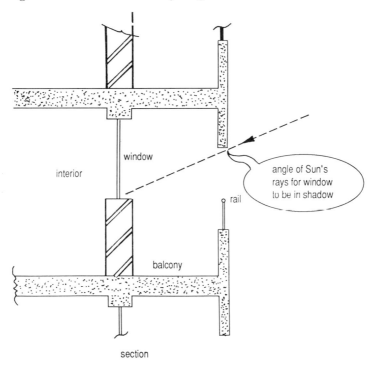

5.2 **Balcony construction to reduce direct solar radiation heat gain**

Internal blinds can give adequate protection against glare and the direct heating effects of sunshine but they are not very effective in reducing solar heat gain. This is because most of the radiant heat is absorbed by the blind and given off as convected heat into the space together with an increase in the MRT. As blinds are of small mass, these effects take place fairly quickly. To give most benefit the blinds should have a white polished surface facing the window to reflect as much radiation as possible. Dark colour blinds and curtains are least effective because they are better absorbers of radiant heat.

Blinds can be placed between the panes of a double glazed windows. In this position the blinds are protected from dirt and mechanical damage. Where a double window, with panes 160–200 mm apart, is installed for the control of external noise this solution might be the most satisfactory, but the MRT can rise because the heat is trapped between the two panes of glass.

Types of glass   More radiation is absorbed by thick glass, double or triple glass. Special heat absorbing glasses are available which absorb infra-red radiation without greatly reducing the transmission of light. The types of glass available transmit about 20 to 60% of solar radiation but about 30% of the absorbed heat is retransmitted as convective heat as well as raising the MRT. The factors for the reduction of heat gain through a single pane of these glasses vary from 0.4 to 0.8. These can be reduced to 0.3 to 0.6 if the heat absorbing glass is used as the external pane of a double glazed window; then the MRT can be reduced considerably.

Walls and roofs   Heat gains through the opaque fabric of walls and roofs are likely to be small compared with gains from other sources, the main exception being the heat gains through a flat roof. To reduce heat gains through flat roofs, attention should be paid to the mass of the structure, and to the thermal insulation. The outer surface should be finished with light colour reflecting material such as white gravel chips. Roofs are also sometimes sprayed with water in countries with high sun altitudes.

Infiltration of outdoor air   Any openable windows should be well fitting to minimize the infiltration of outdoor air. Special attention to the fabric of high rise buildings may be required if excessive infiltration is to be prevented.

Lighting   A number of schemes have used extract air lighting fittings to reduce the heat gain to the space. These fittings are mounted in a false ceiling which may then be used to provide an extract void. Though up to 80% of lamp heat can be removed, the *effective* heat removed is likely to be only approximately 50% since the warm extract air heats the ceiling from which heat is radiated to the space. A system using extract air lighting fittings is often designed so that extracted heat is made available as a source of heat in an application such as air-to-air heat recovery.

AIR DIFFUSION

Air diffusion is defined as the distribution of air in a treated space, using air terminal devices to satisfy certain specified conditions such as air change rate, pressure, cleanliness, temperature, humidity, air velocity and noise level, in a space within the room termed the *occupied zone*. The occupied zone is usually taken to be 1.8 m in height above the floor and 0.3 m from the walls. (These dimensions do not apply in the case of either desk top or local air diffusion.) Air velocities within the occupied zone should be between 0.10 and 0.20 m/s.

Cooling mode

To deal with heat gains, air is introduced into a space in a manner that will not cause draughts or excessive noise levels. For most designs, cooling air is introduced into the room at temperatures well below, and at velocities well above, those which cause a sensation of draught; therefore air must be supplied into a region of the room outside the occupied zone; this region is known as the *mixing zone*. The cold supply air introduced into this zone mixes with room air to form a *total air* stream. The velocities and temperature differences within the envelope of this total air stream reduce so that air can move into the occupied zone with little danger of causing draughts.

These general principles of air diffusion are illustrated in Figure 5.3a for three traditional positions of the supply outlet, ie high level wall grilles, ceiling diffusers and window sill units. Details will vary with different grille types, location, room geometry and use.

Heating mode

To deal with heat losses in winter, air is supplied at a higher temperature than the controlled room condition. The warm supply air, introduced into the mixing zone, mixes with room air to form a total air envelope within which the velocity and supply-to-room temperature difference reduce. There is little danger of producing a warm draught but if the temperature difference between the supply and room air is too great, the total air will remain close to the ceiling due to buoyancy effects. This may lead to temperature gradients from floor to ceiling and stagnant air in the occupied zone. The air distribution patterns for high level wall grilles, ceiling diffusers and window sill are illustrated in Figure 5.3b.

COMMON TERMS USED IN AIR DIFFUSION

Supply air The air flow rate supplied to a terminal device by the upstream duct. With some terminal devices the total air delivered to the treated space by the device may be greater than the supply air because of air induced from the room into the device by virtue of its design.

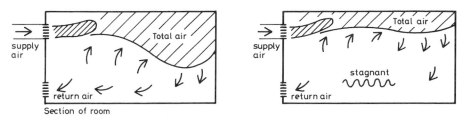

Section of room

(i) High level wall grilles, horizantal projection

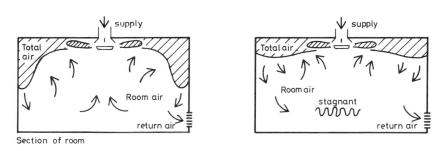

Section of room

(ii) Ceiling diffusers, horizantal projection

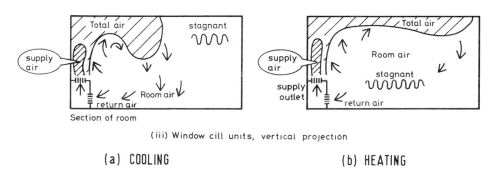

Section of room

(iii) Window cill units, vertical projection

(a) COOLING   (b) HEATING

5.3 Room air diffusion patterns for typical supply outlet positions

Total air The mixture of supply and induced room air which is still under the influence of the supply air.

Throw (for a supply air outlet) The distance between the outlet and a plane which is tangential to a specified envelope of air and perpendicular to the intended direction of flow. (See Figure 5.4.) Outlet performance is often given in terms of the throw, usually with a terminal velocity of about 0.5 m/s; acceptable air movement is achieved from an outlet expressed in terms of throw.

Terminal velocity The velocity of the air stream at the end of its throw. The throw is usually taken as the horizontal distance across the room from the supply outlet. The terminal velocity is an arbitrary value which may range from 0.4 to 1.0 m/s for commercial applications and up to 2 m/s for industrial applications. For a given outlet the throw is usually specified at the same time as the terminal velocity. (See Figure 5.4.)

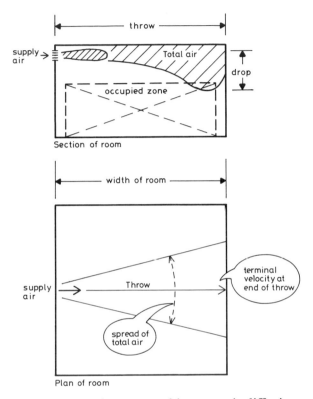

5.4 Definition of terms used in room air diffusion

Overblow occurs when a jet of air from a terminal device meets an opposing wall and is deflected downwards towards the occupied zone.

Drop   For a cold air supply which is dealing with a heat gain, if the room to supply temperature differential is too great, or the throw is inadequate, air may *drop* into the occupied zone, causing draughts. (See Figure 5.4.)

Supply to room temperature differential (Δt)   The difference between the supply air temperature and the mean temperature of the occupied space.

Stagnant region   A local region within the air conditioned/ventilated space in which there is no measurable air velocity (or velocities less than 0.05 m/s).

Convection air currents Variations in surface temperatures in a room give rise to localized heating or cooling of the room air. This sets up localized convection air currents in the space which may be caused by:

– hot/cold windows;
– hot/cold walls;
– occupants;
– lights and equipment;
– radiators;
– chilled panels.

Although difficult to quantify, natural circulation currents can significantly modify the pattern of air introduced through a grille.

Grille noise The *CIBSE Guide*[7] gives recommended maximum grille face velocities for tolerable room noise levels, (see Table 5.2) though these values will be modified by the acoustic properties of the room. However, it will be more normal to use the acoustic data prepared by the grille manufacturer where this is available.

application	outlet velocity
	(m/s)
libraries, sound studios, operating theatres	1.7 - 2.5
churches, residences, hotel bedrooms, hospital rooms and wards, private offices	2.5 - 4.0
banks, theatres, restaurants, classrooms, small shops, general offices, public buildings, ballrooms	4.0 - 5.0
arenas, kitchens, factories, gymnasia, warehouses, department stories, industrial buildings, workshops	5.0 - 7.5

Table 5.2 Air velocities for acceptable noise levels

OUTLET GRILLES AND DIFFUSERS

The five basic types of supply outlets are:

– wall grille;
– slot diffuser;
– ceiling diffuser;
– perforated ceiling panel;
– displacement types.

Each of these has different performance characteristics and should be considered separately. An important factor in the choice of grille type is its ability to entrain and mix room air, a relationship known as the induction ratio. The higher the induction ratio, the more air can be introduced into a room without causing draughts and the more quickly room air is mixed with the supply air. This implies that grilles with high induction ratio can handle larger temperature differences between the supply and room air. Ceiling diffusers have high induction ratios because they diffuse air in all directions. This leads to a short throw and more rapid temperature equalization than slot diffusers. Wall grilles have relatively long throws but low induction ratios.

ASHRAE[8] suggest the use of various outlet types, as given in Table 5.3.

outlet type	air loading l/s per m² of floor area	approx. max. air change per hour (based on 3 m ceiling) (h^{-1})	average room air velocity in occupied space (m/s)
grille	3 - 6	7	0.13 - 0.18
slot	4 - 10	12	0.10 - 0.18
perforated panel	5 - 15	18	0.10 - 0.18
ceiling diffuser	5 - 25	30	0.10 - 0.25
perforated ceiling	5 - 50	60	0.05 - 0.15

Table 5.3 General guide to supply outlet characteristics

GENERAL RULES FOR THE POSITIONING OF GRILLES AND DIFFUSERS

Supply outlets
- If the outlet is to handle both warm and cool air at different times of the year a compromise must be made in the choice of size and position to give satisfactory conditions both in summer and winter. As cold draughts are considered to be more uncomfortable than hot draughts, the designer should incline towards giving more satisfactory room distribution in summer when handling chilled air.
- Avoid overlap between the zones of influence of adjacent supply outlets, as an area of localised increased velocity can occur where two air streams meet, giving rise to uncomfortable draughts.
- When supplying chilled air, an excessive drop is likely to occur if the outlet velocity is below 2 m/s.
- Variable air volume systems are likely to have a variable outlet velocity which will give rise to varying flow patterns in a room. The minimum acceptable outlet velocity is about 2 m/s; variable area outlet diffusers

are available to reduce the outlet area and so maintain constant outlet velocities for satisfactory air diffusion.

For upward vertical projection of supplying air from a floor or window sill mounted outlet, the centre line velocity should be reduced to approximately 0.75 m/s at ceiling level to avoid excessive downward deflection from the ceiling.

– The outlet(s) should be positioned to avoid supply air discharging against ceiling obstructions such as surface mounted light fittings and down-stand beams. Otherwise some of the cold air will be deflected downwards into the occupied space, causing draughts.

Exhaust outlets

Exhaust air is the air which is either extracted or discharged from the conditioned space. This may take one or more of the following forms:

– *Extract* exhaust in which the air is discharged to atmosphere;
– *Relief* exhaust where air is allowed to escape from the treated space should the pressure in the space rise above a specified level;
– *Recirculation* exhaust where air is returned to the air handling system;
– *Transfer* exhaust lets air pass from the treated space to another treated space.

The zone of localized high velocity in association with extract grilles is very close to the grille and therefore the position of the extract grille has little influence upon the overall flow patterns in a room. For this reason there is no need to match every supply outlet with an extract grille. The number of extracts are determined by physical limitations, together with the planning of the control of air movement through various rooms of the building, for example, smoke control in the event of fire, clean and dirty areas of a hospital operating theatre suite of rooms. Extract grilles could be sited successfully at the following places:

– in a stagnant zone;
– close to an excessive heat source, to minimize heat gain to the room;
– close to an excessive cold source to minimize heat loss in the room, eg a window in winter;
– at a point of local low pressure, such as the centre of a circular ceiling diffuser discharging horizontally.

Exhaust grilles should *not* be positioned:

– in the zone of influence of a supply grille so as to prevent conditioned air passing directly into the extract system without having first exchanged heat with the surroundings;
– close to a door or aperture which is frequently opened, so that the exhaust grille does not handle air from another space.

AIR VOLUME FLOW RATE

The supply and extract air volume flow rates required for each room or space in an air conditioned building are determined by a consideration of one or more of the following:

- sensible heat gains;
- latent heat gains;
- sensible heat losses;
- air distribution;
- ventilation;
- air movement control;
- fire precautions.

Surveys of existing air conditioned buildings have shown that fan energy costs account for up to half the total energy costs of the system(s). Therefore, where it is important to reduce energy costs, it is important to design the systems at the lowest flow rates consistent with good design. Lower flow rates will also help to reduce the size of other energy using equipment such as coolers and heaters.

Air supply for summer cooling

Consider the air cooling to offset the design sensible heat gain, q_s. The balance of the heat gain to the cooling effect of the air supply is given by Equation 2.20, ie:

$$q_s = \dot{m}_a c_{pas} (t_R - t_S).$$

The design *mass* air flow rate is therefore given by:

$$\dot{m}_a = \frac{q_s}{c_{pas} \Delta t_c} \tag{5.1}$$

where $\Delta t_c = (t_R - t_S)$, the design cooling air temperature differential.

The supply air volume is obtained by using the specific volume of the supply air condition, v_S:

$$\dot{V} = \dot{m}_a \, v_S. \tag{5.2}$$

Since the room temperature, t_R, is a controlled condition, the flow rate is determined by fixing the supply air temperature, t_S. For economic design, the flow rate should be as small as possible which means having a supply temperature as low as possible without causing draughts. Suitable temperature differentials depend on several factors, some of which have been discussed earlier in the chapter.

Example 5.1 A room is maintained at 21°C and has a design sensible heat gain of 6 kW. Determine the supply air volume flow rate if the room to supply air temperature differential is 8 K.

Solution
The supply air *mass* is obtained using Equation 5.1:

$$\dot{m}_a = \frac{q_s}{c_{pas}\Delta t_c} = \frac{6}{1.02 \times 8} = 0.735 \text{ kg/s}.$$

Using an average value of specific volume at the supply temperature of 13°C, the supply air *volume* is obtained from Equation 5.2:

$$\dot{V} = \dot{m}_a v_s = 0.735 \times 0.82 = 0.603 \text{ m}^3/\text{s}.$$

A number of other considerations are involved in the selection of a suitable temperature differential, some of which are examined with reference to the psychrometric sketch in Figure 5.5.

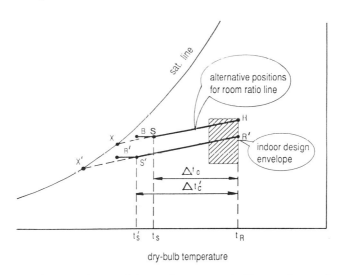

5.5 Alternative positions of room ratio line to determine temperature differential for cooling

Initially a room condition **R** may be taken at the highest enthalpy in the indoor design envelope. From this, the room ratio line, **RX**, (with a relatively small slope for this particular example) is plotted. Any supply condition on this line will maintain the room condition against the design heat gains. The air leaving the cooling coil is at condition **B** and **BS** is a temperature rise due to fan and duct heat gains. Thus the maximum temperature differential, Δt_c, is determined by the position of the room ratio line on the chart, since it is limited by the saturation line and plant operating characteristics such as the contact factor of the coil. By lowering the room condition to R' a lower supply air temperature is possible, resulting in a larger differential, $\Delta t_c'$.

Coil temperature The minimum off-coil air temperature, **B,** may be limited by the cooling coil temperature. Typically, the average lowest average air temperature leaving a coil using chilled water is 8°C.

Removal of latent heat gains

Latent heat gains to a space arise from the following sources:

- infiltration of outdoor air;
- occupants;
- processes.

The total gain from these sources is the *design latent heat gain*, q_1.

To deal with room latent heat gain, air is supplied at a lower moisture content than the room air condition. The balance of the heat gain to the dehumidifying effect of the air supply is given by Equation 2.21, ie:

$$q_1 = \dot{m}_a h_{fg}(g_R - g_S).$$

Where there is a relatively large design latent heat gain compared with sensible heat gain it may be necessary to use the moisture content differential to determine the design flow rate. The design *mass* air flow rate is therefore given by:

$$\dot{m}_a = \frac{q_1}{h_{fg} \Delta g_c} \tag{5.3}$$

where $\Delta g_c = (g_R - g_S)$, the design cooling moisture content differential.

Since the room moisture content, g_R, is the controlled condition, the mass flow rate is determined by fixing the supply air moisture content, g_S, which should be as low as possible, consistent with the performance of the air cooling coil.

Winter heating

For winter heating the system becomes an air heating system, when sensible heat losses from a space arise from the following sources:

- heat transmission through the building fabric due to internal/external temperature differences;
- infiltration of outdoor air.

The total loss from these sources, based on the outdoor air design dry-bulb temperature, is the *design sensible heat loss*, q_s'. The heat balance of the heat loss to the heating effect of the air supply is given by Equation 2.20, ie:

$$q_s' = \dot{m}_a c_{pas}(t_S - t_R).$$

Therefore, if the heat loss is the design heat loss, q_s', then the maximum supply air temperature, t_{Smax}, is given by:

$$t_{Smax} = t_R + \frac{q_s'}{\dot{m}_a c_{pas}}. \tag{5.4}$$

If the supply air temperature is too high for satisfactory room air diffusion, either the design air mass flow rate can be increased or an alternative system of air diffusion used. This is unlikely to be a problem for a *constant air volume* (CAV) system designed for cooling and heating, since the heat gains will be numerically larger than the heat losses. However, with variable air volume system terminal units, which are required to deal with a heating load, excessive supply air temperatures are likely; this problem is examined below.

Total air flow rate

For CAV systems, the total air volume flow rate will normally be the sum of the flow rates to, or from, each outlet in the system.

With *variable air volume* (VAV) systems, the total can be reduced by analysing the variation in cooling load requirements between the different spaces. This is illustrated in Figure 5.6 where the loads of two zones vary through the day and are out of phase with each other. The supply air flow rates for the individual rooms are calculated for the maximum gains to each zone but, since these design flow rates occur at different times, the maximum total of the loads for both zones may be used to determine the fan supply volume.

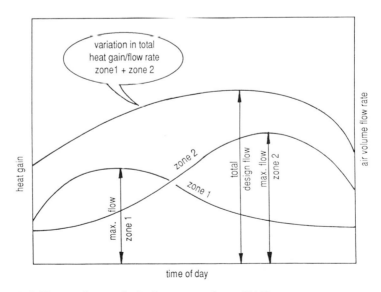

5.6 Heat gains and air flow rates for a VAV system

Variable air volume

To deal with the variations in sensible heat gain to an individual space using a *variable air volume* (VAV) system the flow rate is reduced whilst the supply air temperature remains constant. Since the terminal unit is required to deliver a minimum flow rate to provide ventilation and adequate air diffusion, there is a limit to the reduction in gain that can be dealt with in this way. The minimum to maximum flow rate of a terminal unit is known as the *turn down ratio* (TDR) and for most units this is of the order of 1:4.

Example 5.4 A room is supplied with air from a VAV terminal unit designed for cooling and heating. Determine the supply air temperature when the room is subject to the maximum heat loss for the following design data:

Maximum heat gain	10 kW
Maximum heat loss	4 kW
Room temperature	21°C
Minimum supply air temperature	12°C
Turn down ratio (TDR)	1:4

Solution

The supply air mass is given by Equation 5.1

$$\dot{m}_a = \frac{q_s{}'}{c_{pas}\Delta t_c}$$

$$= \frac{10}{1.02(21-12)} = 1.09 \text{ kg/s.}$$

Therefore the minimum flow rate $=$ TDR

$$\dot{m}_a = 0.25 \times 1.09 = 0.272 \text{ kg/s.}$$

If the terminal unit continues to operate on the minimum flow rate the maximum supply air temperature is given by Equation 5.4:

$$t_{Smax} = t_R + \frac{q_s{}'}{\dot{m}_a c_{pas}}$$

$$= 21 + \frac{4}{0.272 \times 1.02} = 35.4°C.$$

In this last example, the supply air temperature would be considered too high for successful air diffusion and therefore the flow rate should be increased when a maximum supply air temperature is reached. Alternatively, the air can be supplied from another outlet position to promote satisfactory room air diffusion. Such an arrangement is provided in some outlet units, one of which is illustrated in Figure 5.7. Here the air is supplied across the ceiling for cooling but with a rise in supply air temperatures to 2 to 3K above room temperature, the outlet adjusted to give a vertical discharge down the wall.

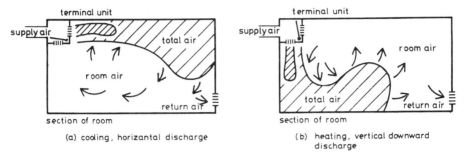

(a) cooling, horizontal discharge   (b) heating, vertical downward discharge

5.7 Air diffusion with a VAV terminal unit for cooling and heating modes of operation

MINIMUM VENTILATION RATES FOR NORMAL OCCUPANCY

Air supply for ventilation purposes is usually taken to be the supply of *outdoor air* into the occupied spaces. Where the building is air conditioned, this is supplied via the ductwork system, the purpose being to maintain a satisfactory indoor air quality by reducing the concentration of air borne contaminants to levels acceptable to the occupants and/or for the activities taking place within the building. A common requirement is the dilution of body odours originating from normal body processes. In the UK there are no statutory requirements with regard to specific minimum ventilation rates. For places of public entertainment, the most quoted authority used by engineers was the old Greater London Council (GLC) Bye-Law which asked for 4 litres per second (l/s) per person, provided that the relative humidity was maintained below 55%, otherwise the minimum should be 6 l/s. The current recommendations in the *CIBSE Guide* are given in Table 5.4, though in the UK it is now usual to take the minimum ventilation rate as 10 l/s; in the USA a figure of 7.5 l/s (15 cfm) is likely to become the accepted standard. Where smoking is permitted the rate should be four to five times these figures.

If the minimum ventilation rate is less than the total air supply rate required to meet the cooling and heating loads of the space, then the difference between the two can usually be recirculated within either the system or the room unit. Where the air supply rate exceeds the minimum ventilation rate and it is not permissible to recirculate air, the total supply air is drawn from outdoors. With VAV systems it is often required to control the minimum ventilation rate rather than maintain a fixed proportion of outdoor to recirculated air (see page 134). It is also worth bearing in mind that, with all-air systems designed for *free cooling*, the system will be supplying considerably more than the minimum ventilation requirement for most of the year (see Chapter 6, pages 125ff).

type of space	smoking	outdoor air supply (1/s)		
		recom- mended	minimum (take greater of two)	
		per person	per person	per m^2 floor area
factories	none	8	5	0.8
offices (open plan)	some	8	5	1.3
shops, department stores and supermarkets	some	8	5	3.0
theatres	some	8	5	-
dance halls	some	12	8	-
hotel bedrooms	heavy	12	8	1.7
laboratories	some	12	8	-
offices (private)	heavy	12	8	1.3
residences (average)	heavy	12	8	-
restaurants (cafeteria)	some	12	8	-
corridors	a per capita basis is not appropriate to these spaces			1.3
kitchens (domestic)				10.0
kitchens (restaurant)				20.0
toilets				10.0

Table 5.4 Recommended outdoor air supply rates for air conditioned spaces

AIR MOVEMENT CONTROL

Air movement control (AMC), sometimes known as *room pressurization*, is the control of air movement within a suite of rooms and within the building complex. This aspect of the design of air handling systems has become increasingly important in recent years. Particular areas of interest are:

- operating room suites in hospitals;
- isolation rooms and intensive care suites;
- clean rooms;
- animal laboratories;
- corridors as escape routes, smoke control zones;
- toilets associated with offices;
- kitchens and serveries.

The general principle is to prevent potentially contaminated air from passing from one area to a clean area by controlling the direction of air movement. To do this it is necessary to plan the direction of flow from a *clean zone* to a relatively *less clean* or *dirty zone*. For air to pass from one zone to another a pressure difference must exist but this can be quite small.

What is important, is that there should be sufficient air flow to minimize the risk of a back-flow from the less clean areas to the areas requiring protection. At the same time, unless the temperatures of each space are controlled at the same level, convection currents can be promoted which might transfer airborne contamination from a less clean area to the protected area.

A hospital operating suite is taken as an example to illustrate the principles involved. The suite of rooms (or spaces) is divided into four zones, *sterile*, *clean*, *transitional* and *dirty*, as given in Table 5.5. This table gives the air supply rates for the dilution of airborne bacterial contaminants which are considered sufficient, provided reasonable mixing of room air occurs. The room pressures are given to help with the sizing and setting-to-work of any pressure relief dampers considered necessary to the overall design; they are not essential to the design solution. Rather, the design procedure[9] deals with the air flows through doorways. When doors are closed, minimizing the movement of contaminated air from a dirty to a clean zone is relatively easy to achieve but once open the problem becomes

class	room	nominal pressure (Pa) (i)	air flow rate for bacterial contamination dilution	
			flow in or supply	flow out or extract
sterile	preparation: (a) lay-up (b) sterile pack	35 25+5	0.20 0.10	– –
	operating room	25	0.65	–
	scrub: (a) open bay (b) semi-open bay	25 25	0 0	0 (ii) 0.10
clean	central sterile pack store anaesthetic scrub	14 14 14	0.10 0.15 –	– 0.15 0.10
transitional	recovery clean corridor access corridor changing plaster	3 3 3 3 3	15ac/h (iv) (iv) 7ac/h 7ac/h	15ac/h (iii) 7ac/h 7ac/h 7ac/h 7ac/h
dirty	disposal corridor disposal room	0 -5 or 0	– –	(v) 0.10

Table 5.5 Hierarchy of cleanliness and recommended air supply and extract rates for dilution of airborne bacterial contaminants in operating room suites (flow rates in m³/s or air changes per hour (ac/h))

Notes to Table 5.5:

(i) Nominal room pressures given in the table are not an essential feature of the design; they are given to facilitate setting up pressure relief dampers, the calculation process and the sizing of air transfer devices.

(ii) The open bay scrub is considered to be part of the operating room; no specific extract is required provided air movement is satisfactory.

(iii) An air change rate of 15ac/h is considered necessary for the control of anaesthetic gas pollution.

(iv) A supply flow rate is necessary to make up 7ac/h after taking account of secondary air from cleaner areas.

(v) No dilution is required; air flow rates required only for temperature control.

room class	doors	room class			
		dirty	transitional	clean	sterile
sterile	single	0.47	0.39	0.28	0 or 0.28
	double	0.95	0.75	0.57	0 or 0.57
clean	single	0.39	0.28	0 or 0.28	–
	double	0.75	0.57	0 or 0.28	–
transitional	single	0.28	0 or 0.28	–	–
	double	0.57	0 or 0.57	–	–
dirty	single	0	–	–	–
	double	0	–	–	–

Table 5.6 Recommended air volume flow rates (m³/s) through a doorway between rooms of different cleanliness to control cross contamination in operating room suites

Notes:

(i) Open door sizes, single – 0.80 × 2.01 m high
    double – 1.80 × 2.01 m high

(ii) The degree of protection required at an open doorway between rooms is dependent upon the degree of difference in cleanliness between them.

(iii) The flow rate required between two rooms of same class tends to zero as class reduces.

(iv) If two rooms are of equal cleanliness, no flow is required and the design of the air movement will assume zero air flow. In certain cases however, interchange is not permitted and a protection air flow rate of 0.28 m³/s is assumed in the design.

more complex. The design of the AMC system assumes that only one door is opened at a time and that the direction and rate of air flow through that doorway is sufficient to prevent any serious back-flow of air to a cleaner area. The recommended flow rates to achieve this are given in Table 5.6.

Based on these criteria AMC control schemes have been developed for several possible operating suite schemes. *Diagrammatic* layouts of two such schemes are shown in Figure 5.8.

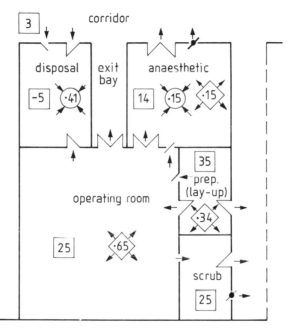

5.8 Diagrammatic layouts of two operating theatre suites[5]
(a) single corridor lay-up

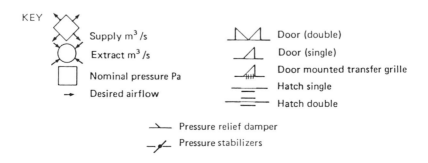

KEY

◇ Supply m³/s
○ Extract m³/s
□ Nominal pressure Pa
→ Desired airflow

⋈ Door (double)
∆ Door (single)
∆ Door mounted transfer grille
═ Hatch single
═ Hatch double

⤬ Pressure relief damper
⤿ Pressure stabilizers

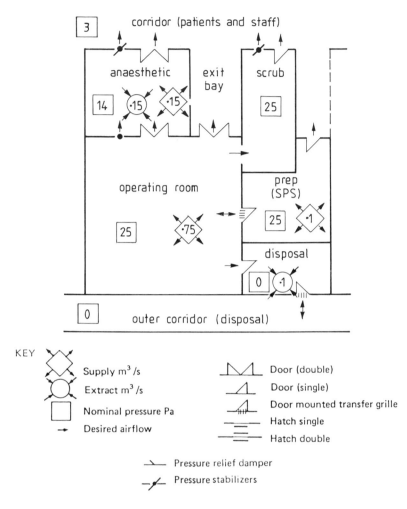

KEY

⬦	Supply m³/s	⌐Ⓜ⌐	Door (double)
⊘	Extract m³/s	⌐△	Door (single)
☐	Nominal pressure Pa	⌐△ⅲ	Door mounted transfer grille
→	Desired airflow	═	Hatch single
		≡	Hatch double

⤙ Pressure relief damper

⤙ Pressure stabilizers

5.8 Diagrammatic layouts of two operating theatre suites[5]
(b) conventional two corridor with sterile pack store

SMOKE CONTROL ESCAPE ROUTES

In some buildings the air conditioning and mechanical ventilation systems must be designed to assist the control of smoke in the event of fire. The aim is to keep escape corridor routes free of smoke, by maintaining a positive pressure there. To achieve this, supply and extract air flow rates are determined based on flow rates through door grilles from the corridor to individual rooms. Separate pressurization systems are provided for staircases.

SYMBOLS

c_{pas} specific heat of humid air

g moisture content

h_{fg} latent heat of evaporation

$\dot{m}_a$ mass flow rate of dry air

q_s sensible heat gain to air conditioned space

q_s' sensible heat loss to air conditioned space

q_l latent heat gain to air conditioned space

t dry-bulb temperature

t_{Smax} maximum supply air temperature

$\dot{V}$ air volume flow rate

v air specific volume

x fraction of supply air volume

y ratio of room sensible to total heat gain

β contact factor

Δg moisture content differential

Δt_c temperature differential for cooling

Subscripts

R   room air condition

S   supply air condition

Abbreviations

AMC   air movement control

RRL   room ratio line

TDR   turn down ratio

6 Systems I

In this chapter a number of air conditioning systems are described to illustrate the principles of design. They are discussed in terms of determining the heating and cooling plant loads together with the operation of the control systems. The plant items give the process requirement, whilst not specifying the heating or cooling medium.

Relatively simple systems are examined in which the air supply rate remains constant; these flow rates are determined as described in the previous chapter. The systems are studied in detail so that the reader may have a clear understanding of the procedures and principles involved. Although these systems are often used in their own right, a thorough understanding ot their operating principles is necessary before going on to consider more complex systems in Chapter 7.

When designing, a block diagram of the system should be built up, together with the seasonal air conditioning processes as complete *cycles* on a psychrometric chart. The air state points should be identified with letters (or numbers) so that the processes on the chart can be related to those on the block diagram.

SYSTEMS USING 100% OUTDOOR AIR

With a system using 100% outdoor air, a decision has already been made that, due to the ventilation requirements, no recirculation of air is allowed.

Summer operation

For summer operation a design procedure would be as follows:
Referring to Figure 6.1:

- decide the room design condition, **R**;
- calculate the design heat gains and determine the room ratio line;
- on the psychrometric chart plot the room condition and the room ratio line;
- determine the supply air condition, **S**, (together with the supply air mass flow rate);
- plot the outdoor design condition, **O**.

The conditions on the chart now show that it is necessary to cool and dehumidify the outdoor air from **O** to **S**, showing that a dehumidifying cooling coil is required. However, the cooling coil process line must cut the saturation line at its ADP, whereas the projection of the line joining **O** and **S** does not. To obtain the off-coil condition **B**, the contact factor of the coil is used to obtain the moisture content of **C**, knowing that the moisture content of the off-coil condition is the same as that of the supply air and that the on-coil moisture content is g_O.

The cooling process line OBC can then be drawn. The air at condition **B** is at a lower temperature than the supply air and the plant therefore requires a heating coil, termed an *after-heater* or *reheater*, to heat the air from **B** to **S**. In practice the temperature rise shown as after-heater load will also include fan and duct heat gains.

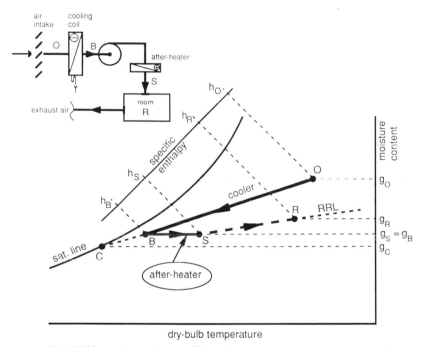

6.1 100% outdoor air conditioning system: summer operation

With the psychrometric process now complete, the cooling coil load is obtained:

$$Q_c = \dot{m}_a(h_O - h_B) \tag{6.1}$$

The cooling coil load, Qc, may be considered as being composed of the following part loads:
fresh air load:

$$Q_{fa} = \dot{m}_a(h_O - h_R) \tag{6.2}$$

room load (total heat gain):

$$Q_{rt} = \dot{m}_a(h_R - h_S) \tag{6.3}$$

after-heat load:

$$Q_{ah} = \dot{m}_a (h_S - h_B).$$ $\qquad$ (6.4)

Such a breakdown of loads illustrates that the refrigeration plant load is not necessarily the total heat gain-cooling load acting on the air conditioned space – it also includes loads due to the need to reheat the air and to cool and dehumidify the outdoor air. For an energy efficient design, the engineer should match plant loads as closely as possible to the building cooling load but this requirement does depend on the type of system chosen. In this case, once the room loads have been determined, the aim should be to minimize the ventilation and after-heater loads.

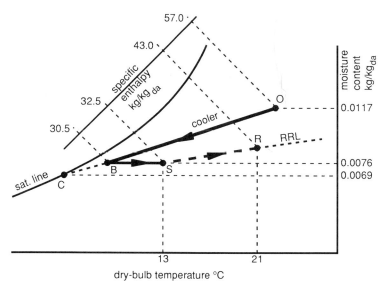

6.2 Psychrometric processes: Example 6.1

Example 6.1 A room, maintained at 21°C dry-bulb temperature and 55%sat, has heat gains of 10 kW sensible and 3 kW latent. For an air conditioning system using 100% outdoor air, determine the plant loads for the following design conditions:

minimum supply air temperature	13°C
outdoor design conditions	27°C dry-bulb
	20°C wet-bulb
contact factor	0.85
specific heat of humid air	1.02 kJ/kg$_{da}$.

Solution
Refer to Figure 6.2.

Supply air mass flow rate is given by Equation 5.1:

$$\dot{m}_a = \frac{10}{1.02(21-13)} = 1.23 \text{ kg/s.}$$

Room ratio line (from Equation 2.23):

$$y = \frac{10}{10+3} = 0.77.$$

On a psychrometric chart:

- plot room condition **R** and room ratio line;
- identify supply air condition **S** on the RRL and obtain moisture content, g_S;
- plot the outdoor condition **O** and obtain moisture content, g_O;
- using the coil contact factor, determine g_C from Equation 2.7:

$$\beta = \frac{g_O - g_S}{g_O - g_C}$$

$$0.85 = \frac{0.0117 - 0.0076}{0.0117 - g_c}$$

$$g_c = 0.0069 \text{ kg/kg}_{da}$$

- plot g_c on the saturation line to give condition **C**;
- draw the straight line **OC**; the intersection of this line with the moisture content line g_S gives the air condition **B** after the cooling coil
- join **B** to **S**.

The psychrometric process is now complete. The cooling coil process is the line **OB**; reading the enthalpies from the chart, the cooling coil load is given by Equation 6.1:

$$Q_c = \dot{m}_a (h_O - h_B)$$
$$= 1.23(57 - 30.5) = 32.6 \text{ kW.}$$

The cooling coil load may be sub-divided as follows:
ventilation load from Equation 6.2:

$$Q_v = \dot{m}_a (h_O - h_R)$$
$$= 1.23(57 - 43) = 17.2 \text{ kW}$$

room load from Equation 6.3:

$$Q_{rt} = \dot{m}_a (h_R - h_S)$$
$$= 1.23(43 - 32.5) = 12.9 \text{ kW.}$$

The after-heat process is the line BS and the load is given by Equation 6.4:

$$Q_{ah} = \dot{m}_a (h_S - h_B)$$
$$= 1.23(32.5 - 30.5) = 2.5 \text{ kW.}$$

Note Enthalpy values measured from the chart do not always produce an exact value compared with table values, hence the small discrepancy between the total room load and that calculated from Equation 6.3.

Winter operation

The operation at the winter design conditions is now considered, using the system already designed for summer operation. The air conditioned space now has a sensible heat loss, q_s' and a latent heat gain, q_l.

Since the air mass flow rate (determined for summer operation) is considered to remain constant, the supply air temperature is given by:

$$t_{Smax} = t_R + \frac{q_s'}{\dot{m}_a c_{pas}}.$$
(6.5)

Referring to Figure 6.3, a procedure similar to that for the summer operation is followed.

- on the psychrometric chart plot the room condition and the room ratio line;
- identify the supply air condition, **S**, on the RRL;
- plot the winter outdoor design condition, **O**.

The conditions on the chart now show that it is necessary to heat and humidify the outdoor air from condition **O** to condition **S**. Since there is already an after-heater in the system, a sensible heating process line may be drawn through **S**. The chart shows that the outdoor air now requires heating and humidifying from **O** to **B**. To achieve this the system can be designed in one of two ways:

- sensible heating followed by adiabatic humidification;
- sensible heating followed by steam humidification.

SYSTEM USING ADIABATIC HUMIDIFICATION

Continue with Figure 6.3 and assume that a sprayed cooling coil with a *humidifying* contact factor, β, is included in the system:

- draw the sensible heating process line through **O**;
- following a line of constant wet-bulb temperature, draw **ABC** such that $\beta = AB/AC$, where:

$$\beta = \frac{g_B - g_A}{g_C - g_A}$$

and all conditions are known except g_C.

With the psychrometric process now complete, the preheater load is given by:

$$Q_{ph} = \dot{m}_a(h_A - h_O)$$
(6.6)

The design after-heater load is given by:

$$Q_{ah} = \dot{m}_a c_{pas}(t_{Smax} - t_B)$$
(6.7)

In practice the temperature rise shown as after-heater load may include a small temperature rise from the fan and duct heat gains.

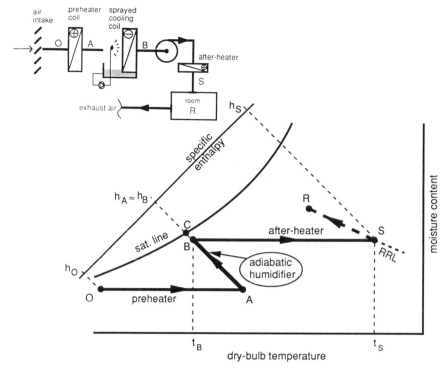

6.3 100% outdoor air conditioning system: winter operation

To summarize: the operation of the plant is as follows. The outdoor air is heated from condition **O** to condition **A** in order that the adiabatic humidification **AB** can take place. Air leaving the humidifier is then at the correct moisture content to maintain the room humidity, the after-heater raising the supply air temperature to **S** to meet the room sensible load requirements.

Example 6.2 The room in Example 6.1 is maintained in winter at 19°C dry-bulb temperature and 45% saturation when the design heat loss is 5 kW sensible and latent heat gain is 2 kW. With the plant now including a preheater and a sprayed cooling coil determine the plant loads for the following design conditions:

outdoor design condition	−4°C dry-bulb, 100%sat.
sprayed coil contact factor	0.75.

Solution

Refer to Figure 6.4.

From Example 6.1, the supply air mass flow rate $= 1.23$ kg/s.

Maximum supply air temperature is given by Equation 6.5:

$$t_{Smax} = t_R + \frac{q_s'}{\dot{m}_a c_{pas}}$$

$$= 19 + \frac{5}{1.02 \times 1.23} = 23°C.$$

Room ratio line:

$$y = \frac{5}{5+2} = 0.71.$$

On the psychrometric chart:

– plot room ratio line;
– identify the supply air condition **S** on the RRL and obtain moisture content, g_S;
– plot the outdoor condition **O** and obtain moisture content, g_O;
– coil contact factor:

$$\beta = \frac{g_S - g_O}{g_C - g_O}$$

$$0.75 = \frac{0.0056 - 0.0027}{g_c - 0.0027}$$

$$g_c = 0.0067 \text{ kg/kg}_{da}$$

– plot g_c on the saturation line to give condition **C**; this gives a wet-bulb temperature of 8°C;
– draw the wet-bulb temperature from **C**; the intersection of this line with the moisture content line g_S gives the air condition **B** after the sprayed cooling coil;
– join **B** to **S**; this gives the after-heater process line;
– draw the preheater process from **O** to cut the 8°C wet-bulb temperature line to give condition **A**.

The psychrometric process is now complete.

The preheater coil load is given by Equation 6.6:

$$Q_{ph} = \dot{m}_a (h_B - h_O) = 1.23 (24 - 2.8) = 26.1 \text{ kW}$$

The design after-heater load is given by Equation 6.7:

$$Q_{ah} = \quad \dot{m}_a c_{pas} (t_{Smax} - t_B) = 1.23 \times 1.02 (23 - 10.5)$$
$$= \quad 15.7 \text{ kW}.$$

Year-round operation of plant

To deal with changes in the outdoor air conditions and variations in room heat gains, it is necessary to regulate the output of the heating and cooling

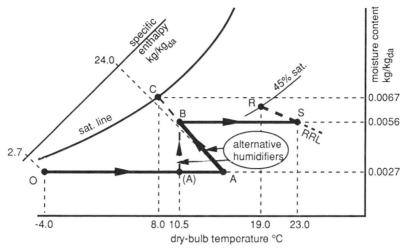

6.4 Psychrometric processes: Examples 6.2 and 6.3

coils. The completed system with controls, together with a summary of the operation on an outdoor condition envelope, is shown in Figure 6.5.

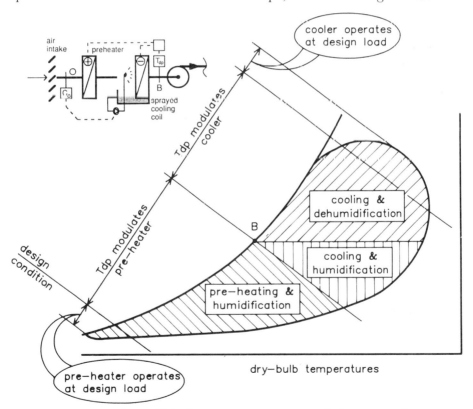

6.5 100% outdoor air conditioning system: summary of plant operation

The system of controls shown operates as follows. A room thermostat, T_R, controls the output of the after-heater to vary the supply air temperature in accordance with the sensible gains to the room. The dry-bulb thermostat, T_{dp}, controls the output of the cooling coil and preheater to maintain the supply air moisture content and hence room humidity. Since the air leaving the cooler is close to the saturation line (and the moisture content of this air condition is the supply air moisture content), this method of controlling humidity is known as *dew-point* control.

When this method of control is used, variations in room humidity can occur from the following:

- contact factor of the sprayed cooling coil;
- proportional band of dew-point controller;
- variations in room latent gain;
- mean air conditions in the duct not sensed by the thermostat T_{dp} under varying loads.

For most systems variations in humidity, caused by these sources of error, will be acceptable in that the room condition can be maintained within the indoor condition envelope. If close control of humidity is required then T_{dp} should be reset by a humidistat mounted either in the room or in the extract air duct.

The system will operate satisfactorily with the humidifier in operation all year-round since, even with water sprayed onto the cooling coil, dehumidification will occur. However, in order to save on pumping energy costs, the humidifier can be switched off by sensor C_o when humidification is not required. Referring to Figure 6.5, the ideal control is one in which this sensor measures either moisture content, vapour pressure or dew-point temperature of condition **B**. Since these ideal sensors are relatively expensive compared with the energy saved, a less satisfactory but cheaper method of switching off the pump with C_o as a dry-bulb thermostat could be considered though it is more usual to exclude it from the system.

SYSTEM USING STEAM HUMIDIFICATION

The second route to achieving the required supply air moisture content is by using a steam humidifier. Refer to Figure 6.6.

Assume that the steam humidifier is placed after the fan. The preheater must raise the air temperature so that the required amount of steam can be added below the saturation line. On a psychrometric chart:

- draw the sensible heating process line through **O**;
- draw the steam humidifier process line to give condition **B**, such that $g_B = g_S$.

With the psychrometric process now complete

The *preheater* load is given by:

$$Q_{ph} = \dot{m}_a c_{pas}(t_A - t_O). \tag{6.8}$$

The *steam humidifier* load is given by:

$$Q_{st} = \dot{m}_a(h_B - h_A). \tag{6.9}$$

The *steam supplied* is given by:

$$\dot{m}_s = \dot{m}_a(g_S - g_O). \tag{6.10}$$

The design after-heater load is given by Equation 6.7.

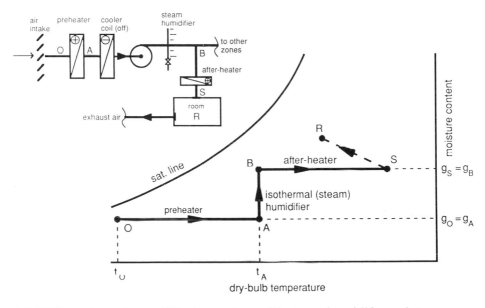

6.6 **100% outdoor air conditioning system with steam humidifier: winter operation**

To summarize the operation of the plant. The outdoor air is heated from condition **O** to condition **A** in order that the steam humidification **AB** can take place. Air leaving the humidifier is at the correct moisture content to maintain the room humidity, the after-heater raising the air to the supply temperature **S** to meet the room sensible load requirements.

Example 6.3 By using relevant data from Example 6.2, determine the plant loads using a steam humidifier in place of an adiabatic humidifier. The temperature after the preheater is to be 10.5°C.

Solution

Refer to Figure 6.4.

On the psychrometric chart the room ratio line, supply and outdoor conditions have been plotted, using data from Example 6.2.

Supply air mass flow rate $= 1.23$ kg/s

Moisture contents:

$$
\begin{aligned}
\text{outdoor air:} \quad & g_O = 0.0027 \text{ kg/kg}_{da} \\
\text{supply air:} \quad & g_S = 0.0056 \text{ kg/kg}_{da}.
\end{aligned}
$$

Preheat load is given by Equation 6.8:

$$
\begin{aligned}
Q_{ph} &= \dot{m}_a c_{pas}(t_A - t_O) \\
&= 1.23 \times 1.02\{10.5 - (-4)\} = 18.2 \text{ kW}.
\end{aligned}
$$

Steam humidifier load is given by Equation 6.9:

$$
\begin{aligned}
Q_{st} &= \dot{m}_a (h_B - h_A) \\
&= 1.23 \times (24.5 - 17.5) = 8.6 \text{ kW}.
\end{aligned}
$$

Steam supply is given by Equation 6.10:

$$
\begin{aligned}
\dot{m}_s &= \dot{m}_a (g_S - g_O) \\
&= 1.23(0.0056 - 0.0027) = 0.00357 \text{ kg/s}.
\end{aligned}
$$

Year-round operation of plant

As with the system using an adiabatic humidifier, it is necessary to regulate the output of the heaters, humidifier and cooling coil. The completed system with controls, together with a summary of its operation on an outdoor condition envelope, is shown in Figure 6.7.

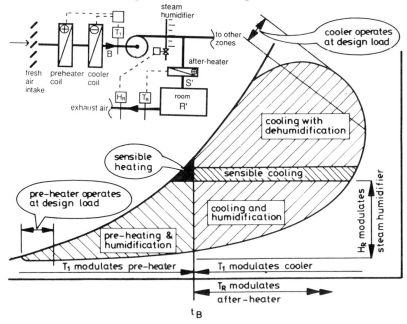

The system of controls operates as follows. The thermostat, T_R, controls the output of the after-heater to vary the supply air temperature in accordance with the sensible gains to the room. The dry-bulb thermostat, T_1, controls the output of preheater and cooling coil in sequence. This thermostat is now only *partially* dew-point control, the cooling coil dehumidifying when the dew-point of the outdoor air is above the controlled condition between the coil and the fan. For the remainder of the year, when the outdoor air dry-bulb temperature is higher than that of the controlled condition t_B, the coil only provides sensible cooling, while the humidistat regulates the output of the steam humidifier to maintain the room humidity.

SYSTEMS USING RECIRCULATED AIR

By recirculating air from the air conditioned space, within the system, the cooling coil load can be reduced and the preheater either omitted or considerably reduced in capacity, compared with using 100% outdoor air. This leads to savings in both capital and energy consumption costs.

The minimum quantity of outdoor air is determined by the fresh air requirements of the room.

SYSTEM USING A FIXED PERCENTAGE OF OUTDOOR AIR

Summer operation

Consider a fixed percentage of outdoor air, determined by the fresh air requirements. Using a design procedure similar to that described for the system using 100% outdoor air, the block diagram and the psychrometric process for summer operation is developed in Figure 6.8. (Note that the water sprays are required for winter operation – where there is no humidifying requirement, as in the tropics, these will be omitted.)

On the psychrometric chart:

- plot the room condition, **R**, and the room ratio line, RRL;
- identify the supply air condition, **S** on the RRL;
- plot the outdoor air design condition, **O**;
- join **O** and **R** and obtain the mixed air condition **M**;
- the cooling coil process line **MB** is drawn for the coil contact factor **MB/MC**;
- the process BS is an after-heater load (including fan and duct heat gains).

◄ 6.7 100% outdoor air conditioning system with steam humidifier: summary of plant operation

With the psychrometric process now complete, the cooling coil load is given by:

$$Q_c = \dot{m}_a(h_M - h_B). \tag{6.11}$$

As before, this load may be broken-down into *room*, *fresh air* and *after-heater* loads. The room load (total heat gain) and after-heat loads are the same as that for the 100% outdoor system, using Equations 6.3 and 6.4. The fresh air load is given by:

$$Q_{fa} = \dot{m}_a(h_M - h_R). \tag{6.12}$$

Therefore, compared with the system using 100% outdoor air, the cooling coil load is reduced by:

$$Q_{diff} = \dot{m}_a(h_O - h_M).$$

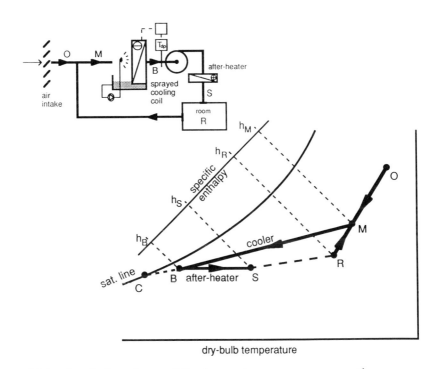

6.8 Recirculation air conditioning system: summer operation

Example 6.4 The air conditioning system in Example 6.1, with the same room heat gains and design conditions, is to be modified so that of the total supply air 75% is recirculated and 25% is fresh air. Calculate the new cooling coil plant load.

Solution
Refer to Figure 6.9.

From Example 6.1 the total air flow rate is 1.23 kg/s and room ratio line, $y = 0.77$.

Using the psychrometric chart:

- plot room condition **R** and the room ratio line. Obtain the room moisture content $g_R = 0.0087$ kg/kg_{da};
- identify the supply air condition **S** on the RRL; moisture content, $g_S = 0.0076$ kg/kg_{da};
- plot the outdoor condition **O** with a moisture content $g_O = 0.0117$ kg/kg_{da}
- join **O** and **R**.

Determine mixed air moisture content from air mixing ratio. Using Equation 2.2:

$$g_M = xg_O + (1-x)g_R$$
$$= 0.25 \times 0.0117 + (1-0.25)\,0.0087$$
$$= 0.00945 \text{ kg/}kg_{da}.$$

Coil contact factor is given by:

$$\beta = \frac{g_M - g_S}{g_M - g_C}$$

$$0.85 = \frac{0.00945 - 0.0076}{0.00945 - g_c}$$

$$g_c = 0.0073 \text{ kg/}kg_{da}.$$

- plot $g_c =$ on the saturation line to give condition **C**;
- draw the straight line MC; the intersection of this line with the moisture content line g_S gives the air condition **B** after the cooling coil.

The cooling coil load is given by Equation 6.11:

$$Q_c = \dot{m}_a(h_M - h_B)$$
$$= 1.23(46.7 - 30.2) = 20.3 \text{ kW}.$$

The ventilation load of the cooling coil is obtained from Equation 6.12:

$$Q_v = \dot{m}_a(h_M - h_R)$$
$$= 1.23(46.7 - 43) = 4.6 \text{ kW}.$$

Compared with the system using 100% outdoor air in Example 6.1, the cooling load has been reduced from 32.6 to 20.3 kW. This reduction has been achieved, in the main, by the reduction of the ventilation load from 17.2 kW to 4.6 kW.

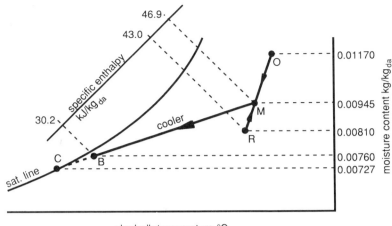

6.9 **Psychrometric processes: Example 6.4**

Winter operation

Consider first the operation at the winter design condition for a system using a relatively *small ratio* of outdoor to recirculated air.

Again, following the design procedure and referring to Figure 6.10:

- plot the room condition **R** and the room ratio line;
- identify the supply air condition, **S**, on the RRL;
- plot the winter outdoor design condition, **O**;
- join **O** and **R** to obtain the mixed air condition **M**;
- draw the after-heat process through **S**;
- the process through the sprayed cooling coil is **MB**.

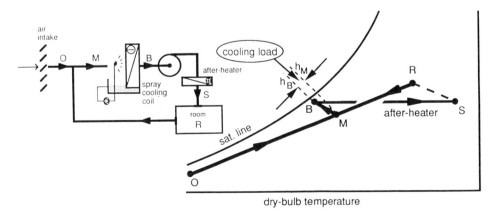

6.10 **Recirculation air conditioning system: winter operation**

The psychrometric cycle is now complete. This cycle shows the enthalpy of the mixed air condition as greater than that of the controlled off-coil condition **B**. Cooling is therefore required in winter, a system operation which is not efficient in energy use. To avoid the use of the refrigeration plant in winter, the air conditioning system may be modified to increase the percentage of outdoor to recirculated air by using modulating dampers, in a *free cooling* system, as described later.

The after-heater load is calculated from Equation 6.4.

Now consider the operation at the winter design condition for a system using a relatively *large ratio* of outdoor to recirculated air. The enthalpy of the mixed air condition would be lower than the controlled off-coil condition **B**. A preheater is therefore required, the load of which is given by:

$$Q_{ph} = \dot{m}_a(h_B - h_M). \tag{6.13}$$

An alternative to placing the preheater in the mixed air stream is to place it in the duct carrying the outdoor air. Though the load of the preheater is the same as that given by Equation 6.13 it may be more conveniently expressed by:

$$Q_{ph} = x\dot{m}_a(h - h_O) \tag{6.14}$$

where x = fraction of outdoor air in the *total* air supply, $\dot{m}_a$.

The advantages of placing the preheater in the fresh air duct are:

- smaller duct size;
- improved heat transfer between primary heat supply and outdoor air. This is particularly important if the preheater is a heat recovery unit using exhausted room air.

RECIRCULATION SYSTEM USING MODULATING DAMPERS – FREE COOLING SYSTEM

With the fixed percentage recirculation system in which a relatively small percentage of outdoor air is used, it was seen that in winter the cooling coil was required to operate at low outdoor enthalpies. A more efficient method would be to use the cold outdoor air instead of the refrigeration plant to achieve the required off-coil condition. Referring to Figure 6.11: by increasing the outdoor air and reducing the recirculated air flow rates, the mixed air condition is such that it coincides with the adiabatic process of the humidifer. To achieve the required changes of flow rates, modulating control dampers are included in the system.

Refer now to block diagram in Figure 6.11. Damper D_1 is fitted in the outdoor air duct and damper D_2 in the recirculation air duct. Damper D_3 is incorporated to ensure that the total flow rate in the system remains approximately constant. These dampers are adjusted by the dew point thermostat T_{dp} to maintain a near-constant, controlled air condition **B**

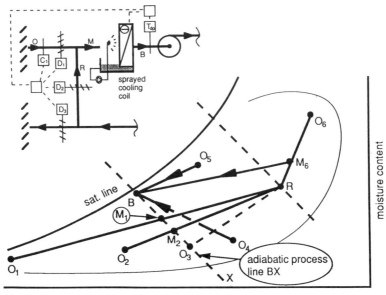

6.11 Psychrometric processes of year-round operation of free cooling system

between the cooling coil and the fan. To illustrate how the system works through the year, a number of processes are shown on Figure 6.11 using representative conditions within the outdoor air condition envelope, progressively increasing in enthalpy.

- *Outdoor air enthalpy* O_1 (which could be the winter design condition). The mixed air condition is M_1 such that it has the same wet-bulb temperature as **B**, taking this to be the adiabatic process through the humidifier, **BX**.
- *Outdoor air enthalpy* O_2. Damper D_1 will open to increase the outdoor air flow rate and damper D_2 will close to reduce the recirculated air flow rate so that the mixed air condition M_2 remains on the line **BX**. Damper D_3 will open to increase the exhaust air flow rate.
- *Outdoor air enthalpy* O_3. Damper D_1 will now be fully open and damper D_2 fully shut. The outdoor air is now the air condition on the line **BX**, with the system operating on 100% outdoor air.
- *Outdoor air enthalpy* O_4. The system will continue to use 100% outdoor air; the cooling coil now comes into operation, with the water sprays providing humidification.
- *Outdoor air enthalpy* O_5. The system continues to use 100% outdoor air with the cooling coil de-humidifying. The washer pump could now be switched off.
- *Outdoor air enthalpy* O_6. The system now operates on minimum outdoor air and maximum recirculation, the original summer design require-

ment. The dampers must therefore be changed over from 100% to minimum outdoor air operation.

The damper change-over is effected by having a sensor and controller C_1 located in the outdoor air duct. For energy efficient operation, C_1 should measure enthalpy with a set-point of the room condition; since the room condition may be subject to small variations (within the indoor condition envelope) another sensor C_2 in the recirculation air duct may be used to reset C_1. This method of damper change-over is known as an *enthalpy control system*.

The use of modulating dampers to vary the proportion of outdoor and recirculation is termed a *free cooling system*. This is because relatively cold outdoor air is used to provide cooling (according to the needs of the system), instead of using the refrigeration plant all year-round.

Year-round operation

The year-round operation of this system is summarized on Figure 6.12. Depending on outdoor conditions, the cooling coil will be required for

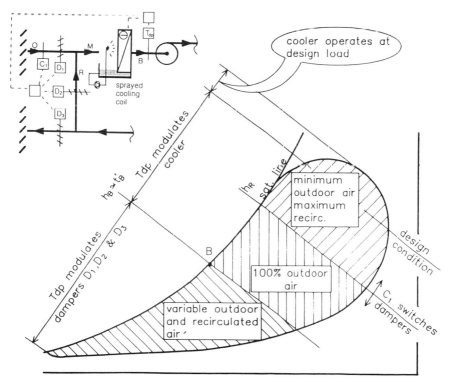

dry—bulb temperature

6.12 Free cooling system with adiabatic humidifier: summary of plant operation with C_1 as an enthalpy sensor

sensible cooling and dehumidification, the water sprays for humidification. The off-coil condition, **B**, is controlled by the dew-point thermostat T_{dp}. The coil and the control dampers D_1, D_2 and D_3 are operated sequentially by the dew-point thermostat T_{dp}. At outdoor air enthalpies below that corresponding to air condition **B**, the ratio of fresh to recirculated air flow rates is varied so that the mixed air condition, **M**, coincides with the adiabatic humidification process line **MB**. The controller C_1, set at the room air enthalpy, effects the damper change-over.

It was shown previously that a preheater is required when the minimum ventilation air flow rate is relatively large. For this recirculation system, modulating dampers can also be incorporated to provide free cooling. In this case the cooling coil, dampers and preheater are controlled in sequence, and the controller C_1, again effecting the damper change-over.

Dry-bulb controller, C_1

Usually an enthalpy signal is generated from a combination of measurements from dry-bulb and relative humidity sensors. At the present time such an enthalpy controller, though theoretically the most energy efficient for system operation, is expensive compared with a dry-bulb thermostat. Also, relative humidity sensors have been found to deteriorate in performance. Therefore, to reduce the capital cost and to obtain reliable performance, it is more usual to use the alternative of a dry-bulb thermostat as the controller at C_1. The operation of such a system is summarized in Figure 6.13.

Some of the features of this figure are as follows:

- As before the cooling coil and dampers are controlled in sequence by T_{dp}.
- The controller C_1 changes over the dampers at the dry-bulb temperature given by the line **GL**, (in this case, set at room temperature):
    on rise in outdoor temperature: dampers go to minimum outdoor air operation;
    on fall in outdoor temperature: dampers go to 100% outdoor air operation.
- The *ideal* changeover line is the enthalpy line **DRK**. The operation of the system in diagram areas **BDRLN** and **GRKP** is as required for efficient operation. Areas **DGR** and **KLR** represent the time when the system is operating *inefficiently*.
- The cooling coil design load is based on the enthalpy of the mixed air condition, shown by the line **EJ**. Therefore when the outdoor air is in the area **EGJ** the cooling coil load will be insufficient to maintain the controlled condition **B**. The area **EFHJ** therefore represents the additional frequency at which the coil operates above the design load.

- The set point of C_1 is shown as the room temperature; this may not be the most energy efficient setting. The objective is to minimize the additional energy consumption in the areas **DGR** and **KLR**. Energy calculation methods are given in Chapter 17.
- To avoid a conflict between the operation of T_{dp} modulating the dampers and C_1 switching them over, line BN should not cross GL. The set points can be obtained by examining their relative positions on an outdoor air design envelope. If there is a danger of *cross-over* one of the controls must be given priority.

An alternative to resetting the dampers with controller C_1 is to use a switch on the cooling coil control valve, set say at half valve stem travel.

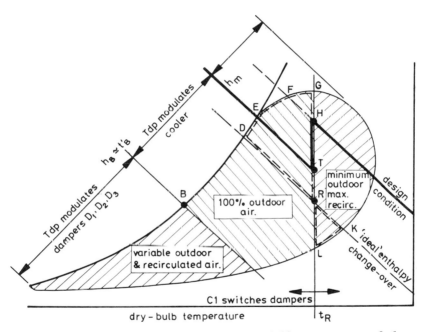

6.13 Free cooling system with adiabatic humidifier: summary of plant operation with C_1 as a dry-bulb thermostat

Recirculation System with no Humidifier

Many of the operating principles described above apply to recirculation systems which do not include an adiabatic humidifier.

A diagram of the central plant is shown in Figure 6.14.

In summer the psychrometric cycle of this system is identical to that shown in Figure 6.1, with the cooling load determined from Equation 6.11. During winter, when the cooling coil is switched off, the dampers are modulated by the thermostat T_1 so that the mixed air dry-bulb temperature is t_R. Since there is no humidifier in the system, the room humidity will vary with the latent heat gains. With no latent gains, the room moisture content will equal that of the outdoor air.

As before, the cooling coil and dampers are controlled in sequence by a dry-bulb thermostat, T_1. The system is only partially dew-point control, that is when the on-coil air dew-point temperature is higher than the coil ADP. During the remainder of the year the system is under temperature control.

The controller C_1 (using either an enthalpy or a dry-bulb temperature sensor) switches the dampers over as described previously; on this figure a dry-bulb thermostat is indicated. In this case, over-cooling and corresponding reheat loads will occur when the cooling coil is operating with the outdoor dry-bulb temperature above t_R. To improve the efficiency of the system the temperature at thermostat T_1 can be reset by an outdoor thermostat according to a schedule determined by an analysis of the room heat gains.

A steam humidifer can be incorporated into the system for full humidity control, similar to that described on page 118ff for the 100% outdoor system.

ZONES

Most air conditioning systems are required to deal with more than one room in a building and to ensure that design conditions are maintained throughout, the building will usually have to be divided into zones. Zoning can be dealt with in a number of ways, depending on the system. The number of zones will affect the amount of plant items, ducting, piping and controls. The more zones there are, the greater the capital costs of the installation will be, and this has to be set against increased efficiency and greater user satisfaction. It may be economical to use separate plants or air handling units to avoid excessively long runs of ducts but, even so, each of these plants may have to supply more than one room. The question arises as to which rooms should be supplied from the same zone heater or terminal unit such as dual duct mixing box. The decision will depend on the tolerance allowed on the optimum air condition and the effect of loads and ventilation requirements.

Heat gains and losses vary according to the time of day and year. If these heat loads are out of phase with one another, the variations between rooms might be such as to warrant separate treatment. Thus usually buildings with different aspects will require zones differentiated by the orientation since the solar heat gains on each of the building faces will be out of phase.

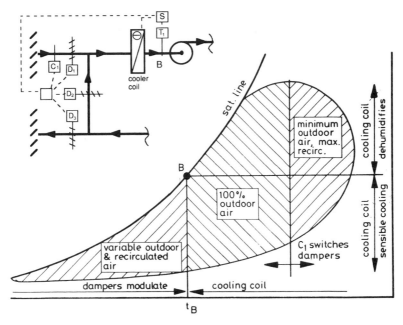

6.14 Free cooling system with no humidifier: summary of plant operation with C_1 as a dry-bulb thermostat

Effect of ventilation requirements

If the supply air rates are based on either the ventilation or air movement control requirements, the ratio of heat gain to air flow rate may differ. The difference between the ratios will indicate whether or not the spaces should be treated separately. This is illustrated by the following example:

Example 6.5 Compare the supply air temperatures for two rooms with design air volume flow rates and heat gains listed below. Both rooms are maintained at 21°C.

room	supply air flow rate $\dot{V}$ (m³/s)	heat gain q_s (kW)
1	1.5	12
2	1.1	4

Take the specific volume of the supply air as 0.82 m³/kg.

Solution

To obtain the supply air temperature use Equation 2.20

$$q_s = \dot{m}_a c_{pas}(t_R - t_S).$$

Room 1

$$12 = (1.5/0.82) \times 1.02(21 - t_S)$$
$$\therefore t_S = 14.6°C.$$

Room 2

$$4 = (1.1/0.82) \times 1.02(21 - t_S)$$
$$\therefore t_S = 18.1°C.$$

With one heater in the supply air to both rooms, the room temperature difference would be 3.5 K. If this difference were considered unacceptable then separate zone heaters would be required.

When more than one room is air conditioned from a central plant the relative humidity in each of the spaces will be only an average condition. If more than one room requires *close* control of humidity either separate plants are required or each room can be provided with its own (zone) steam humidifier.

SYMBOLS

c_{pas}	specific heat of humid air	q_l	latent heat gain to air conditioned space
g	moisture content		
h	specific enthalpy of humid air	t	dry-bulb temperature
		$\dot{V}$	air volume flow rate
$\dot{m}_a$	mass flow rate of dry air	x	fraction of total supply air
$\dot{m}_s$	mass flow rate of steam	y	ratio of room sensible to total heat gain
Q_c	cooling coil load		
Q_{ah}	after-heater load, part load of cooling coil	β	contact factor
Q_{fa}	fresh air load, part load of cooling coil	**Subscripts** (for temperature, moisture content and enthalpy)	
Q_{ph}	pre-heater load	C	air condition on saturation line
Q_{rt}	room total load, part load of cooling coil	M	mixed air condition
Q_{st}	steam humidifier load	O	outdoor air condition
Q_h	heating coil load	R	room condition
q_s	sensible heat gain to air conditioned space	S	supply air condition
q_s'	sensible heat loss to air conditioned space	**Abbreviation**	
		RRL	room ratio line

7 Systems II

In the previous chapter, constant flow rate, zonal reheat systems were described; these are systems in their own right and therefore used for many applications. Certain of their operating characteristics are applicable to the more sophisticated systems used in medium to large buildings such as office blocks, hotels and hospitals.

The most common air conditioning systems are classified within the following categories:

- all air;
- air and water;
- unitary heat pumps;
- self-contained units.

The design and selection of a system should be based on a detailed analysis of the heat gains and losses together with the establishment of flow diagrams and heat balances. For certain periods of the year, different parts of a building may require heating and cooling simultaneously; for overall system efficiency it is desirable to take account of load diversity patterns. The final selection of a system will depend on many factors including an analysis of both the capital and energy costs of the system.

ALL AIR SYSTEMS

VARIABLE AIR VOLUME

Variable air volume (VAV) systems have become increasingly popular in recent years because, compared with constant flow rate/all air systems, they have smaller central plant sizes and a potential for energy saving. The reasons for this are that the total air supplied by the fan deals with the whole building *instantaneous* heat gain, flow rates are reduced in sympathy with the heat gains and there is no reheating of the air as in zonal reheat systems.

The simplest system is for cooling only, suitable for use in:

- internal areas of deep plan buildings, when no heating is required in winter;
- for perimeter zones when cooling is being added to an existing building in which there is an independent heating circuit.

Possible disadvantages may arise from:

– lack of humidity control;
– insufficient supply of fresh air at low flow rates;
– unsatisfactory air distribution at low flow rates.

Air supply flow rates for this system, together with aspects of air diffusion, have been discussed in Chapter 5. Fan selection and fan control for VAV systems are described in Chapter 14.

System arrangement and operation

A diagram of a typical VAV system is shown in Figure 7.1. The central plant consists of:

– control dampers;   – supply fan;
– preheater;   – terminal units.
– cooling coil;

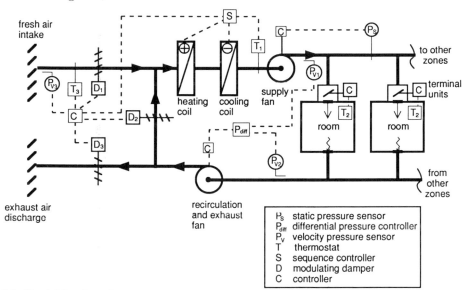

7.1 Variable air volume system

Central plant operation will be similar to that described on page 129f, with the cooling coil, control dampers and preheater controlled in sequence by the thermostat T_1. Damper change-over is from the controller, T_3, sensing the outdoor air condition.

When required to maintain the minimum ventilation rate, controller P_{v3} operates the dampers, if necessary over-riding thermostat T_1. P_{v3} may be a velocity, static pressure or flow rate sensor.

A preheater is required for initial start-up and for normal operation when the minimum ventilation requirements determine that the system operates with relatively large flows of outdoor air.

The flow rates from the terminal units (TU) are controlled through the room thermostats, T_2. It is usual for the TUs to operate with a (minimum) duct static pressure; as they throttle down to deal with reduced room cooling loads, the pressure in the supply air duct will rise. Rather than allow the TUs to operate against this increase in pressure, and at the same time to encourage a reduction in fan operating power consumption, the supply fan flow rate is adjusted, through the static pressure sensor P_s, to maintain the duct base pressure.

The duty of the recirculation/extract fan is matched to the supply fan to maintain a balanced supply and extract within the building. A differential pressure controller, P_{diff}, varies the extract fan flow rate by comparing relevant air velocities (dependent on the duct sizes) with sensors, P_{v1} and P_{v2} in the main supply and extract ducts.

Terminal units

The principle of varying the supply air volume flow rate to deal with changes of heat load was described on page 102. There are various types of terminal units for VAV systems and two of the most common are illustrated in Figure 7.2. In (a) the air flow rate is throttled by varying the outlet area; the device shown here includes a reheater to deal with room heat gains and heat losses when the terminal is operating at the minimum turn-down ratio. In (b) the flow rate is reduced by inflating a bellows device within the unit; the outlet velocity varies as the flow rate changes and the *throw* across the ceiling is maintained by the *coanda* effect of the air as it leaves the aerodynamically designed diffuser. The terminal units will include an acoustic lining to reduce air noise to acceptable levels.

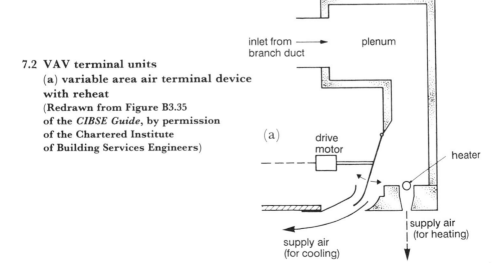

7.2 VAV terminal units
(a) variable area air terminal device
with reheat
(Redrawn from Figure B3.35
of the *CIBSE Guide*, by permission
of the Chartered Institute
of Building Services Engineers)

inlet from ⟶ branch duct

plenum

(a)

drive
motor

heater

supply air
(for heating)

supply air
(for cooling)

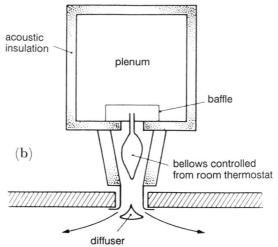

7.2 VAV terminal units

(b) **throttling air terminal device (variable velocity)**
(Redrawn from Figure B3.35 of the *CIBSE Guide*, by permission of the Chartered
Institute of Building Services Engineers)

Psychrometrics

Summer operation
The psychrometric cycle for summer operation is shown in Figure 7.3, for

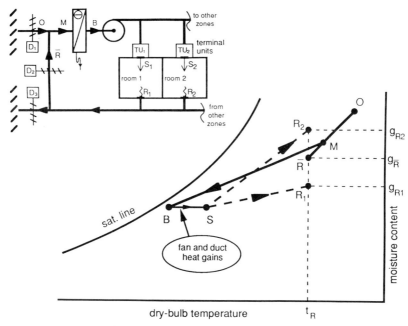

7.3 Psychrometric process for VAV system: summer operation

two representative rooms. The temperature at **B** is controlled by thermostat T_1. Fan and duct heat gains produce the temperature rise **BS** to give the single supply air condition, **S**. Room 1 has relatively small sensible-to-latent heat gains with room ratio line SR_1, whereas Room 2 has relatively large sensible-to-latent heat gains with room ratio line SR_2. From this it is seen that humidity will vary between the individual rooms. The return air, R, is the average value of all room conditions.

The design cooling coil load is given by Equation 6.11; in this equation the supply air mass is determined by analysing the maximum instantaneous heat gain as described on page 101.

Room humidity will also vary within a single space; such variations are demonstrated by the following example.

Example 7.1 In a VAV system, the supply air is maintained at a dry-bulb temperature of 13°C and a moisture content of $0.008 \, kg/kg_{da}$. For the following design criteria, determine the rise in room relative humidity when the sensible heat gain in one of the air conditioned rooms falls to 10 kW, the latent gain remaining constant.

Design heat gains:

sensible	40 kW
latent	5 kW.

Room-to-supply air temperature differential, Δt_c 8 K

Terminal unit turn down ratio	1:4
Humid specific heat, c_{pas}	1.02 kJ/kgK
Latent heat of evaporation, h_{fg}	2450 kJ/kg.

Solution
Refer to Figure 7.4
    Design mass flow rate is given by Equation 5.1:

$$\dot{m}_a = \frac{q_s}{c_{pas}\Delta t_c}$$

$$= \frac{40}{1.02 \times 8} = 4.9 \, kg/s.$$

At design load
The rise in moisture content from supply to room air is given by Equation 2.21:

$$q_l = \dot{m}_a h_{fg}(g_R - g_S)$$

$$\Delta g = (g_R - g_S) = \frac{5}{4.9 \times 2450} = 0.00042 \, kg/kg_{da}.$$

The room moisture content $= g_S + \Delta g$
$$= 0.008 + 0.00042 = 0.00842 \text{ kg/kg}_{da}.$$

At a room temperature of 21°C, this moisture content gives a relative humidity of 54%.

At reduced room load
Since the ratio of reduced room load to design load is 1:4, the terminal unit(s) will also operate at the turn-down ratio of 1:4. The latent heat gain remains constant and therefore the rise in moisture content, $\Delta g'$, will be:

$$\Delta g' = 4 \times 0.00042 = 0.00168 \text{ kg/kg}_{da}.$$

$\therefore$ the room moisture content $= 0.008 + \Delta g'$

$$g_R = 0.008 + 0.00168 = 0.00968 \text{ kg/kg}_{da}.$$

At a room temperature of 21°C this gives a relative humidity of 62%, an increase of 8% above the design load condition.

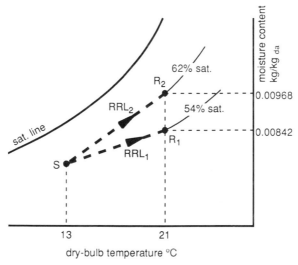

7.4 Psychrometric process for VAV system: Example 7.1

Winter operation
A psychrometric cycle for winter operation is shown in Figure 7.5; the terminal units incorporate heaters for winter heating. The cooling coil is switched off and the temperature at **B** is the mixed air condition, controlled by thermostat T_1 through the modulating dampers. Both rooms have similar sensible heat gains but Room 1 has a relatively large latent heat gain to give room ratio line SR_1; Room 2 has a negligible latent heat gain to give room ratio line SR_2. As with summer operation it is seen that humidity will vary in the different rooms and the return air is the average value of all room conditions.

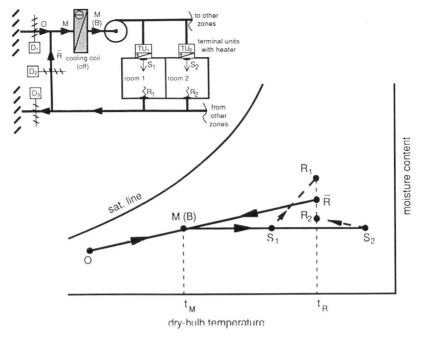

7.5 Psychrometric process for VAV system: winter operation

Ventilation rate

Special consideration should be paid to the supply rate of fresh air if ventilation standards are to be maintained. It has already been shown in Chapter 6 that for most of the year a system designed for free cooling will operate on considerably more than the minimum percentage of outdoor air. However, as the total instantaneous heat gain reduces, the fresh air supply rate will also fall in proportion to the reduction in total supply flow rate. Then, if the air conditioned spaces remain fully occupied, the minimum ventilation requirements will not be met. One solution to this problem is to maintain, through the controller P_1 acting on the dampers (as described above), the minimum flow rate at the fresh air intake. But even if this facility is incorporated into the system, ventilation rates may still be inadequate as is illustrated in the following example:

Example 7.2 Four zones in a building, air conditioned with a VAV system, have design flow rates as given in the table overleaf. The total flow of 15 m³/s is based on an analysis of the total instantaneous cooling load on the building.

The zone ventilation rates are based on maximum occupancy; the total of these determines the minimum flow rate in the central plant of 1.9 m³/s but the designer arbitrarily increases this to 2.25 m³/s to provide 15% of the total flow rate. Compare the ventilation rates of the four zones:

(a) at a ventilation rate of 15% of total supply flow rate
(b) at a ventilation rate of 2.25 m³/s.

system condition	zone				total flow rate
	1	2	3	4	
at design loads:					
maximum supply flow rate	5.50	4.00	6.00	4.00	15.00
minimum ventilation rate	0.55	0.40	0.60	0.40	2.25
at part loads:					
reduced supply flow rate	4.00	2.00	4.00	1.20	11.20
ventilation rate - 15%	0.60	0.30	0.60	0.18	1.68
ventilation rate - 20%	0.80	0.40	0.80	0.24	2.24

Solution
With the system in use the zone cooling loads are out of phase and the zone flow rates are reduced as shown in the table, giving a total supply flow rate of 11.2 m³/s.

(a) *Ventilation flow rate as 15% of total*
All zone supply rates include 15% fresh air. If the zones have the expected occupancy, then the ventilation air for zones 2 and 4 fall below the required level.

(b) *Ventilation flow rate maintained at design flow rate of 2.25 m³/s*
Percentage of total flow rate is now $(2.25/11.2)100 = 20\%$. All zone supply rates include 20% fresh air. If the zones have design occupancy, then the ventilation air for zone 2 is now satisfactory but zone 4 still falls below the required level.

DUAL DUCT SYSTEM

The dual duct system is used where:

– relatively large ventilation/air change rates are required;
– automatic control of individual spaces or varieties of room loads which preclude zoning;
– piped (wet) services cannot be used;
– control of humidity is not required.

Plant arrangement
A diagram of a typical dual duct system is shown in Figure 7.6. The central plant supplies hot and cold air in two separate distribution systems to mixing boxes which serve individual rooms and zones. The major plant items include:

– control dampers;
– supply fan;
– cooling coil in cold duct;
– heating coil in hot duct.

The cooling coil and control dampers are controlled in sequence to maintain the cold duct temperature. The modulating dampers provide *free-cooling*. The heating coil maintains the hot duct temperature.

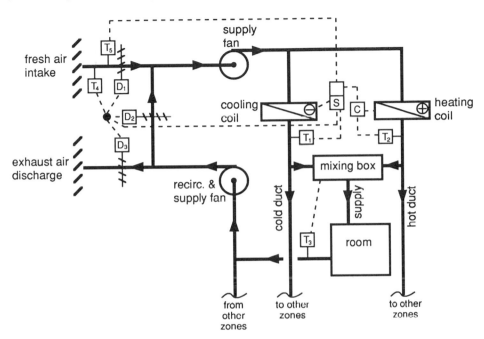

7.6 Dual-duct system: schematic

Mixing boxes

One type of mixing box is shown in Figure 7.7. The mixing damper varies the proportion of cold to hot air while the total supply air is maintained at a constant rate by a system powered, spring-loaded regulator which closes/opens as the pressure at the total pressure sensor rises/falls. All terminal devices include an acoustic attenuating lining.

In recent years, to reduce energy consumption, mixing boxes have been used as variable volume units, particularly for internal zones in buildings where there is no winter heat loss. In these cases the room thermostat reduces the cold duct flow rate to maximum turn down before the hot duct valve begins to operate and bleed-in a proportion of warm air.

Mixing boxes are designed for specific applications, eg mounted under window sill or positioned in the ceiling void; one box can be used to supply a number of outlet grilles or diffusers. These are illustrated in Figure 7.8.

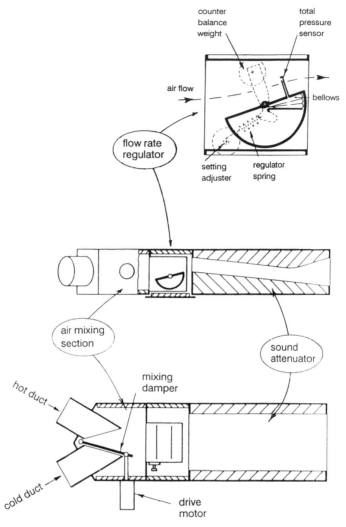

counter
balance
weight

total
pressure
sensor

air flow

bellows

flow rate
regulator

setting
adjuster

regulator
spring

air mixing
section

sound
attenuator

mixing
damper

hot duct

cold duct

drive
motor

7.7 Dual-duct mixing box
(Based on illustrations supplied by courtesy of Trox Brothers Ltd)

Final outlet grilles and diffusers are designed and selected according to the principles of room air diffusion described in Chapter 5.

Scheduling

Mixing *losses* (that is an over-supply of cold/hot air) in the terminal units at off-peak conditions, lead to system inefficiencies. One solution to reducing these losses is to schedule the cold and hot duct air temperatures based on the load and supply air temperature diagrams (see Chapter 17).

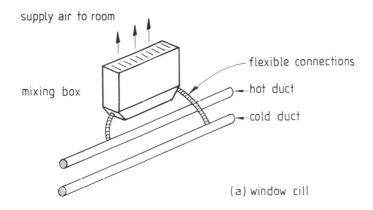

supply air to room

mixing box

flexible connections

hot duct

cold duct

(a) window cill

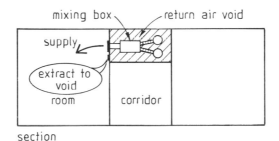

mixing box

return air void

supply

extract to void

room

corridor

section

(b) central corridor false ceiling

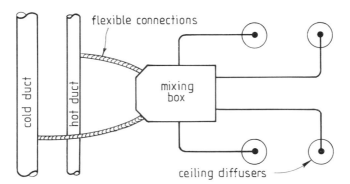

flexible connections

cold duct

hot duct

mixing box

ceiling diffusers

(c) supply to a group of ceiling outlets – plan view

7.8 Alternative locations of mixing boxes

Referring to Figure 7.9 the *cold duct temperature schedule* **LNPQTU** is arrived at as follows. Line **UQB** is the theoretical cold duct temperature to meet the maximum heat gains. Section **BN** of this line cannot be met by the plant so that, instead, the mixed air condition M is used with maximum recirculation (minimum outdoor air) – this is line **LN**. From **N** to **P**, outdoor and recirculated air are varied by modulating the control dampers, bringing the mixed air condition M onto the cold duct temperature line. At **P** the cooler begins to operate, first with the plant working with 100% outdoor air up to **T** and above **T** on maximum recirculation to give the schedule line **PQTU**.

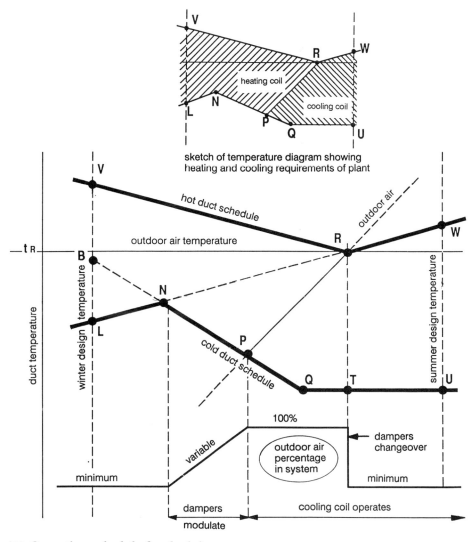

7.9 Operating schedule for dual-duct system

The *hot duct temperature schedule*, **VRW**, is arranged to meet the building fabric transmission losses. To achieve this, air is heated from either the mixed air or the 100% outdoor air, conditions determined from the cold duct schedule. Mixed air is used at outdoor temperatures above the room temperature.

A sketch of the temperature diagram in Figure 7.9 shows the heating and cooling requirements of the plant.

Where appropriate, the cold duct schedule can be modified to take account of:

– fan and duct heat gains;
– the need to prevent excessive room humidities.

To increase operating efficiency further, the hot duct schedule can be depressed to take account of:

– high solar heat gains;
– internal heat gains.

Compared with other systems, the dual-duct system requires more space to accommodate ductwork. To reduce these space requirements, the hot duct may be *undersized*, with reduced air flow rates. If this is done it is necessary to elevate the hot duct temperature schedule to meet the heating loads on the space. The size of the cold duct is based on the maximum (design) *instantaneous* cooling load.

Control

Referring to Figure 7.6, the thermostat T_1, reset by thermostat T_5, maintains the scheduled cold duct temperature by controlling the cooler and modulating dampers in sequence. The control dampers D_1, D_2 and D_3 are switched by T_4 over from 100% outdoor air at the appropriate point in the schedule. Thermostat T_2, also reset by thermostat T_5, maintains the scheduled hot duct temperature by controlling the heater. Thermostat T_3 maintains space temperature by varying the proportions of hot and cold air at the mixing box.

Psychrometric cycles

Summer operation

A room at design heat gain takes all its air from the cold duct and the supply air is at the minimum supply temperature. As the heat gain reduces from the maximum the air from the hot duct will be introduced while the air flow rate from the cold duct reduces, thus raising the supply air temperature.

A psychrometric cycle for summer operation is shown in Figure 7.10, with two representative rooms. Outdoor air **O** mixes with the average return air (room) condition **R** to give the mixed air condition **M**. The heater is not in operation and **M** therefore becomes the hot duct temperature **H**. The temperature of the cold duct at **B** is controlled by the thermostat T_1, to give the cooling coil process **MB**. A temperature rise, due to duct heat gains, gives cold duct temperature **C**. Then air **C** mixes with air **M** to produce various supply air conditions from the mixing boxes; for room 1 with a relatively *large* sensible heat gain, the supply air condition S_1 is close to **C**, with room ratio line S_1R_1. For room 2 with a *small* sensible heat gain, the supply air condition S_2 is closer to **M**, giving the room ratio line S_2R_2.

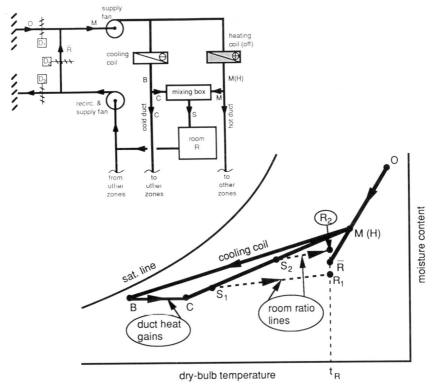

7.10 Psychrometric process for dual-duct system: summer operation

The design cooling coil load is given by Equation 6.11; in this equation the supply air mass is based on the maximum *instantaneous* heat gains.

Winter operation
A typical cycle for winter operation is shown in Figure 7.11. The cooling coil is switched off and the cold duct temperature at **B** is the mixed air condition **M**, maintained by thermostat T_1 through the sequence controller modulating the dampers. The heater is now operating, with the hot duct temperature **H** set by thermostat T_2. Air **H** mixes with air **M** to produce the

various supply conditions at the mixing boxes. Room 1 has a sensible heat loss, a supply air condition S_1 and a room ratio line S_1R_1. Room 2 has a sensible heat gain, a supply condition S_2 and room ratio line S_2R_2.

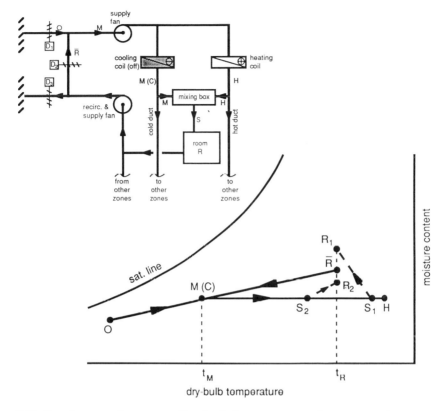

7.11 Psychrometric process for dual-duct system: winter operation

From these examples of summer and winter operation, the room humidity will vary for both cycles. The variations depend on the room latent heat gains, the moisture content in the outdoor air and, in summer, on the relative position of the RRL on the chart.

AIR AND WATER SYSTEMS

Where high air change rates are not required for ventilation and air movement control, air and water systems may be used. Compared with air, the heat carrying capacity of water is very high. Therefore, to reduce the space taken up in the building by the fluid flow networks, heated and cooled water is generated centrally and piped through the building to room terminal units which treat air locally. Fresh air for ventilation may be supplied from a central system or brought in at the unit through the external wall.

Terminal units are usually mounted under the windows. There can be considerable flexibility in the overall designs and the way the units are arranged within the building.

The most common air and water systems are:

- induction unit;
- fan coil.

INDUCTION UNIT SYSTEM

The principle of the induction unit is as follows. A *primary* air supply is generated in a central plant room; this air is sufficient to meet the ventilation requirements of the treated space and is supplied at a relatively high pressure to a header duct in the terminal unit. Here, the primary air, P, is discharged through a set of *nozzles* and this induces *secondary*, (recirculated) air, R, from the room into the unit and over the *secondary coil(s)*. The secondary air is then cooled (and/or heated, depending on the design) to R′, mixed with the primary air to give the supply air condition, S, from the unit to the room.

The induction ratio, I_r, is the ratio of the induced volume air flow rate to the primary air, ie:

$$I_r = \frac{\dot{V}_r}{\dot{V}_p}. \tag{7.1}$$

Without significant error, the induction ratio can also be expressed in terms of the mass flow rates, ie:

$$I_r = \frac{\dot{m}_r}{\dot{m}_p}. \tag{7.2}$$

Induction units have a typical inlet pressure to the nozzle of 250 Pa; dampers are provided to regulate the primary air supply and nozzle pressure. Room temperatures are maintained by a thermostat operating two-port control valves; alternatively a damper may be used to divert room air around the secondary coil. A simple filter is required at the intake to the coil.

There are a number of variations of the induction unit system and these include:

- two-pipe, non-changeover;
- two-pipe, changeover;
- four-pipe, non-changeover.

To illustrate the operating principles, a two-pipe, non-changeover is described below. This system is used in temperate climates where the winter heating load is not excessive. If the climate is such that the winter heat loss is excessive, then either a two-pipe changeover system or a 4-pipe changeover system using two coil units should be considered.

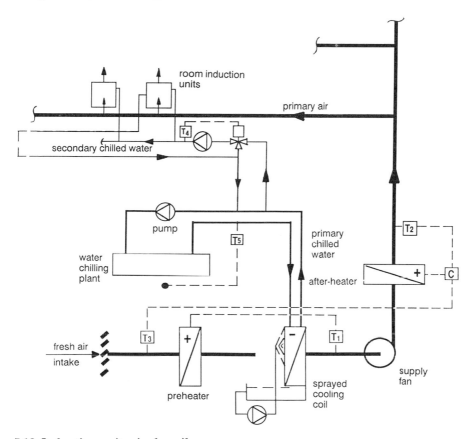

7.12 **Induction unit, single coil**

Two-pipe, non-changeover induction system

A typical arrangement of a two-pipe, non-changeover induction system is shown in Figure 7.13. The secondary chilled water is supplied to the unit coils all the year round. In summer the cooled and dehumidified primary air gives a measure of control of room humidity – these systems are unable to deal with large latent heat gains. In winter the primary air is heated and humidified. The primary air supply temperature takes care of the maximum heating requirements, but since this supplies only 15% to 20% of the total flow rate from the unit, the maximum supply air temperature will need to be considerably higher than for a constant flow, all air system.

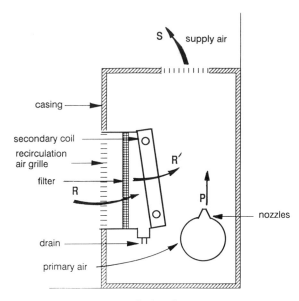

7.13 Non-changeover induction system

For efficient operation a temperature schedule (based on heat analysis of the building heating and cooling loads) is determined for the primary air supply, as shown in Figure 7.14. There is some inefficiency since the primary air is heated to deal with the maximum heat loss and the secondary coil cools the induced air to meet the miscellaneous room sensible heat gains. For more efficient operation, the schedule can be depressed to take account of internal and solar heat gains.

To prevent condensation on the secondary coils the water must be above the room air dew point temperature (the drain pan on the coil caters for condensation of water vapour on start up). To ensure this, the secondary coil water is taken off the piping circuit after the primary cooling coil in the fresh air plant.

Control

The primary air moisture content is determined by the air condition leaving the cooling coil. In summer this is controlled through thermostat T_1. The primary air is heated to its scheduled temperature, with outdoor air thermostat, T_3, which regulates the output of the after-heater. The output of each room unit is usually regulated through a room thermostat acting on a two-port valve.

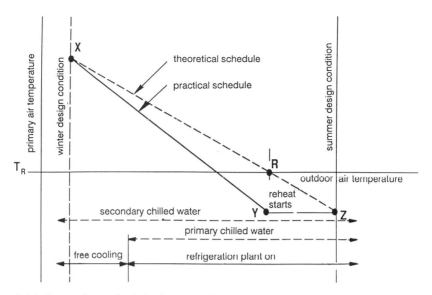

7.14 Operating schedule for non-changeover induction unit system

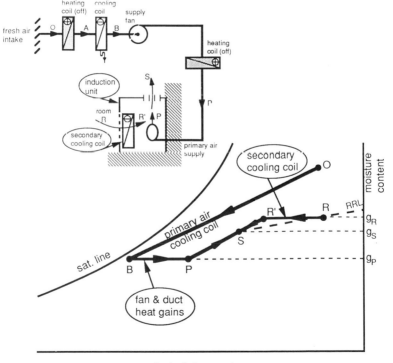

7.15 Psychrometric process for non-changeover induction system: summer operation

Psychrometric cycles

Summer operation

A psychrometric cycle for summer operation is shown in Figure 7.15. Outdoor air **O** is cooled and dehumidified to give the coil process **OB**. A temperature rise from fan and duct heat gains gives primary air temperature **P**. At the induction unit the primary air induces the room air **R** through the secondary coil to cool the air to **R'**. This cooled secondary air then mixes with the primary air **P** to produce the supply air condition, such as **S**, from the terminal unit with a typical room ratio line, **SR**. Some sensible cooling is available from the primary air.

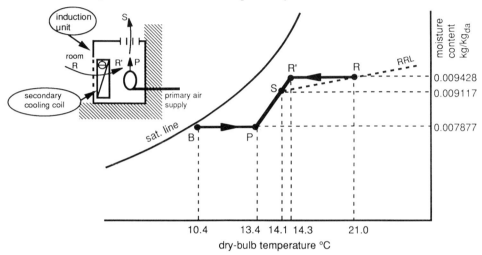

7.16 Psychrometric process: Example 7.3

Example 7.3 An induction unit has an induction ratio of 4:1. For the design conditions given below, determine:

(a) primary air moisture content;
(b) secondary coil load;
(c) temperature of the air off the secondary coil

Data

Room heat gains:		
	sensible	7.5 kW
	latent	0.8 kW
Room condition		21°C, 60%sat
Minimum supply air temperature		14°C
Primary air cooling coil contact factor		1.0
Temperature rise in primary air due to fan and duct heat gains		3 K
Specific heat of humid air		1.02 kJ/kg$_{da}$K.

Solution
Table values of moisture contents are used in the calculations. The results are shown in Figure 7.16.

The supply air mass flow is obtained from Equation 5.1:

$$\dot{m}_a = \frac{7.5}{1.02(21-14)} = 1.05 \text{ kg/s.}$$

Consider latent heat gain; using Equation 2.21:

$$q_l = \dot{m}_a h_{fg}(g_R - g_S).$$

From tables, room moisture content $= 0.009428$ kg/kg$_{da}$

$$0.8 = 1.05 \times 2450 \, (0.009428 - g_S)$$
$$\therefore g_S = 0.009117 \text{ kg/kg}_{da}.$$

Induction ratio, $I_r = \dfrac{4}{1} = \dfrac{g_S - g_P}{g_R - g_S}$

$$4 = \frac{0.009117 - g_P}{0.009428 - 0.009117}$$
$$\therefore g_P = 0.007877 \text{ kg/kg}_{da}.$$

For a primary air cooling coil contact factor of unity, the temperature of the air off the coil $= 10.4°C$.
Primary air supply to the unit $= 10.4 + 3.0 = 13.4°C$.
Primary air mass supply rate $= 0.2 \times 1.05 = 0.21$ kg/s.
Cooling provided by the primary air at the induction unit is given by:

$$Q_p = 0.21 \times 1.02(21 - 13.4) = 1.63 \text{ kW}$$

$$\therefore \text{ secondary coil cooling load} = 7.5 - 1.63 = 5.87 \text{ kW.}$$

Induced air mass supply rate $= 1.05 - 0.21 = 0.84$ kg/s

$\therefore$ temperature of the air off the secondary is given by:

$$5.87 = 0.84 \times 1.02(21 - t_R')$$
$$\therefore t_R' = 14.1°C.$$

The design primary air cooling coil load is given by Equation 6.1. The *total* design load for the secondary coils is based on maximum *instantaneous* cooling load of all units.

Winter operation

A typical cycle for winter operation is shown in Figure 7.17. The primary air is now heated and humidified to condition **B**. The re-heater is operating to give the scheduled primary air temperature **P**. At the induction unit, the primary air induces the room air **R** through the secondary coil to cool the air to **R'**; **R'** mixes with the primary air **P** to produce the supply air condition, **S**, and room ratio line **SR**. The secondary coil cooling load will depend on the level of miscellaneous heat gains offsetting the maximum heat loss; the line **PSR'** varies for each individual room; at maximum heat loss the mixing line to provide the supply condition becomes the line **PR**.

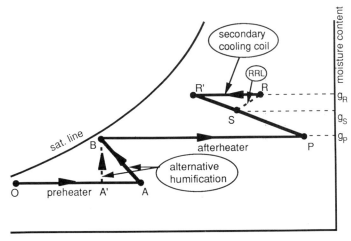

7.17 **Psychrometric process for non-changeover induction system: winter operation**

FAN COIL SYSTEM

The fan coil system is similar to the induction system with changeover, non-changeover and four-pipe systems being used. The room terminal units comprise of a fan, secondary coil and filter (Figure 7.18). Room temperature control is achieved by a thermostat either switching the fan or by a two-port control valve on the secondary coil.

Ventilation air can be drawn through a grille in the external wall as in Figure 7.19. With this arrangement condensation will occur on the secondary coils, for which a drain pipe must be provided. There may also be problems from dirt and noise penetration from the external environment.

A central air conditioning plant for fresh air supply is a more satisfactory solution for ventilation requirements. This will allow more flexibility in positioning the units and at the same time, as with the induction system, give some control over room humidity. The fresh air can be supplied from an outlet which is independent of the fan coil unit, as indicated in Figure 7.20. Though the supply air streams are discharged from separate points in the room, a single psychrometric cycle may be drawn as for the induction unit system in Figures 7.15 and 7.17.

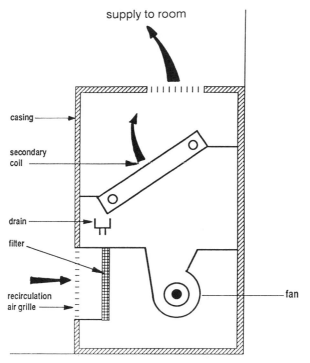

supply to room

casing

secondary
coil

drain

filter

recirculation
air grille

fan

7.18 Fan coil unit

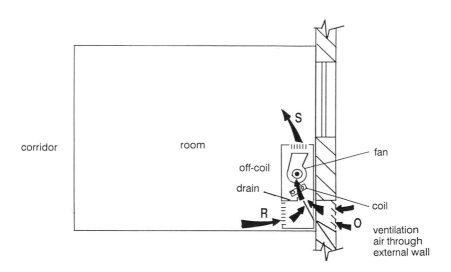

corridor

room

S

off-coil

drain

R

fan

coil

O

ventilation
air through
external wall

7.19 Fan coil unit with ventilation air supply from external wall grille

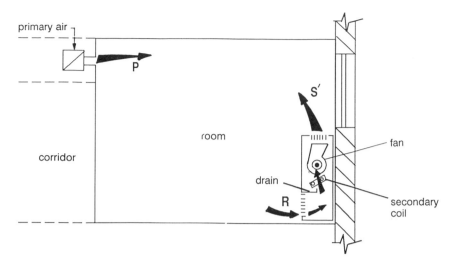

7.20 Fan coil unit with separate ventilation air supply

UNITARY HEAT PUMP SYSTEM

The self-contained room units are reversible heat pumps designed for both cooling and heating. Each unit comprises a small refrigeration unit, a refrigerant/heat exchanger, a refrigerant/air coil and a fan. The heat exchangers in each unit are connected to a water circuit, operating on a constant temperature, as shown in Figure 7.21. This circuit includes a boiler and a cooling tower.

When the unit is required for *cooling* the air coil acts as the evaporator in the refrigeration circuit and the heat exchanger as the condenser; for *heating*, the heat exchanger acts as the evaporator and the air coil as the condenser. This operation is achieved through a reversing valve, a diagram of which is shown in Figure 7.22, controlled by the room thermostat, T_2. The refrigerant flow is reversed through all components except the compressor; the heat exchangers have to be designed for satisfactory operation in either mode.

This unitary heat pump system is sometimes termed a *heat reclaim system*. But this is not strictly true, since the piping circuits connecting the units do not allow the collection and storage of hot water off the room units. It should more correctly be termed a heat balance system, thermal balance being achieved when half the units are heating and the other half are cooling. In this state of operation neither the boiler nor the cooling tower would be operating. When the system is not in balance either the boiler would supply the *out-of-balance* load or the cooling tower would have to reject it.

The system can serve perimeter and central core areas and zoning is not required. Fresh air for ventilation can be supplied either through the

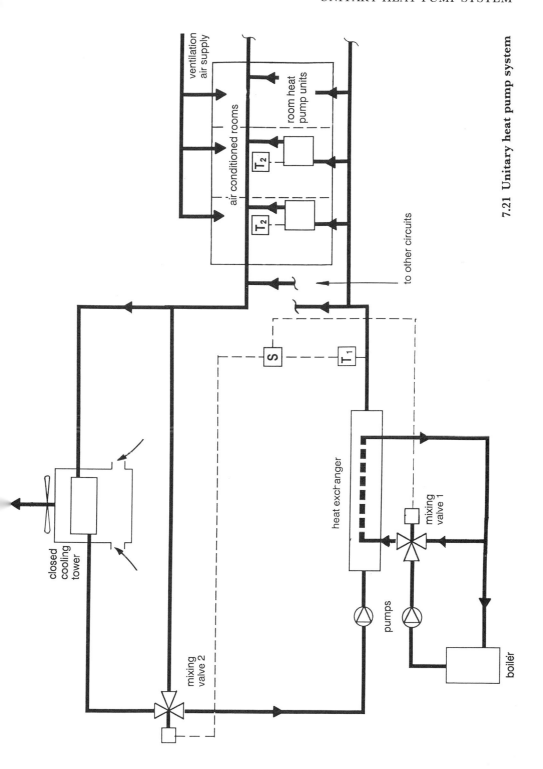

ventilation
air supply

room heat
pump units

air conditioned rooms

T₂

T₂

to other circuits

S

T₁

heat exchanger

closed
cooling
tower

mixing
valve 1

mixing
valve 2

pumps

boiler

7.21 Unitary heat pump system

external wall or from a central plant as described for the induction and fan coil systems.

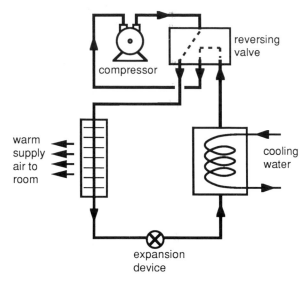

a) heating of room air

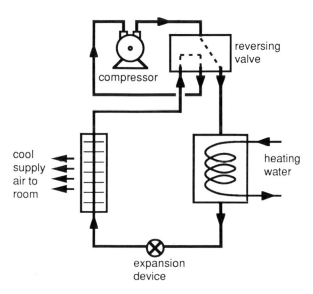

b) cooling of room air

7.22 Operation of reversing valve in room heat pump unit

SELF-CONTAINED AIR CONDITIONING UNITS

Self-contained air conditioning units contain an independent refrigeration circuit for cooling, a fan, a filter and in some cases a heater. They are usually installed under a window, using outdoor air for condenser cooling. Ventilation air is also supplied through the external wall. A drain must be provided to deal with air-side condensation of water vapour. Control is through a room thermostat which cycles the fan on-off or adjusts the fan speed.

Unit sizes have a capacity range of between 1 and 10 kW.

SYMBOLS

c_{pas} specific heat of humid air
g moisture content
h_{fg} latent heat of evaporation
I_r induction ratio
$\dot{m}_a$ mass flow rate of dry air
$\dot{m}_p$ mass flow rate of primary air
$\dot{m}_r$ mass flow rate of recirculated air
Q_p cooling load provided by primary air
q_s sensible heat gain to air conditioned space
q_s' sensible heat loss from air conditioned space
q_l latent heat gain to air conditioned space
t dry-bulb temperature

$\dot{V}_p$ volume flow rate of primary air
$\dot{V}_r$ volume flow rate of recirculated air
Δt_c temperature differential for cooling
Δ_g moisture content difference

Subscripts (for temperature and moisture content)
P   primary air condition
R   room air condition
S   supply air condition

Abbreviations
RRL   room ratio line
VAV   variable air volume

Learning Resources
Centre

8 Refrigeration and Heat Pump Systems

(by R W James)

Air conditioning usually implies that mechanical cooling is required, necessitating the provision of refrigeration systems. The cooling coils in the air conditioning plant may be supplied with either refrigerant or chilled water. Whichever is appropriate, the design of the refrigeration system requires detailed consideration if it is to match the air conditioning demands for cooling and achieve energy efficient operation. At the same time some of the heating requirements can be met using heat rejected by the refrigeration systems, or the system may be used purely for heating, as in a heat pump.

With power cycles such as those involving boilers, steam turbines and condensers, heat is received by the working fluid at a high temperature and rejected at a low temperature, while a net amount of work is done by the fluid. The prime purpose of a refrigerator is to extract heat from a space, whereas a heat pump's principal purpose is to supply heat at an elevated temperature. The refrigerator and heat pump are identical in principle and one machine is sometimes used to fulfil both functions.

PERFORMANCE CRITERIA

An adiabatic process is one in which there is no heat transfer between the fluid and its surroundings, a reversible process is an ideal process in which the fluid is in equilibrium at all times and is analogous to a frictionless process in machines. A process which is reversible and adiabatic is said to be isentropic.

A cycle is reversible if it consists only of reversible processes. The original concept of this was introduced by SADI CARNOT, who conceived a cycle in which all heat enters a system from a source at constant temperature, and all heat leaving it is rejected to a sink at constant temperature. It follows that the processes in which heat is exchanged with the surroundings must be isothermal, the temperature of the fluid never differing by more than an infinitesimal amount from the fixed temperature of the source or sink. When reversed, the cycle gives the highest possible efficiency for a refrigerator and is used as a basis for comparison.

A reversed CARNOT cycle (Figure 8.1), using a wet vapour as a working fluid, has the following processes:

– vapour is compressed isentropically from a low pressure and temperature to a higher pressure and temperature;

– the vapour is then passed through a condenser in which it is condensed at constant pressure;
– the fluid is expanded isentropically to its original pressure;
– finally, the fluid is evaporated at constant pressure to its original state.

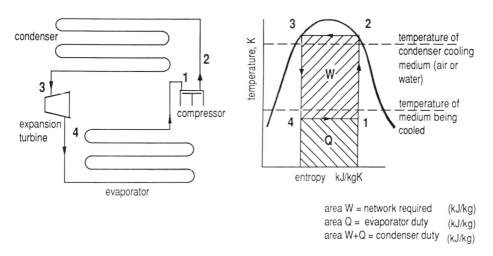

area W = network required   (kJ/kg)
area Q = evaporator duty   (kJ/kg)
area W+Q = condenser duty   (kJ/kg)

8.1 Reversed Carnot refrigeration or heat pump cycle

The criterion of performance of the cycle, expressed as the ratio output/input, depends upon what is regarded as the output. In a refrigerator the *coefficient of performance*, COP, is defined as:

$$COP = \frac{\text{heat extracted in the evaporator}}{\text{net work done on the fluid}}. \tag{8.1}$$

For a heat pump:

$$COP = \frac{\text{heat rejected in condenser}}{\text{net work done on the fluid}}. \tag{8.2}$$

From Equation 8.2 it is evident that the COP of a heat pump is the reciprocal of the efficiency of a power cycle. A reversed Carnot cycle operating between an upper temperature T_a and a lower temperature T_b will therefore have a COP of $T_a/(T_a - T_b)$ for a heat pump and $T_b/(T_a - T_b)$ for a refrigerator. The Carnot COP cannot be achieved in practice but forms a convenient basis for comparison between cycles.

PRACTICAL REFRIGERATION CYCLES

It is possible to show that all reversible cycles operating between the same two reservoirs have the same coefficient of performance, ie that of the reversed Carnot cycle. However, practical refrigeration systems operate on a different cycle which has a lower ideal COP but is more suitable in other respects. The expansion machine is replaced by a simple throttle valve referred to as the expansion valve, the expansion process is then one of constant enthalpy instead of being isentropic. The compression process is carried out in the superheated region. Finally the condensed liquid is often cooled below its saturation temperature (sub-cooled) in the condenser or in the pipe connecting the condenser to the throttle valve.

The cycle is best demonstrated by using a pressure/enthalpy chart, which is illustrated in Figure 8.2. Since pressure is one of the co-ordinates of this diagram, the main family of curves comprises those of constant temperature. In the liquid region these isothermals are nearly vertical because the effect of pressure on enthalpy is negligible; they run horizontally across to the saturated vapour line since, in the wet region of the chart, they are also lines of constant pressure. In the superheated region of the chart they fall increasingly steeply. Also on the chart lines of constant volume are added in the wet and superheated regions, lines of constant dryness fraction in the wet region and lines of constant entropy in the superheated region. As with the psychrometric chart a knowledge of two properties allows the others to be determined.

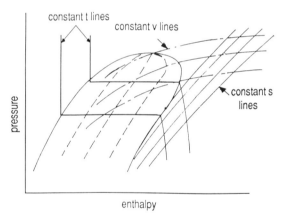

8.2 Pressure enthalpy diagram

BASIC CALCULATIONS

Figure 8.3 shows the saturation cycle with isentropic compression on the ph chart. Simple mass and energy balances give the following equations:

Refrigeration effect (kJ/kg):

$$Q_e = h_1 - h_3. \tag{8.3}$$

Compressor work done (kJ/kg):

$$W_D = h_2 - h_1. \tag{8.4}$$

Condenser duty (kJ/kg):

$$Q_c = h_2 - h_3$$
$$= Q_e + W_D. \quad (8.5)$$

Refrigerant mass flow rate (kg/s):

$$\dot{m}_r = Q_e/(h_1 - h_3). \quad (8.6)$$

Compressor intake volume (m^3/s):

$$V_i = \dot{m}_r \, v_i. \quad (8.7)$$

Compressor isentropic power (kW):

$$W_i = \dot{m}_r(h_2 - h_1). \quad (8.8)$$

Heat rejected in condenser (kW):

$$q_c = \dot{m}(h_2 - h_3). \quad (8.9)$$

Refrigeration duty (kW):

$$q_e = \dot{m}(h_1 - h_3). \quad (8.10)$$

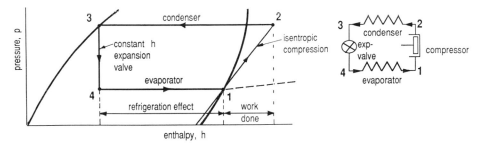

8.3 Saturation cycle with isentropic compression

Example 8.1 The system shown in Figure 8.3 operates on Refrigerant 12 (R12) with an evaporating pressure of 5 bar and a condensing pressure of 15 bar. The cooling coil load is 200 kW. Calculate the following:

(a) the COP when operating as a heat pump;
(b) the COP when operating as a refrigerator;
(c) the refrigerant mass flow rate;
(d) compressor intake volume flow rate;
(e) compressor isentropic power.

Solution
The system shown in Figure 8.3 assumes isentropic compression, saturated liquid leaves the condenser and dry saturated vapour leaves the evaporator.

From the pressure/enthalpy diagram (Figure 8.4):

$$h_1 = 258.0 \text{ kJ/kg}$$
$$h_2 = 277.4 \text{ kJ/kg}$$
$$h_3 = 158.6 \text{ kJ/kg}$$
$$h_4 = h_3$$
$$v_{i1} = 0.0344 \text{ m}^3/\text{kg}$$

The enthalpy values will vary depending on the actual chart used but this will not affect the answers.

The refrigeration effect is given by Equation 8.3:

$$Q_e = h_1 - h_3$$
$$258 - 158.6 = 99.4 \text{ kJ/kg}$$

The condenser duty is given by Equation 8.5:

$$Q_c = h_2 - h_3$$
$$277.4 - 158.6 = 118.8 \text{ kJ/kg}$$

The work done by the compressor is given by Equation 8.4:

$$W_D = h_2 - h_1$$
$$= 277.4 - 258 = 19.4 \text{ kJ/kg}$$

(a) The COP for cooling is given by Equation 8.1:

$$COP = \frac{\text{heat extracted in the evaporator}}{\text{net work done on the fluid}}$$

$$= \frac{Q_e}{W_D} = \frac{99.4}{19.4} = 5.1$$

(b) The COP for heating (heat pump) is given by Equation 8.2:

$$COP = \frac{\text{heat rejected in condenser}}{\text{net work done on the fluid}}$$

$$= \frac{Q_c}{W_D} = \frac{118.8}{19.4} = 6.1$$

(c) The refrigerant mass flow rate is given by Equation 8.6:

$$\dot{m}_r = Q_e/(h_1 - h_3)$$
$$= 99.4/(258 - 158.6) = 2.01 \text{ kg/s}$$

(d) The compressor intake volume is given by Equation 8.7:

$$V_i = \dot{m}_r \, v_i$$
$$= 2.01 \times 0.0344 = 0.0692 \text{ m}^3/\text{s}$$

(e) The compressor isentropic power is given by Equation 8.8:

$$W_i = \dot{m}_r \, (h_2 - h_1)$$
$$2.01 \, (277.4 - 258) = 39.1 \text{ kW.}$$

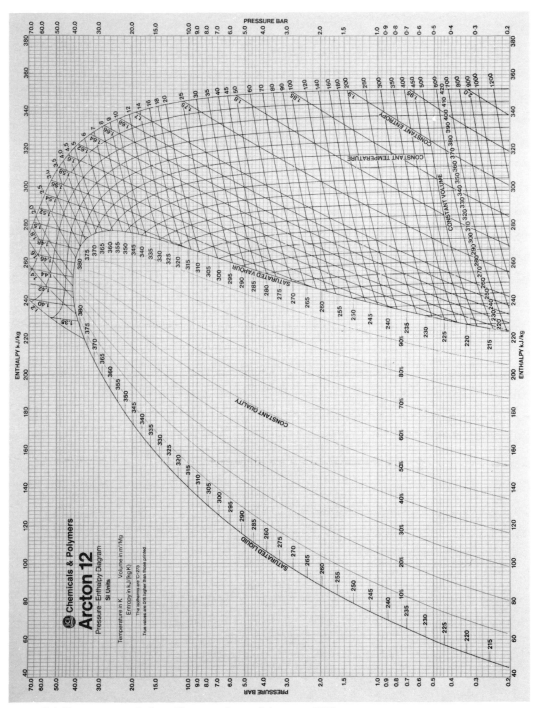

8.4 Pressure enthalpy diagram for refrigerant R12
(Reproduced by permission of ICI Chemicals and Polymers Ltd)

Effect of liquid sub-cooling and vapour superheating

Figure 8.5 shows the effect of liquid sub-cooling in the condenser and vapour superheating in the evaporator. Liquid sub-cooling increases the refrigeration effect and therefore decreases the refrigerant mass flow rate, compressor induced volume and power for the same refrigeration duty. Vapour superheating in the evaporator increases the refrigeration effect and decreases the refrigerant mass flow rate for the same refrigeration duty but there is little change in the compressor induced volume or power. However, when the vapour is heated in the tubes between the evaporator and compressor, the compressor induced volume and power is increased.

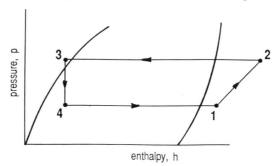

8.5 **Effect of liquid sub-cooling and vapour superheating**

Example 8.2 Repeat Example 8.1 assuming that the vapour entering the compressor is superheated by 10 K:

(a) in the evaporator;
(b) in the tubes between the evaporator and compressor.

Assume that all other conditions remain the same.

Solution
(a) From a pressure/enthalpy diagram
$h_1 = 264.5 \text{ kJ/kg}$
$h_2 = 284.9 \text{ kJ/kg}$
h_3 and h_4 are as in Example 8.1
$v_{i1} = 0.0366 \text{ m}^3/\text{kg}$

Using the appropriate equations:

Refrigeration effect	$Q_e = 105.9 \text{ kJ/kg}$	(from Eq. 8.3)
Condenser duty	$Q_c = 126.3 \text{ kJ/kg}$	(from Eq. 8.5)
Compressor work done	$W_D = 20.4 \text{ kJ/kg}$	(from Eq. 8.4)
COP (cooling)	$COP = 5.2$	(from Eq. 8.1)
COP (heating)	$COP = 6.2$	(from Eq. 8.2)
Refrigerant mass flow rate	$\dot{m}_r = 1.89 \text{ kg/s}$	(from Eq. 8.6)
Compressor intake volume	$V_i = 0.069 \text{ m}^3/\text{s}$	(from Eq. 8.7)
Compressor isentropic power	$W_i = 38.6 \text{ kW}$	(from Eq. 8.8)

(b) h_1, h_2, h_3, h_4, and v_{i1} are as in part (a) but there is now an additional point on the ph chart, that of the saturated vapour leaving the evaporator; let this be denoted as condition 5. Then h_5 will have the same value as h_4 in Example 8.1, ie, 258.0 kJ/kg.

Using the appropriate equations:

Refrigeration effect	$Q_e = 99.4$ kJ/kg	(from Eq. 8.3)
Condenser duty	$Q_c = 126.3$ kJ/kg – as in part (a)	
Compressor work done	$W_D = 20.4$ kJ/kg – as in part (a)	
COP (cooling)	COP = 4.9	(from Eq. 8.1)
COP (heating)	COP = 6.2	(from Eq. 8.2)
Refrigerant mass flow rate	$\dot{m}_r = 2.01$ kg/s	(from Eq. 8.6)
Compressor intake volume	$V_i = 0.073$ m^3/s	(from Eq. 8.7)
Compressor isentropic power	$W_i = 41.1$ kW	(from Eq. 8.8)

Examples 1 and 2 show that, in theory, the influence of suction superheat is small if the vapour is superheated in the evaporator and provides useful refrigeration duty. In practice small liquid droplets are entrained in the suction vapour reducing the refrigeration effect; their influence decreases as suction vapour superheat increases.

SUCTION/LIQUID LINE HEAT EXCHANGERS

Suction/liquid line heat exchangers sub-cool the liquid refrigerant leaving the condenser by heating the vapour leaving the evaporator. Their influence on induced volume and power is usually small but their inclusion in the system provides other advantages; if friction or change of fluid elevation of the evaporator over the condenser reduces the pressure of the liquid in the pipes between the condenser and expansion valve, it will boil and cause problems. Throttle valves are not designed to cope with vapour/liquid mixtures at entry. The heat exchanger prevents boiling by sub-cooling the liquid.

SYSTEM DESIGN AND RELEVANT PLANT COMPONENTS

The diversity of systems make it impractical to discuss their design or the plant components fully. The emphasis here is therefore one of presenting an overview appropriate to the air conditioning industry. Figures 8.6 and 8.7 show the essential components of systems with flooded and dry expansion evaporators respectively.

The low pressure region of the system in Figure 8.6 is to the left of the dotted line and the high pressure region to the right. If this system were a boiler and steam turbine set the essential components would be the same except that the expansion valve would be replaced by a liquid pump and the liquid receiver would be called a *hotwell*. The significant difference is in the high and low pressure regions which would be reversed. This provides a practical insight into the reason why refrigerators and heat pumps are the same machines and why, when reversed, the cycle becomes that of an engine.

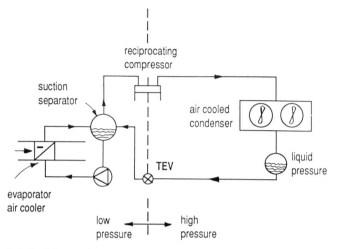

8.6 Refrigeration system with flooded evaporator

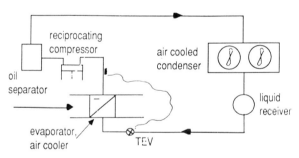

8.7 Refrigeration system with dry expansion evaporator

Compressors

Reciprocating compressors are widely used for air conditioning applications but for large systems screw or centrifugal are often selected. Rotary vane type compressors have improved considerably in recent years and can now be expected to make inroads into the air conditioning market.

The volume of vapour to be removed can be calculated as previously, the determination of the compressor swept volume requires a knowledge of its volumetric efficiency. For a reciprocating compressor this is influenced by such things as clearance volume, valve size, valve inertia, valve spring tension, and effects of cylinder heating. The volumetric efficiency, η, can be expressed as a function of the ratio of condenser pressure to evaporator pressure, r. This may take the form:

$$\eta = A - Br^y \qquad\qquad (8.11)$$

where the constants A, B and index y vary with compressor design and operating speed.

Previously it was shown that the COP improves if the difference between the condensing and evaporating temperatures is reduced. Equation 11 further shows that the lower the pressure ratio, the higher will be the compressor volumetric efficiency and therefore the smaller will be the compressor size. These are important fundamentals for the system design.

As the condensing pressure rises the compressor is required to do more work per kilowatt of refrigeration duty and therefore the capacity of a particular system is reduced. As the evaporator (and therefore compressor suction) pressure rises, less work is required for the same refrigeration effect. However, for a particular system the increased suction pressure results in an increase in the density of the refrigerant and hence an increase in mass flow rate and capacity. The combined characteristic is shown in Figure 8.8.

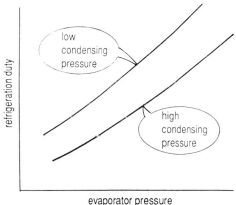

8.8 Variation of refrigeration capacity with compressor suction and discharge pressures

The capacity of a reciprocating compressor is controlled by manipulating the number of cyclinders operating or varying its speed. Other methods are available such as suction gas throttling but these are not energy efficient.

Thermostatic expansion valve

Thermostatic expansion valves are designed to adjust the flow of refrigerant into the evaporator so that no liquid leaves, but the proportion of the evaporator containing a liquid/vapour mixture is a maximum. This flow controller is widely used and responds indirectly to vapour superheat at evaporator outlet. In its most usual form its sensor consists of a small bulb or phial, containing the same refrigerant as the plant and in thermal contact with the evaporator outlet tube. It is connected to a flow regulator by a small bore tube. The regulator responds to the difference between the evaporator and phial pressures which is related to the vapour superheat at the position where the phial is clamped to the pipe.

Recently electronic flow controllers have become available, these have two temperature sensors; one at evaporator inlet for measuring the saturation temperature, and one at evaporator outlet for measuring the temperature of the superheated vapour. The refrigerant flow is controlled to maintain a constant difference between these readings.

Packaged water chillers

Packaged water chilling units, typically as shown in Figure 8.9, are normally located on the roof of buildings. The size of these units range from 30 kW to in excess of 1000 kW. Chilled water pipes require insulating to prevent frost damage. Large roof top units will have the refrigeration components outside the plant room which is a disadvantage for servicing and maintenance. Their location within the plant room would require a large floor area and an additional cost for the supply and extract air ducts.

8.9 Packaged liquid chilling plant with air cooled condenser
(Courtesy York International)

An alternative to the packaged water chilling unit is to use a separate air cooled condenser and liquid receiver. Then a packaged condensing unit would comprise a compressor, evaporator for cooling water, expansion valve and associated piping and controls. Such a unit is usually located inside the plantroom. These units require connecting pipes from the compressor discharge to the condenser, and from the condenser to the expansion valve via the liquid receiver.

Package condensing units

These are available for direct coupling to cooling coils, they usually comprise a compressor, air or water cooled condenser, liquid receiver and safety controls mounted on a base.

CONDENSING SYSTEM DESIGN

When the compressor capacity is reduced the refrigerant velocity in the pipework is also reduced. This is unlikely to effect the expansion valve feed; but low velocities in the compressor discharge line can result in problems transferring oil carried over from the compressor to the condenser and from condensation of refrigerant if the pipework is not designed correctly. A check valve (one-way valve) fitted in this line at entry to the condenser will prevent condensation of refrigerant. Double risers are sometimes used as shown in Figure 8.10.

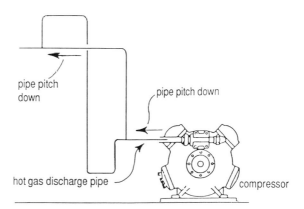

8.10 Double riser from compressor to condenser

When the capacity is substantially reduced, the larger tube traps oil, and the refrigerant is then routed through the smaller tube. When the capacity is increased, the trapped oil is carried into the condenser allowing both pipes to function normally. A check valve is installed at the condenser entry to prevent vapour condensing in these pipes when the compressor is switched off.

The pipe connecting the condenser to the liquid receiver must be sized to ensure good draining. The flow should be part liquid and part vapour so that vapour can pass in either direction. The liquid velocity should not exceed 0.5 m/s.

Liquid refrigerant lines connecting the receiver and expansion valve must be free of vapour. A pressure reduction caused by friction or a change in static head will result in vapour being formed unless the liquid is sub-cooled. If the liquid receiver is mounted above the expansion valve and the pipes are correctly sized the problem should not occur; if there is a problem, a suction/liquid line heat exchanger should be fitted.

Suction lines

With any system in which the evaporators are remote from the compressor, the design of the piping between the evaporator and compressor is critical. A large pressure drop in this pipe reduces the system COP, whereas low refrigerant velocities allow the oil to be trapped. Generally the suction vapour velocity should always exceed 6 m/s when the machine is operating at minimum capacity; if the compressor is located higher than the evaporator the piping should be designed in a similar manner to that shown in Figure 8.10.

Oil separators

The use of a good oil separator reduces the problems encountered but even if this has a high efficiency some oil will still escape into the system. The design must cater for efficient oil circulation.

CAPACITY CONTROL

The capacity of a refrigeration system must be manipulated to match the load. Part-load efficiencies are particularly important for air conditioning applications in a temperate climate. The use of on/off control results in temperature fluctuations which are unacceptable for many applications and on larger machines frequent starting and stopping will cause damage to the system. Other capacity control option available are:

- two speed compressor;
- variable speed compressor;
- cylinder unloading;
- back pressure throttling;
- hot gas bypass.

Two speed compressors provide energy efficient capacity control but with large temperature fluctuations. Varying the compressor speed using an AC frequency inverter provides energy efficient control and temperature accuracy but the capital cost is higher. Cylinder unloading is usually achieved using a mechanism that prevents the compressor suction valve from closing, this method is energy efficient but step changes in capacity result in temperature fluctuations.

Back-pressure throttling is achieved by restricting the refrigerant flow at compressor inlet. The energy consumption is reduced as the capacity is reduced but the system is much less efficient than with the control systems previously discussed. The system is simple and cheap but the range through which the capacity can be varied is limited.

Hot gas bypass is a system by which some of the vapour discharged by the compressor is metered into the evaporator. The system provides the evaporator with an artificial load. Smooth capacity control from 0% to 100% load is possible and high temperature accuracy can be achieved. However the energy consumption is almost unaffected by capacity changes and the system is therefore inefficient.

COOLING TOWERS AND WATER COOLED CONDENSERS

An alternative design of the condensing system is to use a cooling tower and a water cooled condenser; the condenser is then much smaller and can be located in the plantroom. The condensing pressure is generally much lower because the cooling tower performance depends on the ambient wet-bulb temperature. The net result is reduced energy costs but increased capital and maintenance costs.

Since public awareness that Legionnaires' disease is often traced to cooling towers, there has been an increase in the use of dry or sensible water coolers for condensers. This has resulted in bulkier condensing equipment, higher condensing temperatures and reduced COP. A development for air conditioning applications with winter cooling requirements such as computer suites has been to use a glycol solution for condenser cooling, which can also be used directly in a cooling coil during winter. This saves compressor energy during such periods.

In water cooled condensers, the refrigerant should condense at 5 to 10 K above the water inlet temperature, a low value results in a higher COP but also higher initial costs. With open type cooling towers, the water should cool to within 3 to 6 K of the air wet-bulb temperature and with closed cooling towers the water should be cooled to within 8 to 12 K of the air wet-bulb temperature. (Cooling towers are described in Chapter 9.)

CHOICE OF COOLING SYSTEM

Design engineers often have to choose between direct expansion (DX) evaporators (Figure 8.7) and a water chilling system (Figures 8.11 and 8.12). For systems employing a single air cooler, a direct expansion evaporator would normally be used whereas when cooling is required in many locations, and it is not practical to use smaller multiple units, a water chilling system is required. For many applications the choice is difficult and the final system may owe much to familiarity with particular equipment. Also water chillers are often chosen to allow the design to be processed easily and quickly without the involvement of a refrigeration company.

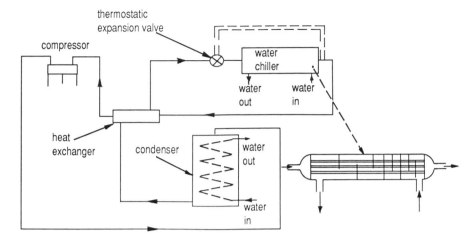

8.11 Water chilling system with a dry expansion evaporator

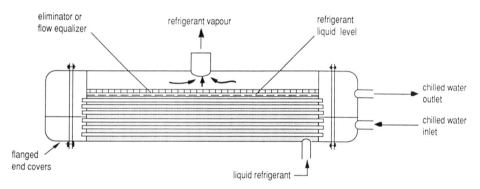

8.12 Alternative flooded evaporator for water chilling

The choice between a direct expansion system and the use of a heat transfer medium should be made by considering the various factors particular to each application. The case for and against direct expansion systems compared with water chilling systems deserves consideration of the following:

For direct expansion cooling coils:
– there is less pipework and equipment since there are no cold water circuits;
– higher coefficients of performance are possible;
– potential for reverse cycles or using the condenser to reheat air;
– avoids problems associated with valve authority and valve characteristics in chilled water systems;
– total system capital cost is often reduced;

- energy efficient capacity control over a wide range is possible with recently developed technology;
- water chillers require protection against freezing during operation and during shut down;
- heat gains to chilled water in large piped distribution systems;
- lower off-coil air temperature can be achieved.

Against direct expansion cooling coils:
- potential for refrigerant leakage into the occupied space;
- analogue control arguably more difficult than the use of a three-way valve on a water/air heat exchanger;
- distribution of refrigerant piping to locations remote from the plant can cause problems, such as excessive pressure drop or leakage of refrigerant, or oil could be trapped;
- control of multiple coils through wide capacity ranges is more difficult;
- frost on coil surfaces is more likely to occur;
- water can be chilled without running the compressor for much of the year;
- increases the quantity of CFC refrigerant used;
- noise control is simplified by using a central water chilling plant;
- variable air volume flow rates through the coil require special consideration.

The potential for refrigerant leakage from direct expansion evaporators has increased with the introduction of thin wall copper tubing into their construction. Their use in public buildings was illegal under the former Greater London Council regulations.

Water chillers with flooded evaporators (Figure 8.12) can be designed to operate on a thermo-syphon system to provide free cooling when the ambient wet-bulb temperature is low enough.

SYSTEM COMPARISON

It is informative to make a theoretical comparison between systems that can perform the same air conditioning process on the basis of refrigeration duty, required compressor swept volume, work done and power requirements.

Consider air being cooled from 23°C with a moisture content of 0.012 kg/kg to 15°C with a moisture content of 0.01 kg/kg and an air mass flow rate of 7.4 kg/s. The air cooling coil load for these conditions is 100 kW. The following systems are considered to perform this air conditioning process.

System 1 An R22 direct expansion evaporator cooling air, in which the refrigerant evaporates at 8°C and leaves superheated at 14°C.

System 2 An R22 water chilling plant with a dry expansion evaporator and an air/water heat exchanger.

For the dry expansion water chiller, it is assumed that the water enters the chiller at 14°C and leaves at 7°C, and that the refrigerant is evaporating at 2°C and leaves the chiller at 8°C.

The refrigerant is assumed to condense at 35°C and to leave the condenser at 33°C with both systems. To simplify the calculations the heat transfer between all pipes and their surroundings, the influence of pressure drop in pipework and heat exchangers and water pumping power for System 2 is neglected.

Calculations on the basic cycle, using Equations 8.1 to 8.10, then give the results in Table 8.1.

parameter	system	
	1	2
refrigerating effect (kJ/kg)	172.2	171.6
refrigerant flow rate (kg/s)	0.581	0.583
compressor displacement (m^3/s)	0.0218	0.0269
isentropic work (kJ/kg)	19.8	24.4
isentropic power (kW)	11.5	14.2

Table 8.1 Comparison of refrigeration systems 1 and 2

These calculations show that for indirect cooling there is an increase of approximately 24% in both compressor size and power when compared with the direct expansion system.

Systems incorporating dry expansion evaporators can benefit from suction/liquid line heat transfer. This applies to the system with the direct expansion evaporator of Figure 8.7 and to the dry expansion liquid chiller of Figure 8.11. Up to 40% of a dry expansion evaporator may contain a mixture of liquid droplets and superheated vapour with a consequent poor heat transfer coefficient.

If suction/liquid line heat transfer is used, a greater proportion of the evaporator will contain liquid in contact with the tube wall; this is achieved by sensing superheat between the heat exchanger and compressor to control the evaporator feed. This arrangement increases the risk of liquid refrigerant entering the compressor. By using computer control and an electronic expansion valve, superheat can be sensed at the heat exchanger outlet during normal operation and at evaporator outlet if the systems transient behaviour results in risk to the compressor.

WATER CHILLING SYSTEMS

The water circuits for a water chilling system with condenser heat recovery are shown in Figure 8.13. The heating and cooling requirements are rarely matched. In winter, more heating duty is required and in summer, more cooling duty is required. The system capacity is normally determined by the cooling requirements and excess heat is dissipated in a cooling tower, additional heat is sometimes required and this is usually provided by a boiler.

The system shown in Figure 8.13 can be controlled to match the heating requirements and for this an additional evaporator is installed in the ambient air to collect additional heat when required. The cooling tower should be of the closed type so that any contaminants picked up are not circulated through the system.

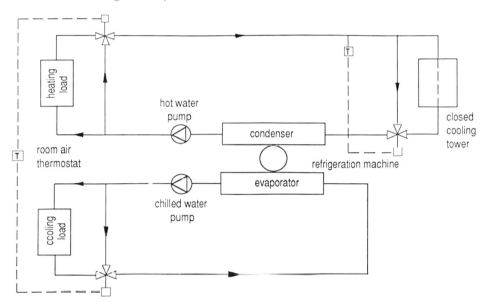

8.13 Chilled water circuit with energy recovery

Thermal storage can be introduced as shown in Figure 8.14, to cope with peak loads, thus reducing the liquid chiller size.

An open cooling tower can be used in conjunction with a double bungle condenser as shown in Figure 8.15. Some of the condenser tubes are connected to the cooling tower and the remainder to the load.

When a water chilling system is the correct choice, design and installa-tion costs can be reduced by purchasing packaged liquid chillers. This type of plant poses few problems to the system designer but when more than one is employed the running costs at part-load may be higher than necessary. This results from some of the available evaporation and condensation surface areas not being utilized when a compressor stops.

An integrated system design could utilize all the heat exchanger surface area available at part-load reducing the temperature difference and increasing the refrigeration system COP. Nevertheless, separate systems would be preferred with hermetic compressors, since a motor burnout would pollute all parts of the system including the remaining compressors.

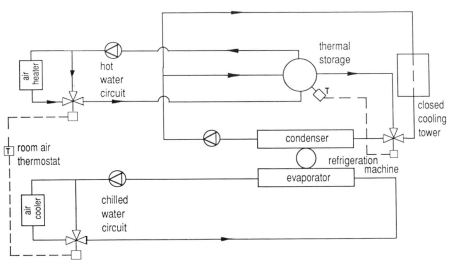

8.14 Water circuits with thermal storage for condenser heat

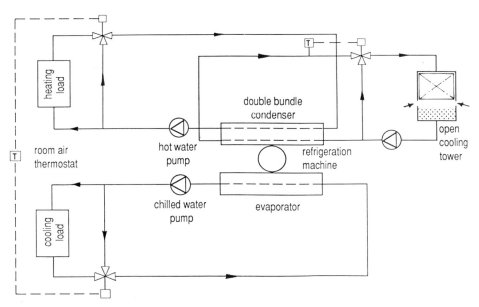

8.15 Heat recovery system with double bundle condenser

REFRIGERANTS

The ideal refrigerant would be non-toxic and should have:

- a large refrigeration effect requiring a small mass flow rate;
- a small amount of theoretical work to be done during compression;
- a small vapour specific volume.

These characteristics would result in a small compressor and a low power requirement. Additionally for reliable plant operation:

- the temperature of the discharge superheated vapour from the compressor should be low to avoid oil breakdown;
- evaporating pressure should be above atmospheric and the condensing pressure should be low; this reduces leakage problems and gives a lower pressure ratio.

Refrigerant thermodynamic properties are also important, for example a low liquid specific heat results in less vapour being formed when the pressure is reduced.

Table 8.2 shows the boiling point at atmospheric pressure (1013 mbar) for commonly used refrigerants:

refrigerant	chemical structure	boiling point at 1013 mbar $^\circ$C
R11	CCl_3F_2	23.8
R12	CCl_2F_2	-29.8
R22	$CHClF_2$	-40.8
R502	$CHClF_2/$ C_2ClF_5	-45.6

Table 8.2 Common refrigerants

The use of refrigerants R11, R12 and R502 will decline on account of ozone layer depletion. R22 is suitable for air conditioning applications; it minimizes ozone depletion, and is less harmful. New refrigerants are being developed which are environment friendly.

SYMBOLS

COP   coefficient of performance
Q_e refrigeration effect
Q_c condenser duty
h   enthalpy
$\dot{m}_r$ refrigerant mass flow rate
q_c heat rejected in condenser
q_e refrigerant duty
T   absolute temperature
V_i compressor intake volume
v_i specific volume
W_D work done
W_i compressor isentropic power
x   dryness fraction
η volumetric efficiency

Subscripts
1, 2, 3 and 4 refer to specific conditions in the refrigeration cycle

9 Humidifiers and Cooling Towers

The analysis of the humidity requirements of the occupants, the building contents and fabric and manufacturing/industrial processes in the air conditioned space will determine the need for a humidifier in the system. The psychrometric processes associated with the different types of humidifier have been given in Chapter 2. This chapter describes in more detail the humidifying equipment available, their operating characteristics and guidance on practical application.

For many years, mains water was used as a cooling medium for industrial applications, as well as for condenser cooling requirements of refrigeration plants. However, advancing technology and man's increasing expectancy of his domestic, social and working conditions have resulted in heightened demands for water and this has meant that clean water is now relatively expensive. Consequently, where there is a cooling requirement there are sound financial reasons to practice water conservation by using a recirculatory cooling water system. The most efficient plant available for this is the evaporative cooling tower.

HUMIDIFIERS

Spray-type air washer

The expression washer is a misnomer, the name originating from the days when this piece of equipment was used as an air cleaning device in mechanical ventilation systems. In fact it was found to be inefficient for air filtration requirements but was found to be an efficient adiabatic humidifier. Over many years the water spray humidifier was developed into a sophisticated piece of air conditioning equipment in which various psychrometric processes could be achieved. Some of these processes are described on page 44–48. By using three banks of sprays, contact factors (defined in Chapter 2) of up to 0.96 can be obtained.

Figure 9.1 shows a typical spray-type washer. It consists of a chamber containing a set of nozzles which generate the fine spray of water, a tank for the collection of unevaporated spray water and an eliminator section at the discharge end removes entrained droplets of water from the airstream which would otherwise pass into the ductwork system. A pump recirculates water to the nozzles at a rate which is up to 50 times more than the evaporation rate.

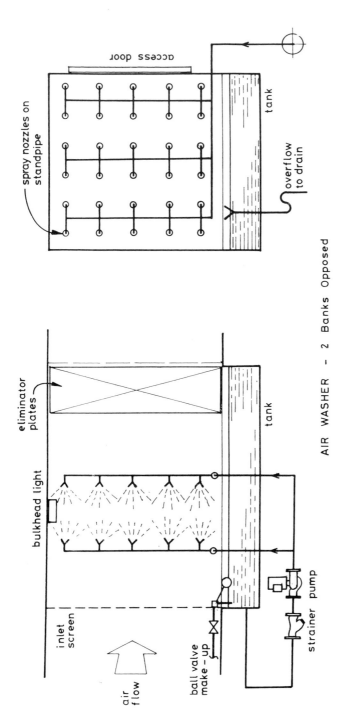

AIR WASHER — 2 Banks Opposed

9.1 Air washer with two banks of nozzles

The simplest design has a single bank of spray nozzles mounted in a casing usually 1 to 2 metres long. Two or more spray banks are generally used when a high contact factor is required and for cooling and dehumidification applications which require chilled water. The nozzles are spaced to give uniform coverage across the airway. The casing, tank and eliminator plates may be constructed from various alternative materials, but most usually sheet steel galvanized after manufacture, for relatively small sizes; builders work construction is usual for larger equipment. The eliminator plates consist of a series of vertical plates, 25 mm apart, with a number of bends and lips to promote the separation of water droplets from the air. An inlet screen may be provided to ensure even air flow through the washer. Other accessories include access and inspection doors, ball valve for water supply make-up, *quick-fill* for rapid filling after maintenance, trapped overflow, drain and bulkhead light.

Washers are commonly available from 1 to 100m³/s capacity, though there is no standardization of design parameters such as water-spray density and spray pressure. Typical design values are:

face velocity*	2.5 m/s
water circulation rate	1.0 l/s per 1.0 m³/s air
pump pressure	200 kPa
air pressure drop	100 Pa

Sprayed cooling coil

With the sprayed cooling coil humidifier, the water for humidification is sprayed through a bank of nozzles onto the air side of a chilled water coil or direct-expansion coil. This arrangement avoids some of the problems associated with open type air washers eg scaling on the chilled water side of refrigeration evaporator tubes.

A diagram of a typical arrangement is shown in Figure 9.2. The large wetted surface of the coil allows close contact between air and water, giving humidifying contact factors of up to 0.8. With no coolant being supplied to the cooling coil the humidification process will be adiabatic. As explained in Chapter 2, when the coil comes into operation with a cooling load, the off-coil condition will depend on the ADP and the contact factor of the coil.

---

*face velocity is defined as the average velocity entering the equipment, ie the air volume flow rate divided by the cross section area.

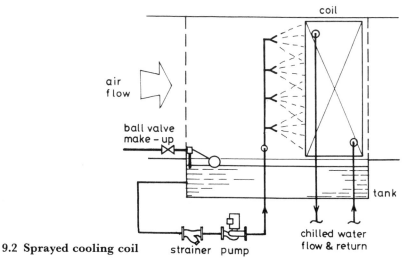

9.2 Sprayed cooling coil

There are no standard design parameters; typical design values are:

face velocity	2.5 m/s
water circulation rate	0.5 l/s per m³/s air
pump pressure	100 kPa
air pressure drop –	depends on cooling coil pressure drop, typically 200 Pa

Capillary washer

A diagram of a typical arrangement of a capillary washer is shown in Figure 9.3. A bank of capillary cells is fitted across the airway, each cell

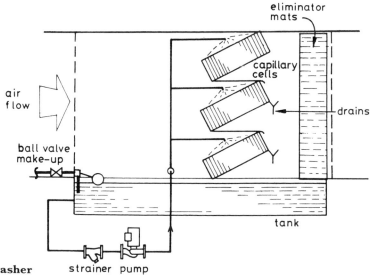

9.3 Capillary washer

comprising an open-ended box of galvanized sheet steel closely packed with glass fibres. Water sprayed onto one face of the bank of cells passes through to the other side, wetting the glass fibres. Air passing through the cells comes into close proximity with the large wetted surface area of the fibres, giving a high contact factor of up to 0.95. The water may be sprayed onto the upstream or downstream faces of the cells to give parallel or counter flow heat and mass transfer.

The capillary washer also acts as an efficient air filter and as the cells accumulate dirt, the pressure drop increases and the contact factor deteriorates. It is therefore necessary to fit pre-filters to prevent excessive accumulation of deposits in the body of the cells. Even with filters, glass fibres tend to clog up and if this type of humidifier is used today it is usual for a material such as open-mesh polyurethane foam to be used as packing.

Design values are similar to the sprayed cooling coil. Typical capillary cell dimensions are 600×600 mm across the face and 200 mm deep.

Spinning disc humidifier

A schematic drawing of a spinning disc humidifier is shown in Figure 9.4. The unit consists of a 0.2 kW induction motor to the end plate of which is bolted a spun copper dish, on whose rim is fixed a ring of stainless steel teeth. The stainless steel shaft of the motor is fitted with two conical brass hubs, which locate the stainless steel disc by means of a locking nut at the end of the shaft.

Water is fed onto the inner conical brass hub through the nozzle. This water is then forced by centrifugal action to the tip of the disc and onto the teeth, generating a fine mist of water droplets. The quantity of water generated as this mist compared with the quantity of water supplied, is known as the atomizing (decimal) efficiency and with good design this can be as high as 0.7. However, the humidifying contact factor itself can be as high as 0.85.

The disc units are incorporated into a plant casing as shown in Figure 9.5, the number of units being dependent on the total air flow rate; the ancillary control gear is mounted on the outside of the duct. Typical design values are:

air flow rate, one disc	$2.0 \text{ m}^3/\text{s}$
face velocity	2.5 m/s
pressure drop	50 Pa

Steam humidifiers

This group of humidifiers use steam either from a central source, such as a boiler, or from a local generator close to the point of injection into the air duct. The advantages of these units compared with water spray types are:

- smaller air pressure drop;
- duct air velocity can remain high;

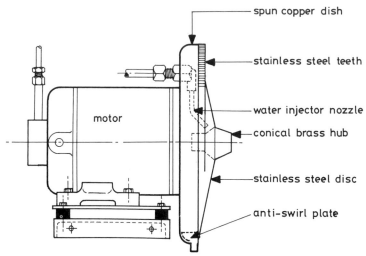

spun copper dish

stainless steel teeth

water injector nozzle

motor

conical brass hub

stainless steel disc

anti-swirl plate

9.4 Spinning disc humidifier – four disc unit

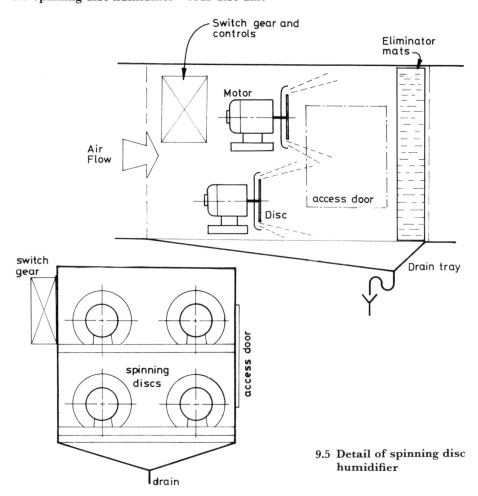

Switch gear and controls

Eliminator mats

Motor

Air Flow

access door

Disc

Drain tray

switch gear

spinning discs

access door

drain

9.5 Detail of spinning disc humidifier

– eliminator plates or mats are not required;
– less risk of corrosion;
– no risk of bacterial contamination from standing water.

The steam injected into the air stream should be clean and dry. Drying the steam will minimize any odour that may be present in the steam supply from the boiler.

Figure 9.6 shows a typical unit, which is designed to dry the steam as well as to control its flow rate. The steam supply, reduced to the operating pressure, feeds into a separating chamber allowing any condensate to drain to a trap at the base of the unit. The steam flow rate is controlled by the valve at the top of the unit, before passing into a drying chamber which is completely jacketed by the separating chamber. From here the dried steam passes into a distribution manifold which is itself jacketed by the incoming steam. The top of the drying chamber contains metal swarf to act as sound absorber. The steam capacity of these units range from 0.15 to 72 g/s with a maximum operating pressure of 1.4 bar.

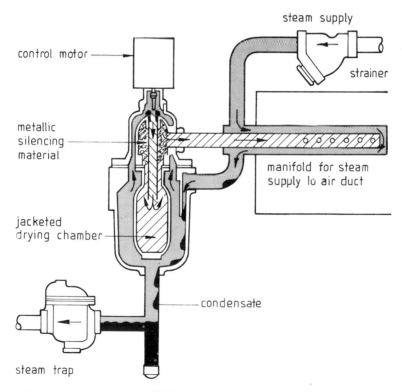

9.6 Jacketed steam humidifier (for mains steam)

A typical unit for local steam generation is shown in Figure 9.7. Water in a small cylinder is brought to the boil with electrical resistance elements selected to ensure constant output irrespective of the conductivity of the water. Primary heating can also be provided by boiler steam or high pressure hot water. The fresh steam so produced passes over a series of baffles before being injected into the air stream via a stainless steel tray mounted in the duct. These units are available to provide a range of steam outputs, from 0.5 to 30 g/s.

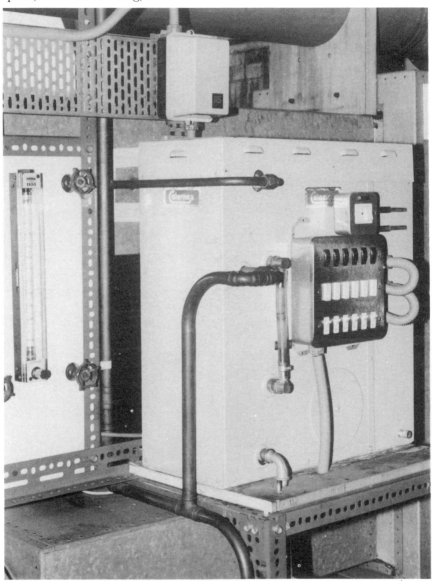

9.7 Steam humidifier, local generation

Steam humidifiers are suitable for systems with air supply rates up to $10 \, m^3/s$. They take up less plant space than a water spray type and, if required, can more easily be incorporated into branches supplying different zones of the building. To avoid condensation on duct walls the system should be designed and controlled to ensure that the air condition leaving the humidifier is less than 90% saturation. Where the humidifier is controlled from a room humidistat, a duct mounted high limit humidistat should be placed in the duct downstream of the humidifier as a further precaution against condensation.

COOLING TOWERS

A basic cooling tower is illustrated in Figure 9.8. The principal components and their functions are as follows:

- a *case* encloses the cooling process and provides a structure for supporting the other items;
- the *packing* is a structure designed to provide the maximum area for the water to flow over; this maximizes the area of water–air surface for heat and mass transfer. The materials for the packing used to be wood or metal slats but these are now usually of plastic;
- a *fan* moves the required amount of air through the water covered packing;
- *eliminators* are designed to minimize the carry-over and drift of water droplets from being carried away from the tower by the wind;
- a *water distribution system* ensures that the circulating water has maximum contact with the air. Nozzles produce fine water droplets and spread them uniformly over the packing;
- a *water reservoir* is an integral part of the casing, collecting the cooled water;
- a *pump* is required to provide the necessary pressure to circulate the water through the system; the water gravitates through the packing back into the reservoir.

Evaporation accounts for most of the heat transfer and therefore the process is referred to as *evaporative* cooling. In addition, depending upon the relative temperatures of the air and water, convection and conduction account for approximately a quarter of the heat transferred.

Open cooling towers

The type of tower shown in Figure 9.8 is a *mechanical, induced draught, contra-flow* tower; it is named in this way because the fan draws a steady flow of air through the packing in a flow direction opposite to that of the water which falls by gravity. All the water used in the system is open, at some time, to the atmosphere; hence the term *open* tower, to contrast with a *closed* tower where the condenser water is in a closed piped water circuit.

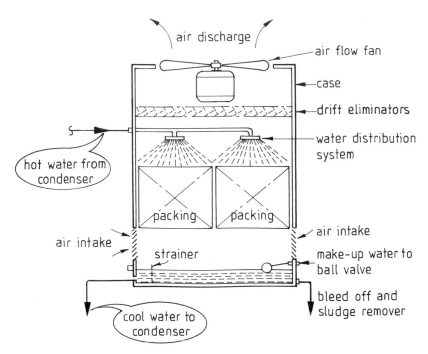

9.8 Principal elements of a cooling tower

Another type of contra-flow tower is the *forced draught* tower, two patterns of which are shown in Figures 9.9 and 9.10. In both these cases the air is discharged by the fan into the packing but otherwise the principles of operation are the same as for the induced pattern. Note the use of the centrifugal fan in Figure 9.9 and the axial flow fan in Figure 9.10. Both types of fan are suitable for forced draught application but, because of its configuration, the centrifugal fan is not normally suitable for an induced draught tower.

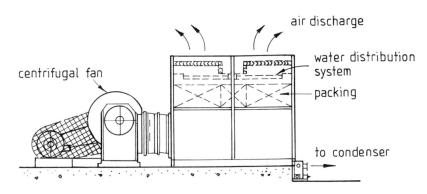

9.9 Forced draught contra-flow tower with centrifugal fan

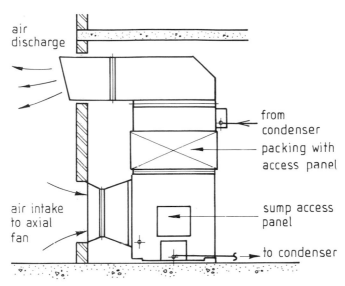

9.10 Forced draught contra-flow tower with axial flow fan

Cross-flow towers can be forced or induced draught. The cooling principles are similar to contra-flow types but the air flows horizontally through the packing causing a cross-flow through the falling water. The compact form of *induced* cross-flow tower involves twin packs, one on either side of the fan as illustrated in Figure 9.11. Because of the larger drift eliminators, there is less risk of carry-over with this configuration compared with a vertical tower and low air velocities result in lower fan power and reduced noise levels. Heights are little more than that of the packing; consequently a tower of this type is suitable for roof locations where there is an architectural need for a low profile.

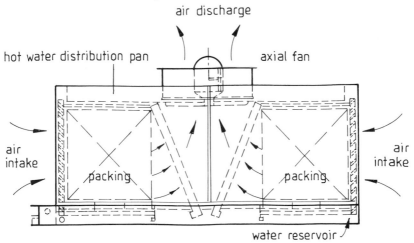

9.11 Twin pack, induced draught, cross-flow tower

Another advantage of the cross-flow tower is its simple water distribution system which consists of a shallow tank having a perforated case. Hot water from the refrigeration system condenser is directed from an open pipe into the tank situated immediately above the packing. This is in contrast to the contra-flow towers where the air has to pass through the water distribution system consisting of spray nozzles on a trough and weir arrangement.

Closed cooling towers

Closed cooling towers are constructed on similar lines to open towers. The major difference is that the condenser water does not come into direct contact with the ambient air. Instead, a pipe coil is used through which the condenser cooling (primary) water is circulated. Secondary water, for evaporative cooling, is then sprayed onto the outside of the coil as shown in Figure 9.12. For most applications, primary water can be cooled to within approximately 10 K of ambient wet-bulb temperature compared with approximately 5 K from an open cooling tower.

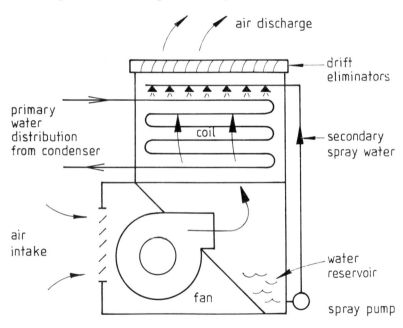

9.12 **Diagram of closed cooling tower**

LEGIONNAIRES' DISEASE

Legionella Pneumophila occur naturally in streams and stagnant pools and can be transmitted into building systems by way of the public water distribution network. Figure 9.13 shows the typical ranges of temperature of water in various building services together with its effect on the bacteria.

The organism is harmless in low concentrations and at low temperatures but will multiply rapidly if ideal breeding conditions exist in the temperature range 25 to 45°C. As the temperature increases above about 47° the bacteria is killed at progressively increasing rates until at 70° they die instantly.

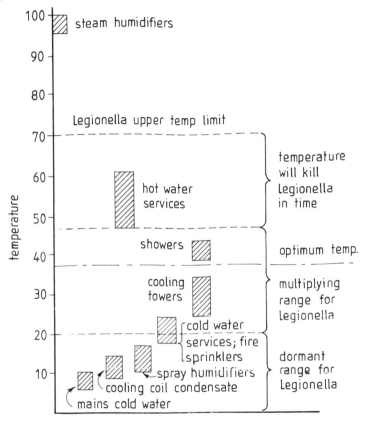

9.13 **Typical temperature ranges for water used in building engineering services, in relation to the Legionella bacteria**[1]

Transmission of the bacteria requires aerosols to be generated from a contaminated water source. The most hazardous aerosol size is 5 μm, water droplets larger than this will settle out of the airstream while smaller sizes are much less likely to carry the bacteria. Occasionally, inadequate maintenance and ineffective drift eliminators on some cooling tower installations have caused the dispersal of Legionella bacteria (in aerosol form) into the atmosphere resulting in outbreaks of Legionnaires' disease. The adverse publicity surrounding these outbreaks has inevitably led to doubts about cooling tower technology and caused system engineers to reconsider the alternatives. One of these is the air cooled condenser in which no water is used. This equipment is certainly hygienically safe but other considerations make it a poor substitute for the evaporative cooling tower. By using only air *sensible* cooling results in an increase of approximately 30% of compressor power compared with a cooling tower.

With closed towers, the secondary spray water only is exposed to the atmosphere (a much smaller quantity of water than the primary water); it therefore presents a lower potential for dispersal of the bacteria than an open tower.

In the past, cooling towers have sometimes been *under-rated* for the condensing requirements. The result of this is that the temperature of the water returning to the tower, from the condensers, will often be well above 25°C, a condition conducive to the multiplication of the Legionella bacteria. It is important that all spray type towers should be sized so that the water in contact with the atmosphere will not exceed, or rarely exceed, this critical temperature.

There have been no reports of water spray humidifiers causing an outbreak of Legionnaires' disease. In a ducted air system the aerosols should have dried out before reaching the conditioned space; and the aerosol size, from such units as spinning discs are smaller than the critical size.

It is important to reiterate that eliminator plates on all water spray equipment should be designed to minimize carry-over of aerosols.

Water treatment

With all recirculatory-type spray equipment, bacterial growth (and possibly algae) will occur in the water tank or reservoir. Bacteria may also breed on cooling coil surfaces when these are wet due to dehumidification. As well as being a potential health hazard this growth will reduce overall plant efficiency. Also, the build up of salts in the recirculated water due to continued evaporation will affect heat and mass transfer and cause corrosion. Hence, water treatment is usually necessary to prevent the increase in bacteria and scale formation. Continuous bleed-off, or periodic discharge (blow-down), together with regular cleaning of the water tank and air-side surfaces is also vital.

The water treatment programme must be comprehensive and monitored by sample analysis, together with visual inspections as to general cleanliness.

A hot water (pasteurization) system has been developed recently which destroys the Legionella organism; it is hoped that this system will be used to make cooling towers, and other water supplies, safer.[2]

In many areas, the normal supply water to a *non-storage adiabatic humidifier* contains a relatively high proportion of carbonates and when the aerosols generated by the humidifier evaporate, the solids remain in the airstream as a fine dust. This dust should be removed either by filtration after the humidifier or by water treatment which, to be effective, must remove the solids; simply to change their chemical composition is not sufficient.

10 Exhaust Air Heat Recovery

A system which makes optimum use of recirculated air is the most efficient method of recovering heat from the exhaust air. However, if for ventilation purposes, the system has to be designed for either 100% outdoor air or a large percentage of outdoor air, then consideration should be given to the incorporation of a heat recovery unit (HRU) for exhaust air heat recovery. Generally the heat recovery units described below refer to the exhaust air heating the supply air (the normal economic requirement in cold or temperate climates) but in most cases the units can be arranged to provide some cooling during summer operation.

EFFICIENCY

The efficiency of an air-to-air heat recovery unit may be defined as follows. Referring to Figure 10.1:

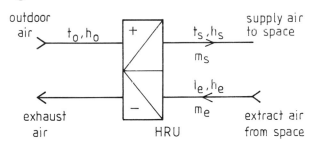

10.1 Schematic of an air-to-air heat recovery unit: definition of efficiency

For sensible heat exchange:

$$\eta = \frac{\dot{m}_S(t_S - t_O)}{\dot{m}_E(t_E - t_O)} 100 \qquad\qquad (10.1)$$

For total heat exchange:

$$\eta = \frac{\dot{m}_S(h_S - h_O)}{\dot{m}_E(h_E - h_O)} 100 \qquad\qquad (10.2)$$

where $\dot{m}_S$ = air mass flow rate in the supply duct
$\quad\ \dot{m}_E$ = air mass flow rate in the extract duct.

THERMAL WHEEL (REGENERATOR)

The thermal wheel comprises a framework, like a thick cartwheel, filled with a suitable matrix of large surface area through which the air can pass. The unit is installed between the two counter flowing airstreams as illustrated in Figure 10.2. The wheel is rotated slowly, driven by a small electric motor with a maximum speed of 10 rpm. The part of the wheel in the exhaust air is warmed up and this in turn heats the incoming air as the wheel revolves. The rate of heat transfer is controlled by varying the speed of rotation of the wheel.

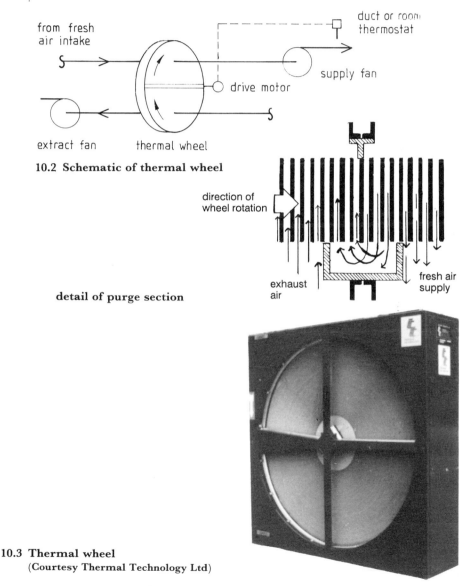

10.2 Schematic of thermal wheel

detail of purge section

10.3 Thermal wheel
    (Courtesy Thermal Technology Ltd)

Hygroscopic wheels are available which transfer latent heat as well as sensible heat and these will be particularly suitable for use in spaces which have high humidity. The psychrometric processes that occur across a hygroscopic wheel are shown in Figure 10.4.

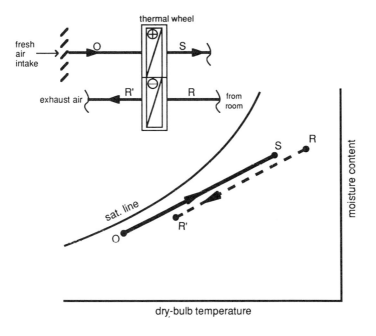

10.4 **Psychrometric process: hydroscopic thermal wheel**

Exhaust air will not carry over to the supply air provided the correct pressure differentials are observed. To avoid leakage of the exhaust air into the supply air stream, the air duct pressure should be positive in the supply with respect to the exhaust, with the leakage path as indicated in Figure 10.5. This can be ensured by placing the supply fan on the upstream side of the wheel; but an excessive pressure differential across the unit will result in a flow of outdoor air to the exhaust with a reduction in overall efficiency. Where air flows from exhaust to supply, the efficiency is increased but with the risk of the exhaust air contaminating the supply air. A *purge* fitting, illustrated in Figure 10.3b, can be installed at the interface of the two air streams to minimize this transfer. Typical design values are:

size, diameter	0.5 to 4.5 m
air flow rate, range	0.2 to 70 m^3/s
pressure drops	60 to 250 Pa
efficiency of heat transfer	70% to 90%

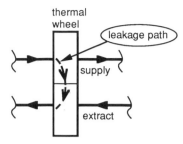

10.5 Leakage path across a thermal wheel

Some advantages of the thermal wheel are that efficiency remains high whatever the thermal load, there are no problems with bacterial growth and that frost/ice does not build up at sub-zero outdoor air temperatures on *sensible* heat exchangers (frosting can occur on hygroscopic wheels). There may be difficulties in arranging the plant to obtain the required pressure differentials and there is the need to bring exhaust duct close to the supply.

HEAT PIPES

Figure 10.6 illustrates the operating principle of a heat pipe. A length of pipe, up to 3 m long and 50 mm diameter and sealed at both ends, contains a tightly fitting sleeve of porous material around a hollow core together with a charge of refrigerant. With one end in the warm air stream, (**A**), and the other in the cold air stream, (**B**), refrigerant evaporates at (**A**) absorbing heat. The gas passes along the inside to condense at (**B**), giving out heat and the refrigerant travels back by capillary action through the wick to complete the cycle.

In an air conditioning system, a bank of these heat pipes is used. For efficiency of heat transfer, the bank of pipes will be finned, the whole unit acting as a pair of conventional heat transfer coils. Efficiency is improved by tilting the tubes and, for conventional systems, this tilt is about 6° with the warmer end in the high temperature air stream. Control of output from this unit may be achieved by face and bypass dampers in the supply air duct as shown in Figure 10.6(b). The heat pipe works in reverse for summer operation. The principal advantage of this unit is its relative simplicity resulting in low maintenance costs.

Typical design values are:

air flow rate, range		up to 5 m³/s.
face velocity		2.5 m/s
efficiency of heat transfer:		
	4 row coil	50%
	8 row coil	70%
pressure drop		
	4 row coil	70 Pa
	8 row coil	140 Pa

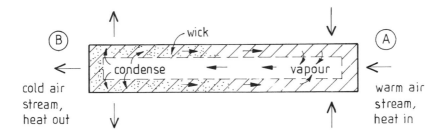

(a) detail of single heat pipe

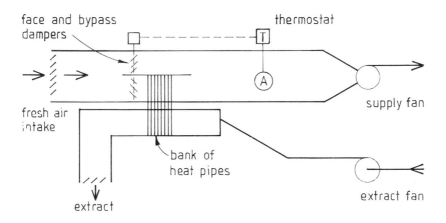

(b) arrangement of heat pipe unit in system

10.6 Heat pipe heat recovery unit

PARALLEL PLATE HEAT EXCHANGER

The principle of operation of the parallel plate heat exchanger is illustrated in Figure 10.7. A basic unit comprises an open ended box with a series of thin plates of metal, plastic or glass. These form narrow linear passages, alternate rows of which carry the supply air, the remainder the exhaust air; the plates can be finned to enhance heat transfer. Some latent heat may also be recovered when the outdoor temperature is sufficiently low to condense moisture on the exhaust air-side of the plates.

Units are obtainable in a large range of sizes, able to deal with air flow rates of up to 2 m³/s; modules can be bolted together in parallel if higher system flow rates are required. Individual units are sometimes incorporated into package air conditioners as in Figure 10.8. Efficiencies are in the range 60 to 75% and the output can be controlled with a modulating damper system as shown in Figure 10.7(b).

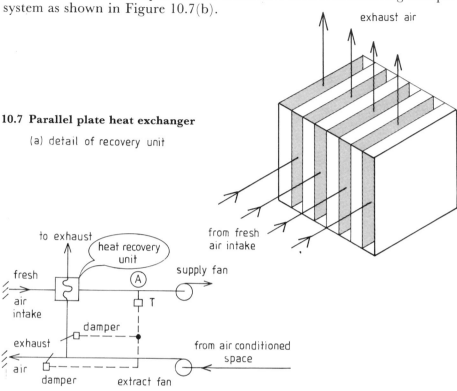

10.7 Parallel plate heat exchanger

(a) detail of recovery unit

(b) heat recovery unit in ventilation system. thermostat T modulates dampers to maintain temperature at (A)

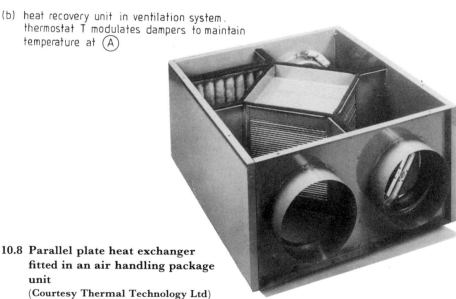

10.8 Parallel plate heat exchanger fitted in an air handling package unit
(Courtesy Thermal Technology Ltd)

The advantages of this unit are that little maintenance is required and there is no possibility of cross contamination between the two airstreams. Apart from the need to bring the supply and extract ducts close together, disadvantages centre around condensation within the unit. In winter this can cause the unit to *ice-up* with consequent loss of efficiency and reduced flow rates. This type of HRU may therefore require frost protection, usually in the form of a preheater. If the unit is arranged for summer cooling, condensation can occur on the supply air side and in these circumstances there is a risk of bacterial contamination.

RUN-AROUND SYSTEM

Two pipe-coil heat exchangers, one in the supply duct and one in the exhaust duct are connected with a closed, pumped/piping system filled with water dosed with anti-freeze. The exchanger in the exhaust air duct transfers heat to the water which is circulated to the exchanger in the supply duct. A typical arrangement is shown in Figure 10.9, the control of the supply duct temperature being achieved by a thermostat and three-way valve. The back-up that may be necessary can be obtained by a heater in the piping system, rather than by a second air heating coil.

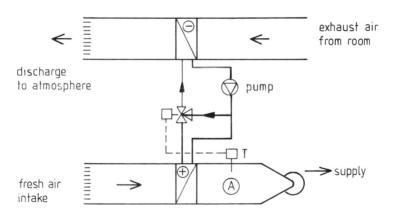

10.9 Two coil run-around system

The efficiency of heat transfer will usually be in the range 40 to 70% depending on the design of the system. The pipework and fittings should be insulated to maintain the maximum efficiency. In very cold weather the liquid leaving the supply air coil can be cold enough to form ice on the exhaust coil and when this is the case a protective heating circuit will be necessary.

This method is used where the supply and exhaust ducts cannot easily be run close together. Generally it is the most appropriate scheme for use where existing systems are to be upgraded. A number of coils may be incorporated, making this method a more flexible way of recovering heat from the exhaust air compared with other methods of energy conservation.

General

To save fan power, the unit can be bypassed when it is not providing heat to the system. For this to be effective the flow rate handled by the fan should also be regulated.

In some systems, where the HRU is unable to meet the whole of the heating load, back-up heaters will be required. If the air conditioning system is switched off, whether at night or at weekends, a conventionally sized heater will be required for start-up of the system. As with all heat exchange equipment, a filter should be placed upstream of the unit in both supply and exhaust airstreams. A preheater can be used to prevent condensation within the heat recovery unit where this could cause problems.

ECONOMIC ANALYSIS

When analysing the economics of an air-to-air heat recovery unit the following cost differences should be considered, compared with a system designed without the unit:

Capital costs
- heat recovery unit;
- associated controls;
- ductwork connections;
- filter in exhaust duct;
- increase in fan size to deal with the additional pressure drops, supply and extract;
- reduced heater and cooler battery sizes;
- reduced boiler and refrigeration plant sizes.

Annual costs
- savings in energy for heating and cooling plant;
- additional fan energy consumption;
- energy consumption of ancillary equipment, eg motor for thermal wheel, pump for run-around system;
- maintenance costs.

Most economic studies of units using 100% outdoor air show a pay-back period of between one and five years. The economics are improved with higher exhaust air temperatures (when heating) and for systems with longer hours of operation. As well as comparing the different methods, it will usually be worth investigating two or three unit sizes since an HRU of the smallest size for a given air flow rate will not necessarily produce the shortest pay-back period.

11 Air Filters

Outdoor and recirculated air used in air conditioning and mechanical ventilation contain impurities and contaminants which need to be removed for the following reasons:

- to maintain an acceptable level of air purity for occupants, process work and building equipment;
- to protect the air conditioning plant;
- to prevent staining on interior decorations and furniture.

In addition, airborne contaminant arising from any process occurring within the treated space may have to be removed from the exhaust air. Examples of this include grease in kitchen extract systems and wood dust in saw mills.

Atmospheric contaminants are classified as solid, liquid, gaseous or organic and usually, within these categories, they are ordered roughly according to particle size and type. The particle size is relatively small and the unit of measurement commonly used is the micrometre ($\mu m = 10^{-6}$ m). Most of the staining in buildings is caused by particles with a size range of between 0.5 and 5 μm. Although invisible to the human eye, these particles have little mass but are very numerous. For comfort air conditioning applications it is usually considered necessary to include a filter with an efficiency high enough to remove most of the particles in that range.

FILTER PERFORMANCE

Efficiency

The most important characteristic of filter performance is its efficiency. Efficiency is a measure of the ability of the filter to remove dust from the air, expressed in terms of the dust concentration upstream and downstream of the filter; it is given by the equation:

$$\eta = \frac{C_1 - C_2}{C_1} 100 \qquad (11.1)$$

where C_1 = concentration upstream
C_2 = concentration downstream.

Concentration can be expressed in terms of: either mass of dust, number of particles or staining power per unit of surface area. Where the concentrations are expressed as a mass, the efficiency is known as the *arrestance*, A.

There are a number of ways of determining filter efficiency using standard tests based on the way concentration is expressed and, as these give different values, it is important to state the test method used. For example most filters whose efficiency is based on the mass of particles removed will have efficiencies of more than 90%, as the heavier, larger particles are more easily trapped by the filter medium; whereas based on a staining power test, the efficiency may be as low as, say, 20%.

There is no one test method suitable for all types of filters and results from different methods cannot be directly compared. To compare the performance of various filters, the same test method must be used together with the same test dust or aerosol, eg the test dust should have the same particle size distribution.

Tests of filters for general purposes

Tests for arrestance and dust spot efficiency are covered by BS 6540: part 1[1], ASHRAE Standard 52–76[2] and Eurovent 4/5.[3]

The gravimetric tests use a standardized, synthetic dust with the specification given in Table 11.1. The test is operated by feeding a measured, controlled quantity of test dust into the air stream, giving the value of C_1 in Table 11.1. The mass of dust leaving the filter under test is determined by passing all the air through a high efficiency filter. The gain in mass of this second filter determines the value of C_2.

proportion by mass	composition
72%	natural earth dust, graded to give the following particle size by mass: 0 to 5 m – 39% 5 to 10 m – 18% 10 to 20 m – 16% tolerance +3% 10 to 40 m – 18% 10 to 80 m – 9% typical bulk density: 1057 kg/m^3
23%	Molocco (carbon) black; a sub-micro particle material forming fluffy aggregages typical bulk density: 272 kg/m^3
5%	No.7 cotton linters, ground and sieved through a 4 mm screen

Table 11.1 Synthetic dust for testing filters (using standards ASHRAE 52-76 and Eurovent 4/5)

Dust spot efficiency

Dust spot efficiency, or blackness, test uses the airborne contaminants already present in the atmosphere. Air is sampled on either side of the filter via target filter papers using equal flow rates. The sample drawn through the downstream target is continuous, whilst that through the upstream is intermittent with a measured time interval such that the opacity of both target papers are comparable. The efficiency is then a function of the total quantity of air passing through each target combined with its light transmission.

Dust holding capacity

The dust holding capacity of a filter is the amount of dust that it can hold whilst maintaining its specified efficiency, or within its rated pressure drop rise, from *clean* to *dirty*; its value is obtained at the same time as the efficiency tests. For most filters efficiency and dust loading are inter-related and therefore the efficiency is obtained a number of times during the course of a test; the dust holding capacity is the integrated amount of dust measured at each part of the test. A different technique is used for self-renewable filters such as the automatic roll filter.

High efficiency filters

High efficiency absolute (HEPA) filters are tested using the sodium chloride test which is covered by BS 3928:1969[4] and by Eurovent 4/4[5]. The test is a discolouration test using small salt crystals which are carried by the air stream. Samples of air upstream and downstream of the filter are passed through a flame of burning hydrogen gas which becomes bright yellow when exposed to sodium chloride. The intensity of the colour relates directly to the concentration of sodium chloride particles, the brightness of flame being measured with a photosensitive cell.

In the USA the DOP penetration test is used. In this test a vapour of di-octyl-phthalate is generated with a particle size of approximately $0.3 \, \mu m$ with a cloud concentration of $80 \, mg/m^3$. Light scattering techniques are used to obtain the upstream and downstream concentrations. The performance of high efficiency filters is often expressed as the *penetration*, P, which is given by:

$$P = 100 - \eta. \tag{11.2}$$

A DOP standard test is also available for on-site testing.

Face velocity

The face velocity, v_f, is the mean velocity of the air (m/s) entering the effective face area of the filter.

$$v_f = \dot{V}/Ad \tag{11.3}$$

where $\dot{V}$ = volumetric flow rate (m^3/s)
A_d = cross sectional area of duct connection of filter (m^2)

Usually the maximum velocity is recommended by the manufacturer, although this is not necessarily consistent with the face velocity of a filter sized for economic operation.

Pressure drop

The pressure drop, Δp, across a filter is related to the face velocity of the following equation:

$$\Delta p = b_{vfn}$$

where $1 < n < 2$.

(11.4)

For most filters the value of the constant b will rise depending on the amount of dust it is holding.

In much of the literature on filters the pressure drop is termed, erroneously, the *resistance*. (The reader is referred to page 233 for the preferred definition of resistance for components in a ducted air system.)

FILTER TYPES

Brief descriptions of the various filters used in air conditioning systems are given below; typical design and operating characteristics are set out in Table 11.2.

filter type	face velocity	pressure drop		arrestance	dust spot or sodium flame efficiency
		initial	final		
	(m/s)	(Pa)	(Pa)	(%)	(%)
dry fabric panel	2.0	70	100	70 – 95	–
bag:					
low efficiency	2.5	70	300	90 – 95	45
medium efficiency	2.0	140	600	98	80 – 85
high efficiency	1.8	200	600	99.5	90 – 95
automatic roll	2.5	80	160	80 – 90	–
absolute:					
low efficiency	2.5	250	500	–	95
medium efficiency	1.2	250	500	–	99.5
high efficiency	1.2	250	500	–	99.997
viscous panel	2.0	80	–	65 – 80	–
automatic	2.0	100	–	65 – 80	–
electrostatic	2.5	100	–	–	90

Table 11.2 Typical design and operating characteristics of air filters used in air conditioning and mechanical ventilation systems

Dry fabric

Dry fabric filters use materials such as cotton wool, glass fibre, cotton fabric and pleated papers, as the filter medium. The efficiency of the filter depends largely on the area of the medium offered to the air stream. The different types of dry fabric filter include the following:

Panel or cell The filter material is mounted in panels up to 600 mm square. Panels are fitted into a common frame and are either placed at 90° to the air flow or as a series of oblique cells as in Figure 11.1. The framework is either part of an air handling packaged unit or a separate plant item connected into the ductwork system. For maintenance, small units are usually side withdrawal, larger units front withdrawal; the dirty cells are discarded and replaced with new.

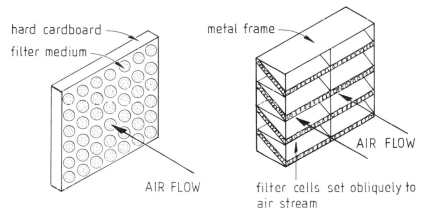

11.1 Panel filters

Bag A set of bags, made of the filter material, is mounted in a frame, with the open ends of the bags facing the air stream as in Figure 11.2. This provides a large area of material which makes the filter suitable for large volume flow rates at relatively low pressure drops. Filter material is used to provide low, medium and high efficiencies; the last is often referred to as a semi-absolute filter. The framework and maintenance is similar to that described for panel filters.

Automatic-roll In this filter the medium is stretched between two rollers, one of which is driven by a motor, as in Figure 11.3. As the filter becomes dirtier the pressure drop across it increases and the differential pressure switch starts the motor which rolls on a clean area of filter until the pressure has dropped to the low limit of the switch. Some media are cleanable but most are discarded once the whole roll is dirty. Currently this filter type is rarely used in new installations.

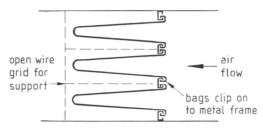

open wire
grid for
support

air
flow

bags clip on
to metal frame

(a) diagrammatic arrangement

11.2 Bag filter (b) bank of filters (Courtesy Vokes Ltd)

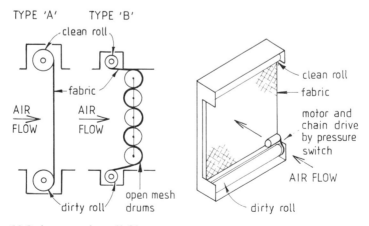

TYPE 'A'   TYPE 'B'

clean roll

fabric

AIR
FLOW

AIR
FLOW

open mesh
drums

dirty roll

clean roll

fabric

motor and
chain drive
by pressure
switch

AIR FLOW

dirty roll

11.3 Automatic roll filter

Absolute High efficiency particulate absolute (HEPA) filters use densely packed pleated filter material as shown in Figure 11.4. Filters are graded under low, medium and high efficiencies, the latter up to 99.99%. Some HEPA filters are available to cope with adverse environmental conditions.

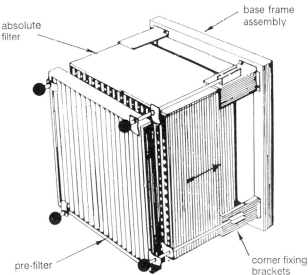

11.4 Absolute filter (Courtesy Vokes Ltd)
    (a) **typical filter cell**
    (b) **fixing arrangements of a single cell filter unit with front access (cells bolt together to form multiple unit)**

The dense filter material results in a high pressure drop for the rated air flow and for economic operation it is often necessary to select a larger filter to give a lower pressure drop, rather than install a unit at the upper limit of at its maximum capacity. A pre-filter should be used to remove the relatively large particles from the incoming air and so prolong the life of the absolute filter cells. When dirty the cells are discarded.

Viscous impingement

Viscous impingement filters have a large dust holding capacity and are used where there are high levels of atmospheric contaminants of large particle size. The filter medium is coated with non-inflammable, non-toxic and odourless oil to which contaminants adhere as they pass through the filter. The different types of filters in this class are:

Cell type In which the cells are arranged in banks, often in 'V' formation; each cell consists of an open ended box in which the medium is formed from either wire mesh or metal swarf coated in oil. Dirty cells can be cleaned and re-used.

Automatic type Consists of hinged plates on open mesh screen which moves under control of a time switch. The plates pass through an oil bath which cleans the oil (Figure 11.5). Sludge should be removed and the oil replaced on a regular basis.

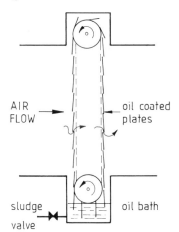

11.5 Viscous impingement filters

Electrostatic

An electrostatic filter consists of two main sections as shown in Figure 11.6. The first section is the *ionizing section*; it consists of a series of fine wires charged to a voltage of up to 13 kV, placed alternately with earthed rods. This sets up a corona discharge and as the airborne particles pass through the ionizing field, they receive a positive electrostatic charge. The second section through which the air passes is the *collector section*; this consists of a series of parallel, vertical metal plates with a potential difference of from 6 to 7 kV between adjacent plates. The ionized dust particles are attracted towards these plates to which they adhere. The plates are sometimes coated with oil to help dust retention.

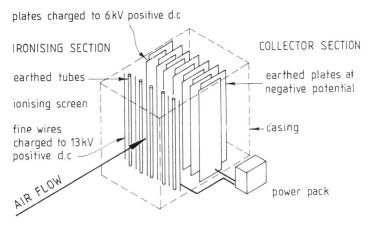

plates charged to 6 kV positive d.c

IRONISING SECTION

earthed tubes

ionising screen

fine wires charged to 13 kV positive d.c

AIR FLOW

COLLECTOR SECTION

earthed plates at negative potential

casing

power pack

11.6 Electrostatic filter

The filters are cleaned automatically by washing with high pressure water.

Adsorption

Adsorption filters are used for the removal of odours, tobacco smoke and some poisonous gases such as SO_2. The air is passed across a large surface area of activated carbon. The contaminating gas is attracted to the carbon which eventually becomes saturated with the gas. It can be re-activated by removing and heating the carbon to an appropriate temperature.

FILTER SELECTION

The primary requirement is for the filters to remove airborne contaminants at the required efficiency for the application, whilst retaining the dust removed from the air stream. Other considerations are as follows:

Environmental conditions Some conditions, eg high humidity and temperature, can adversely affect the filter and hence appropriate media should be chosen.

Maintenance With rising labour costs, the requirement for minimum maintenance becomes more important. Servicing methods include:

– replacing complete filter;
– renewing filter media;
– reconditioning or cleaning;
– automatic self-cleaning – eg electrostatic filters.

The frequency of filter replacement or cleaning should be considered; in some cases it may be reduced by selecting a different type of filter or by using a pre-filter.

Clogged filters are a common cause of failure of systems to supply sufficient air. Suitable indicators and alarms must be provided to ensure filters are maintained correctly and adequate access should be allowed for replacement and cleaning.

Pressure drop   Manufacturers aim to design the filter fabric in such a way as to minimize pressure drop for a given velocity. Nevertheless, the filter should be chosen with as low a pressure drop as possible, consistent with required efficiency and initial costs. This is particularly useful with high efficiency absolute filters which have high pressure drops at the maximum recommended face velocities. A derated filter will usually give increased capital costs for builders work and/or ductwork, which has to be off-set against the fan energy savings.

Pre-filters   The life of expensive high efficiency absolute filters can be extended considerably by providing pre-filters to remove the larger particles. Care must be taken in striking the right economic balance. Pre-filters are provided as an integral part of an electrostatic filter system.

Costs   One type of filter may have low initial costs but high operating costs whereas another type may have high capital cost with low running costs. The initial cost will depend on the filter type, associated ductwork and any ancillary equipment. The operating cost of the filter is made up of the replacement cost of the media at the end of its useful life, the cost of replacing and/or cleaning the filter media, the energy and cleaning material costs and the average fan power required to meet the filter pressure drop (the mean value between clean and dirty operation).

There is no reliable up-to-date cost data available on which to make such an economic selection. An interim solution to this problem is to calculate what may be termed the *owner benefit index* (OBI). This index is the ratio of the filter efficiency to the total owning and operating costs. Thus for a given total cost the OBI increases with the efficiency and the higher the OBI the more benefit to the client/building owner in terms of:

– reduced maintenance, decorating and housekeeping;
– protection of contents and products;
– protection of high efficiency (absolute) filters.

SYSTEM DESIGN

Position of the filter in the system   As it is essential to prevent the build up of contaminants on air conditioning plant items the filter will normally be the first plant item in the system.

For special areas where high standards of air cleanliness are required an additional filter may need to be provided as the last plant item. This should be fitted after the fan to cope with any ingress of air which may occur in the plant. Alternatively each room outlet may be fitted with a terminal filter.

Recirculated air   A case can be made for cleaning contaminated exhaust air to allow it to be recirculated. The use of potentially contaminated air can be an emotive subject, eg recirculation in hospital operating rooms. Even so, techniques for air cleaning are available and may prove to be the cheapest and most appropriate plant arrangement. The choice will usually depend on the quality of the installation and maintenance to ensure acceptable standards of air cleanliness.

Changes in air flow rate   The change in air flow rates resulting from increased pressure drop across a dirty filter can be examined by referring to the system and fan characteristics. If the reduction in air flow rate is unacceptable then some method of controlling near constant flow rate can be achieved by using one of the following methods:

- hand reset damper;
- automatic damper;
- inlet guide vanes operated with a static pressure controller;
- variable speed fan.

Protection against fog and frost   Where the filter is to be used in a system employing 100% outdoor air, or where the filter is placed in the door fresh air duct, fog can sometimes saturate and degrade the filter media; freezing fog can block it completely. Where these conditions are likely it will be advisable to provide a protective heater upstream of the filter under the control of a low limit thermostat.

Fresh air intake   The position of the intake should be sited away from local sources of contamination, and at least 2 metres off the ground but preferably at roof level; this precaution will reduce the load on the filter. A bird and/or insect screen should be fitted. For buildings which are located in areas liable to sand storms, grilles should be fitted to trap any heavy airborne particles.

Installation   Care should be taken to provide adequate seals between the filter units and the holding frames; this is especially important with absolute filters.

12 Fluid Flow – General Principles

This chapter considers the general principles of air and water flow in ducts and pipes, both fluids being treated here as incompressible; the relevant equations are given and the general characteristics of flow are described. The methods of calculating pressure losses in straight ducts and in fittings are illustrated, together with the pressure distributions. The concept of resistance is explained and the chapter concludes by giving some relevant methods for measuring flow rates. An understanding of these topics is important before going on, in subsequent chapters, to consider the design and sizing of the ductwork systems, fan selection and on-site balancing procedures.

The emphasis is on ducted air systems, though many of the principles are also applicable to piped water systems.

FLOW CONTINUITY

If an incompressible fluid flows steadily in a closed duct (or pipe), the *mass* flow rate remains constant:

$$\dot{m} = \rho A \bar{v} = \text{constant} \tag{12.1}$$

If the density remains unchanged, the *volume* flow rate remains constant:

$$\dot{V} = A\bar{v} = \text{constant}. \tag{12.2}$$

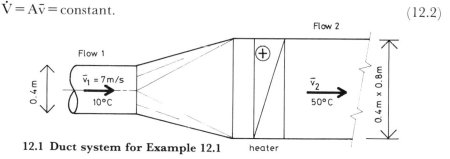

12.1 Duct system for Example 12.1

Example 12.1 The mean air velocity in a 0.4 m diameter duct is 7 m/s, the temperature being 10°C. The duct expands to a rectangular cross section of 0.4×0.8 m to accommodate a heater which raises the air temperature to 50°C. Determine the mean velocity of the air in the rectangular duct after the heater.

Solution

Referring to Figure 12.1.

Air densities (corrected for temperature):

circular duct: $\rho_1 = 1.2\dfrac{293}{(273+10)} = 1.242 \text{ kg/m}^3$

rectangular duct (after heater): $\rho_2 = 1.2\dfrac{293}{(273+50)} = 1.089 \text{ kg/m}^3$

Duct areas:

circular duct: $A_1 = \dfrac{\pi 0.4^2}{4} = 0.126 \text{ m}^2$

rectangular duct, $A_2 = 0.4 \times 0.8 = 0.32 \text{ m}^2$

Using Equation 12.1:

$$\dot{m} = \rho_1 A_1 \bar{v}_1 = \rho_2 A_2 \bar{v}_2$$
$$1.242 \times 0.126 \times 7 = 1.089 \times 0.32 \times \bar{v}_2$$
$$\therefore \bar{v}_2 = 3.14 \text{ m/s}.$$

CONSERVATION OF ENERGY

In any fluid, the total energy is the sum of the potential, pressure and kinetic energies. (This is summarized by BERNOULIS Equation which is given in text books on fluid mechanics.) The potential energy, due to the elevation of the fluid above a datum, is assumed constant for an air flow system. Hence the total energy of the fluid may be expressed in terms of its static (internal) pressure, p_s and its velocity (dynamic) pressure, p_v. Static pressure acts in all directions at a point in a fluid, whereas velocity pressure can only act in the direction of the flow. This is illustrated in Figure 12.2.

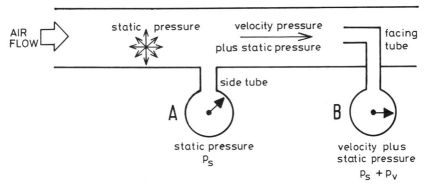

12.2 Pressures in a duct

A pressure gauge A, connected flush with the duct wall will read the static pressure and gauge B, connected to a tube facing into the air stream, will read the sum of the velocity and static pressures, which is the total pressure, p_t, at the point of measurement; ie:

$$p_t = p_s + p_v \qquad\qquad (12.3)$$

The velocity pressure, p_v, is given by the relationship:

$$p_v = 1/2\rho v^2. \qquad\qquad (12.4)$$

The pressure relationship of Equation 12.3 is illustrated by the Pitot-static tube, an instrument used for measuring air pressures in a duct. A diagram of this instrument is shown in Figure 12.3. The *Pitot* tube is the inner tube, the head of which is parallel to the duct axis facing directly into the air stream; this measures the *total* pressure. The *static* (outer) tube with holes at right angles to the air stream senses only static pressure. By making appropriate connections to a manometer, the total, velocity and static pressures can be measured.

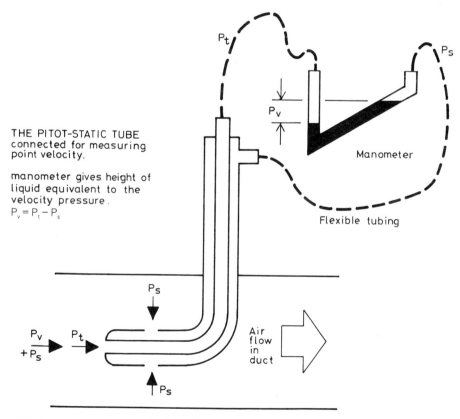

THE PITOT-STATIC TUBE connected for measuring point velocity.

manometer gives height of liquid equivalent to the velocity pressure.
$P_v = P_t - P_s$

Manometer

Flexible tubing

P_t

P_s

P_v

P_s

P_s

$\dfrac{P_v}{+P_s}$ P_t

Air flow in duct

12.3 Pressures measured with a Pitot-static tube

Confusion sometimes arises when adding or subtracting pressures because of the difficulty in deciding whether manometer readings are *plus* or *minus*. Manometer readings always give the *difference* between two pressures. However, in the case where a Pitot-static tube is used to read velocity pressure, the two pressure readings taken are total and static, and the manometer reads velocity pressure direct, this always having a positive value. Care must be taken to allocate the correct sign to readings of total and static pressure, which depend on whether the readings are taken on the suction or discharge side of the fan. In each case, using a manometer, the pressure measured is relative to a datum of atmospheric pressure, hence the *plus* and *minus*. If absolute pressures are considered (or measured) the confusion does not occur, as may be seen in the following examples.

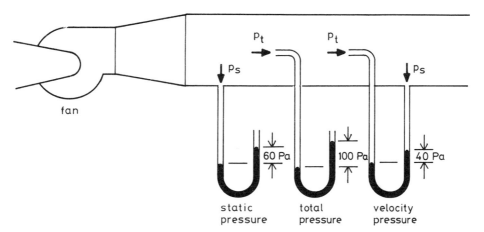

12.4 Pressures measured on the discharge side of a fan

Example 12.2 In a duct on the discharge side of a fan the total pressure is measured as 100 Pa and the static pressure as 60 Pa, relative to atmospheric pressure of 1000 mbar. If the air temperature is 30°C, determine the air velocity.

Solution

Referring to Figure 12.4.

$$\text{Air density} = 1.2 \frac{P_{at}}{1013} \frac{(273+20)}{(273+t_a)} = 1.2 \frac{1000}{1013} \frac{293}{(273+30)}$$

$$= 1.125 \text{ kg/m}^3$$

Since the pressures are measured on the discharge side of the fan, both are *positive* with respect to atmospheric pressure.

Using Equation 12.3 and remembering that 1 mbar = 100 Pa:

Relative to atmospheric
pressure

$p_t = p_s + p_v$
$100 = 60 + p_v$

Absolute pressures (Pa)

$p_t = 100000 + 100 = 100100$
$p_s = 100000 + 60 = 100060$

∴ $p_v = 40$ Pa

p_v 40

Using Equation 12.4:

$$p_v = 0.5\ \rho v^2$$
$$40 = 0.5 \times 1.125\ v^2$$

$$\therefore v = \sqrt{\frac{40}{0.5 \times 1.125}} = 8.43\ \text{m/s}.$$

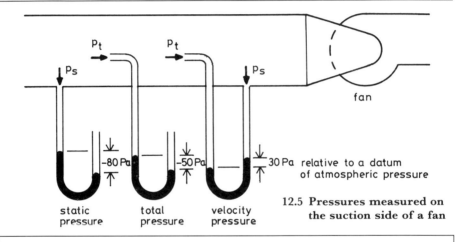

12.5 **Pressures measured on the suction side of a fan**

static pressure   total pressure   velocity pressure

Example 12.3 On the suction side of the fan in Example 12.2, the total pressure is measured as 50 Pa and the static pressure as 80 Pa, relative to atmospheric pressure of 1000 mbar. Determine the air velocity.

Solution

Referring to Figure 12.5. Since the pressures are measured on the suction side of the fan, both are *negative* with respect to atmospheric pressure. Using Equation 12.3:

Relative to atmospheric
pressure

$p_t = p_s + p_v$
$-50 = -80 + p_v$

Absolute pressures (Pa)

$p_t = 100000 - 50 = 999950$
$p_s = 100000 - 80 = 999920$

∴ $p_v = 30$ Pa

p_v 30

Using Equation 12.4:

$$p_v = 0.5 \ \rho v^2$$
$$30 = 0.5 \times 1.125 \ v^2$$

$$\therefore \ v = \sqrt{\frac{30}{0.5 \times 1.125}} = 7.30 \ \text{m/s}.$$

REYNOLDS NUMBER

The ratio of inertia force to viscous force of a fluid flowing in a closed duct or pipe is known as the REYNOLDS number, Re. For systems to have the same flow regimes at different flow conditions, the REYNOLDS numbers must be equal.

$$Re_D = \frac{vD}{\nu}$$

$$(12.5)$$

where ν = kinematic viscosity.

For air at temperature t_a:

$$\nu = (1.32 + 0.0092 \ t_a) \times 10^{-5} \ \text{m}^2/\text{s} \qquad (12.6)$$

$\therefore$ for air at 20°C, $\nu = 1.5 \times 10^{-5} \ \text{m}^2/\text{s}$

For water there is no simple relationship relating kinematic viscosity to temperature and reference should be made to published table values, eg the *CIBSE Guide*.[1]

For water at: 20°C, $\nu = 1.00 \times 10^{-6} \ \text{m}^2/\text{s}$
80°C, $\nu = 0.36 \times 10^{-6} \ \text{m}^2/\text{s}$

The diameter, D, is a characteristic dimension of the duct through which the fluid is flowing. When the duct is other than circular, the *equivalent* diameter, D_e, is used; for a rectangular duct this is known as the *hydraulic mean diameter* which is given by:

$$D_e = \frac{2A}{(b + W)} \qquad (12.7)$$

Example 12.4 Determine the REYNOLDS number of the air flowing in the rectangular duct in Example 12.1.

Solution

Using the data from Example 12.1:

Kinematic viscosity of air at temperature of 50°C is given by Equation 12.6:

$$v = (1.32 + 0.0092t_a) \times 10^{-5} = (1.32 + 0.0092 \times 50) \times 10^{-5}$$
$$= 1.78 \times 10^{-5} \, \text{m}^2/\text{s}.$$

Hydraulic diameter is given by Equation 12.7:

$$D_e = \frac{2A}{(b+W)} = \frac{2 \times 0.4 \times 0.8}{(0.4 + 0.8)} = 0.533 \, \text{m}.$$

Mean duct velocity in the rectangular duct = 3.14 m/s.
The REYNOLDS number is given by Equation 12.5:

$$Re_D = \frac{vD}{v} = \frac{3.14 \times 0.533}{1.78 \times 10^{-5}} = 0.94 \times 10^5.$$

FLOW CHARACTERISTICS

When a fluid such as air is introduced into a circular, straight duct with a suitably shaped inlet the velocities near the inlet will be equal at all points in a transverse section up to the vicinity of the duct wall. At this point the boundary layer is beginning to form and, as the fluid progresses down the duct, wall friction acts on the fluid the layer becomes thicker. A symmetrical velocity profile develops; beyond a certain distance the velocity profile remains the same, and this flow is known as *fully developed pipe flow*. The progressive development of the velocity profile along such a duct is shown in Figure 12.6, though the actual shape of the profile will depend on the REYNOLDS number, the nature of the flow (which can be either laminar or turbulent) and the roughness of the duct walls.

By introducing a suitable tracer into a fluid it can be shown that at low speeds no mixing occurs, but when the speed has increased beyond a critical value, the tracer becomes rapidly diffused in the fluid due to small and rapid velocity fluctuations. The flow without mixing is known as *laminar* flow and that with mixing as *turbulent* flow. The transition between the two types of flow occurs at a REYNOLDS number approximately in the range 2300 to 4000. Above Re = 4000 the flow will be turbulent and below Re = 2300 the flow will be laminar. The speed at which the change occurs between the two types of flow is known as the *critical speed*, v_{crit}, ie:

$$v_{crit} < \frac{3000v}{D}$$

The velocities encountered in ducted air and piped water systems are usually well above the critical speed. The lengths of duct required after the entrance section to achieve *fully developed, turbulent pipe flow* depends on the intake configuration and the amount of disturbance in the entering fluid. There is no general relationship to predict this length; with a rounded entrance the distance is more than 30 diameters of duct length but with an abrupt entry perhaps half this distance.

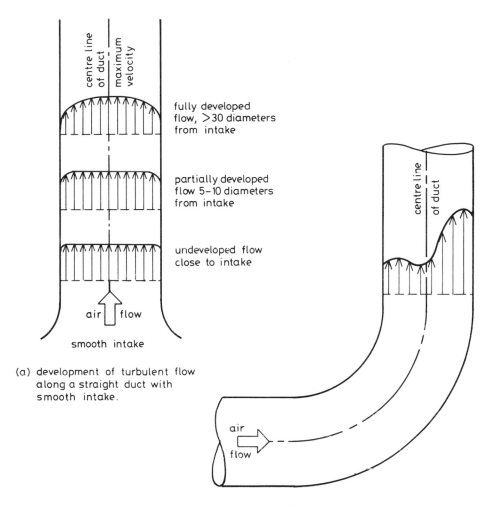

fully developed
flow, >30 diameters
from intake

partially developed
flow 5–10 diameters
from intake

undeveloped flow
close to intake

air flow

smooth intake

(a) development of turbulent flow
    along a straight duct with
    smooth intake.

air
flow

12.6 Flow profiles in a duct   (b) asymmetrical flow following a bend.

Asymmetrical velocity profiles, also illustrated in Figure 12.6, can be produced at the beginning of a straight duct when a certain type of entrance is used, such as a bend or an abrupt constriction. This profile will continue down the duct but with a reducing amount of asymmetry until, ultimately, fully developed pipe flow is established. To achieve this condition of flow a straight length of between 50 and 100 equivalent diameters is likely to be required.

The majority of published investigations of fluid flow in closed conduits have been in circular ducts or pipes. Flows in non-circular ducts have not been investigated to the same extent though, in the essentials of development and type of flow, it will be similar to that in a circular duct. In addition to the normal flow characteristics, secondary flows are set up, superimposed on the main flow. These secondary flows have the effect of feeding more air into the corner of the duct, thus increasing the velocities in these areas above that which would exist without the effect. However, though not without importance, it is unlikely that these secondary flows have a significant influence on topics of interest to the building services engineer, eg the accuracy of on-site measurement of flow rate.

PRESSURE LOSSES IN STRAIGHT DUCTS

For fully developed, turbulent pipe-flow the pressure loss in a straight duct is given by:

$$\Delta p_f = \frac{fL}{D} \frac{\rho \bar{v}^2}{2} \tag{12.8}$$

or:

$$\Delta p_f = K_f p_v \tag{12.9}$$

where:

$$K_f = \frac{fL}{D} \tag{12.10}$$

K_f is termed the *straight duct loss coefficient*.

The friction factor, f, varies with duct diameter, the duct wall roughness, k, and the REYNOLDS number. These variations are shown on the MOODY chart, Figure 12.7.

The friction factor may be calculated from the following equation:

$$f = \frac{0.25}{[\log_{10}(k/3.7D + 5.74/Re_D^{0.9})]^2} \tag{12.11}$$

Example 12.5   A 12 metre length of 0.3 m diameter galvanized sheet steel duct carries an air flow rate of 0.4 m³/s. If the average roughness of the duct is 0.15 mm and the air density is 1.2 kg/m³ at 20°C, determine the pressure loss due to friction.

Solution

Duct cross section area = 0.0707 m².
Mean air velocity = 0.4/0.0707 = 5.66 m/s.

12.7 Moody chart – friction coefficient versus Reynolds number
(Based on Figure 8.1, D S Miller, *Internal Flow Systems*, (2nd edition), 1990, BHRA, Cranfield, UK, by permission)

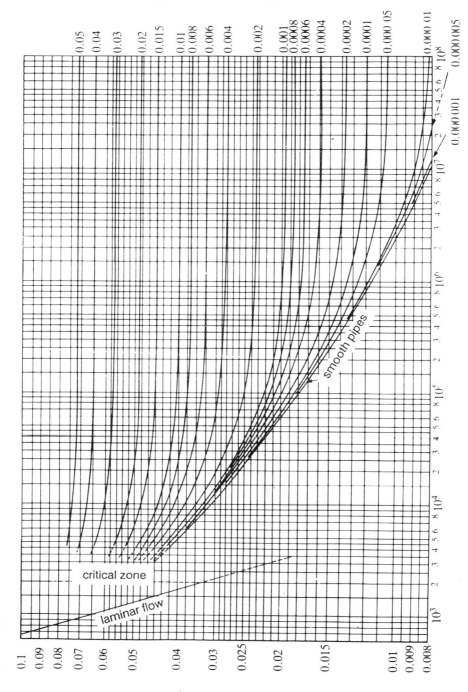

relative roughness, k/D

Reynolds number, Re

friction coefficient, f

smooth pipes

critical zone

laminar flow

The REYNOLDS number is given by Equation 12.5:

$$Re = \frac{\bar{v}D}{v} = \frac{5.66 \times 0.3}{1.5 \times 10^{-5}} = 1.13 \times 10^5.$$

The friction factor is given by Equation 12.11:

$$f = \frac{0.25}{[\log_{10}(k/3.7D + 5.74/Re_D^{0.9})]^2}$$

$$f = \frac{0.25}{\left[\log_{10}\left(\dfrac{0.00015}{3.7 \times 0.3} + \dfrac{5.74}{113000^{0.9}}\right)\right]^2}$$

$$= 0.201.$$

(This result may be confirmed from Figure 12.7. The relative roughness $= k/D = 0.00015/0.3 = 0.0005$ and at a REYNOLDS number of 1.13×10^5 the friction factor is 0.0201, a figure which agrees with the calculated value.)

The equivalent pressure loss coefficient for the duct is given by Equation 12.10, ie

$$K_f = \frac{fL}{D} = \frac{0.0201 \times 12}{0.3} = 0.804.$$

The velocity pressure is given by Equation 12.4:

$$p_v = \frac{1.2 \times 5.66^2}{2} = 19.2 \text{ Pa}$$

$\therefore$ pressure loss due to friction, using Equation 12.9:

$$\Delta p_f = K_f p_v = 0.804 \times 19.2 = 15.4 \text{ Pa}.$$

For manual calculations it is usual to use a *friction chart* as described in Chapter 13, leaving the use of Equations 12.8 and 12.11 for use in computer programs.

PRESSURE DISTRIBUTION

The total energy in a fluid will remain constant unless changed by external forces. Thus the total pressure will be constant unless frictional or dynamic pressure losses are present. Therefore if the velocity changes in a fluid, the static pressure will also change to ensure constant total pressure in a *no-loss* system. Since all components in a system do have pressure losses, the difference in total pressures measured at two points in a system will be equal to the pressure losses between the two points and the change in static pressure will be determined by using Equation 12.3. *Relative to a datum of atmospheric pressure*, velocity pressure is always positive; total and static pressures may be either negative or positive depending on whether the duct is on the suction or discharge side of the fan, as previously explained.

Straight duct pressure distribution

With a straight duct of constant cross-sectional area the velocity pressure remains constant along the length of the duct. Therefore, the static pressure drop along the duct is equal to the pressure loss due to friction.

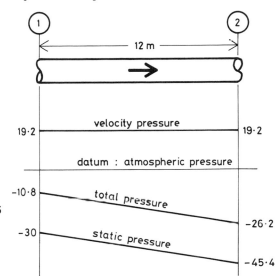

12.8 Pressure distribution in a straight duct: Example 12.6 (all pressures in pascals)

Example 12.6 Plot the pressure distributions for the 0.3 m diameter duct in Example 12.5. The duct is on the suction side of the fan and the static pressure at the entrance to the duct is 30 Pa, relative to atmospheric pressure.

Solution

The pressures are plotted in Figure 12.8.
   From Example 12.5:
      duct velocity $= 5.66$ m/s
      air density $= 1.2$ kg/m^3
      friction pressure drop $= 15.5$ Pa
      velocity pressure $= 19.2$ Pa.
   Static pressure at end of duct, $p_{s2} = (-30) - 15.4 = -45.4$ Pa.
   Using Equation 12.3 the total pressure at duct entrance:

$$p_{t1} = p_{s1} + p_v = (-30) + 19.2 = -10.8 \text{ Pa}$$

total pressure at duct exit:

$$p_{t2} = p_{s2} + p_v = (-45.4) + 19.2 = -26.2 \text{ Pa}.$$

PRESSURE LOSSES IN FITTINGS

Fittings in ductwork systems include bends, expansions, contractions, branches, duct inlets from large spaces, discharges to large spaces and constrictions such as orifice plates. The pressure loss across such fittings is calculated from the product of the velocity pressure, p_v, and a pressure loss coefficient, K, ie:

$$\Delta p = K p_v \qquad (12.12)$$

where p_v is based on the mean duct velocity, $\bar{v}$, as determined from Equation 12.2 and K is related to the fitting.

Example 12.7 Determine the pressure loss in a bend which follows the straight length of duct in Example 12.5, and which has a discharge duct of the same cross section. The bend loss coefficient, K_b, is given as 0.24.

Solution

From Example 12.5:

velocity pressure, $p_v = 19.2$ Pa.

The pressure loss across the bend is given by Equation 12.12:

$$\Delta p = K_b p_v = 0.24 \times 19.2 = 4.61 \text{ Pa.}$$

Determination of loss coefficients

Pressure loss coefficients for duct fittings, defined by Equation 12.12, are obtained from experimental data. In experiments to determine these coefficients, the standard test arrangement is, initially, a straight duct sufficiently long to produce fully developed pipe flow at the upstream face of the fitting; another straight duct downstream of the fitting allows redevelopment of the flow after the disturbance caused by the fitting. The pressure gradient due to friction losses along the duct will be uniform with no fitting present, but with a fitting installed, additional losses occur. These are made up of dynamic and friction losses in the fitting itself and additional friction losses in the downstream duct due to turbulence and velocity gradients generated by the fitting. For calculation purposes these additional losses are attributed to the fitting, it being assumed that the fitting under test has *no duct length*. The general relationship between pressure losses in the straight duct and in a fitting are illustrated in Figure 12.9.

Tables of loss coefficients for the various fittings, drawn from a number of sources, are given in the following chapter.

Loss coefficients may have some dependence on REYNOLDS number: the correction factors relating to bends are given in Chapter 13, (Figure 13.7). However, there is little reliable data for the majority of fittings and it is not

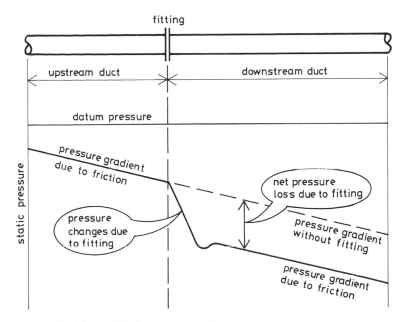

12.9 Definition of fitting pressure loss

usual to allow for REYNOLDS number effects in duct sizing calculations in commercial systems.

Interaction effects

In ductwork systems installed in buildings, it is often the case that the inlet and outlet duct lengths are too short to allow fully developed pipe flow at the entrance to a fitting and redevelopment of the flow after the disturbance. The fully developed pipe flow arrangement described above provides a datum from which departures due to interference between components can be investigated. Interference between two fittings may occur when the length of straight duct between them is insufficient to allow complete redevelopment of the flow after the first item. Where there is no straight duct between fittings, or where the spacer piece is relatively short, the static pressure distributions in the fittings often interact, causing effects far different from those encountered with long lengths of straight duct upstream and downstream of the fitting. Consequently pressure losses will differ from the sum of losses in the mutually independent fittings. MILLER[2] has investigated and published the interaction effects for a number of typical pairs of fittings, for example various arrangements of bends, bend-diffuser, and diffuser-bend combinations. Using his data, the net pressure loss across a pair of fittings may be calculated by using a correction factor applied to the sum of their individual loss coefficients based on the inlet velocity pressure to the first component. Some interaction factors for bends are given in Table 13.8 page 261.

PRESSURE DISTRIBUTION IN DUCT FITTINGS

Duct expansion
For a duct expansion the outlet velocity pressure is lower than inlet velocity

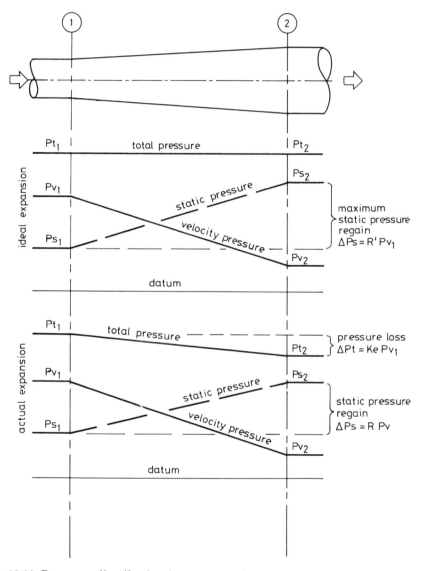

12.10 Pressure distribution in an expansion

pressure. In a *ideal* expansion there will be no pressure loss and the total pressure at the outlet will be the equal to that at the inlet. It follows, therefore, that there will be a *rise* in static pressure corresponding to the drop in velocity pressure. Static pressure regain (or recovery) is an important characteristic of an expansion.

Referring to Figure 12.10 and using Equation 12.3:

$$p_{t1} = p_{t2} = p_{s1} + p_{v1} = p_{s2} + p_{v2}$$
$$p_{s2} - p_{s1} = p_{v1} - p_{v2}.$$

The static pressure regain coefficient, R', for *maximum* pressure recovery is defined by:

$$R'p_{v1} = p_{v1} - p_{v2}$$

$$\therefore R' = 1 - \left(\frac{\bar{v}_2}{\bar{v}_1}\right)^2 \tag{12.13}$$

For a *real* expansion the pressure loss is given by:

$$\Delta p = K_e p_{v1}$$

The *actual* static pressure recovery with regain coefficient, R, is therefore given by:

$$\Delta p_{rs} = R p_{v1} \tag{12.14}$$

$$R p_{v1} = R' p_{v1} - K_e p_{v1}$$

$$\therefore R = R' - K_e \tag{12.15}$$

Example 12.8 Plot the pressure distribution for an expansion with an inlet velocity of 10 m/s and an outlet velocity of 5 m/s. The total angle of expansion is 30° and the inlet static pressure 30 Pa relative to atmosphere on the discharge side of the fan.

Solution

Referring to Figure 12.11:

Using a standard air density, the velocity pressures are:

$$\begin{array}{ll} \textit{inlet:} & p_{v1} = 0.6 \times 10^2 = 60 \text{ Pa} \\ \textit{outlet:} & p_{v2} = 0.6 \times 5^2 = 15 \text{ Pa.} \end{array}$$

For a *total* expansion angle of 30°:

$$K_e = 0.17 \text{ (from Table 13.9)}$$

total pressure loss across expansion is given by Equation 12.12:

$$\Delta p_e = K_e p_{v1} = 0.17 \times 60 = 10.2 \text{ Pa.}$$

The theoretical pressure recovery coefficient is given by Equation 12.13:

$$R' = 1 - \left(\frac{\bar{v}_2}{\bar{v}_1}\right)^2 = 1 - 0.5^2 = 0.75.$$

The pressure regain coefficient is given by Equation 12.15:

$$R = R' - K_e = 0.75 - 0.17 = 0.58.$$

The static pressure recovery is given by Equation 12.14:

$$Rp_{v1} = 0.58 \times 60 = 34.8 \text{ Pa}.$$

Static pressure at outlet $= 30 + 34.8 = 64.8$ Pa.
Total pressure from Equation 12.3:

inlet	$p_{t1} = p_{s1} + p_{v1} = 30 + 60 = 90$ Pa	
outlet	$p_{t2} = p_{s2} + p_{v2} = 64.8 + 15 = 79.8$ Pa.	

Difference in total pressures $= 90 - 79.8 = 10.2$ Pa, the pressure loss calculated above.

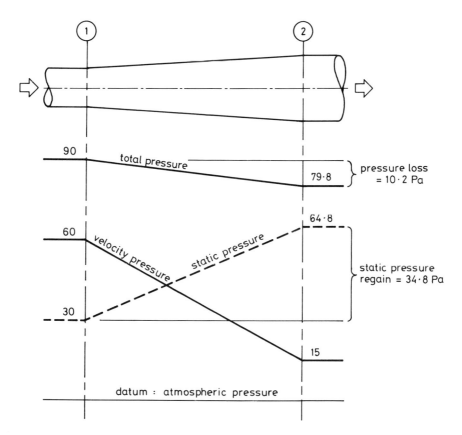

**12.11 Pressure changes in an expansion: Example 12.8
(all pressures in pascals)**

Contraction

With a contraction fitting the velocity increases and this increase results in a *depression* of the static pressure. For *gradual* contractions there is only a small pressure loss and therefore the static depression is approximately equal to the increase in velocity pressure. Here again, as with a duct expansion, the pressure distribution can be plotted graphically by using the pressure relationship in Equation 12.3 and illustrated in Figure 12.12.

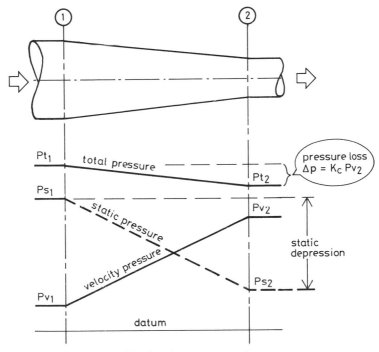

12.12 **Pressure distribution in a contraction**

System intake and discharge

The pressure distribution at an intake to, and discharge from, a system are illustrated in Figures 12.13 and 12.14.

If there are no losses at the intake, then the static pressure will fall below atmospheric pressure to a value equal to the velocity pressure, thus maintaining zero total pressure. In practice the static pressure will be lower than the numerical value of p_v by an amount equal to the net pressure loss in the intake. (Use is made of this phenomenon in an inlet to duct-work systems for measuring flow rates, eg conical inlets described on page 238–9.)

At the discharge from the system the total pressure drops to ambient pressure, (usually atmospheric pressure), of the space into which the air is being discharged. At the face of the outlet the system static pressure is at the datum (atmospheric) pressure and the pressure loss will be the velocity pressure at the outlet.

12.13 Pressure distribution at a suction inlet

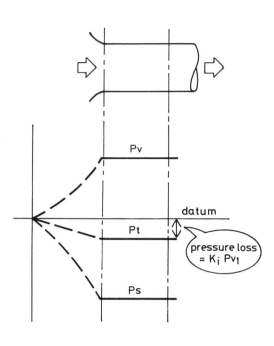

12.14 Pressure distribution at a discharge

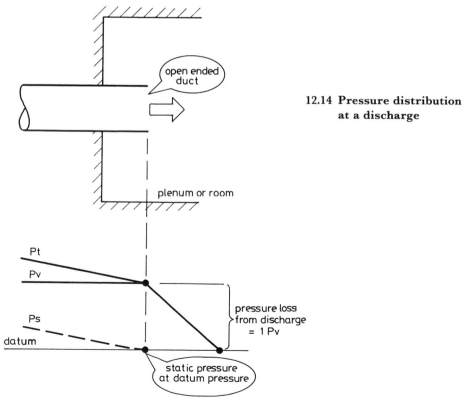

Complete system

For the pressure distribution in a complete system it is necessary to calculate the total pressure losses in each of the system components such as straight ducts, fittings and plant items. These losses will determine the total pressure gradients. Then, knowing the velocity pressures in each part of the system, the static pressures at the entrances and discharges from each of the system components are found following the rules already outlined above. The pressure distributions through a complete system may then be plotted; those for a ducted fan are given in Figure 14.18 of Chapter 14.

RESISTANCE

The concept of resistance is useful in solving various problems associated with fluid flow networks with fully developed turbulent pipe-flow. The assumption is that the straight duct friction and the duct fitting velocity pressure coefficients remain constant, ie they are independent of REYNOLDS number. The resistance of a length of duct/pipe, fitting or plant item in a system is defined by what is known as the *square law*; that is:

$$\Delta p = r \dot{V}^2 \tag{12.16}$$

Resistance in series

Figure 12.15 shows three fittings with resistances r_1, r_2 and r_3 connected in series. The flow rate is the same for each item and the total pressure drop is the sum of the pressure drops across the individual items.

$$\Delta p_t = \Delta p_1 + \Delta p_2 + \Delta p_3$$

Using Equation 12.16 for each of these pressure drops:

$$r_t \dot{V} = r_1 \dot{V}^2 + r_2 \dot{V}^2 + r_3 \dot{V}^2$$
$$r_t = r_1 + r_2 + r_3 \tag{12.17}$$

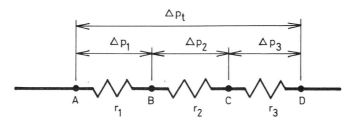

12.15 Resistances in series

Resistance in parallel

Figure 12.16 shows three fittings with resistances r_1, r_2 and r_3 connected in parallel. The total flow rate divides into the separate branches, the pressure drop across the nodal points **A** and **B** being the same for each branch.

$$\dot{V}_t = \dot{V}_1 + \dot{V}_2 + \dot{V}_3$$

Using Equation 12.16 for each of these flow rates:

$$\sqrt{\frac{\Delta p_t}{r_t}} = \sqrt{\frac{\Delta p_1}{r_1}} + \sqrt{\frac{\Delta p_2}{r_2}} + \sqrt{\frac{\Delta p_3}{r_3}}$$

Since

$$\Delta p_t = \Delta p_1 = \Delta p_2 = \Delta p_3$$

$$\frac{1}{\sqrt{r_t}} = \frac{1}{\sqrt{r_1}} + \frac{1}{\sqrt{r_2}} + \frac{1}{\sqrt{r_3}} \tag{12.18}$$

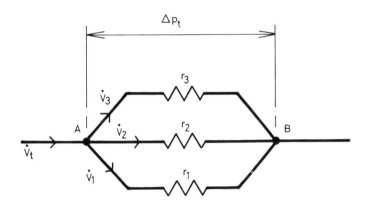

12.16 Resistances in parallel

Example 12.9 Figure 12.17 shows part of a system required to handle outdoor air in varying proportions. When operating on 100% outdoor air, the total flow rate is 3 m³/s, 1.2 m³/s passing through the preheater. When in this mode of operation the following pressure drops occur:

A to **B** – 20 Pa
B to **C** – 45 Pa
C to **D** – 12 Pa.

If **A** is at atmospheric pressure and the pressure at **D** remains constant, determine the percentage change in flow rate through the preheater when the damper is fully shut.

Solution

Referring to a schematic of the system in Figure 12.18 which shows resistances in place of the plant items in Figure 12.17; for the flow rates with the damper fully open the relevant resistances are calculated using Equation 12.16.

$$\Delta p = r \dot{V}^2$$

$$r_1 = \frac{20}{3^2} = 2.22$$

$$r_2 = \frac{45}{1.2^2} = 31.2$$

$$r_4 = \frac{12}{3^2} = 1.33.$$

With the damper fully shut the resistances r_1, r_2 and r_4 are in series. Using Equation 12.17, the total resistance is obtained:

$$r_t = r_1 + r_2 + r_4$$
$$= 2.22 + 31.2 + 1.33 = 34.75.$$

The pressure at point **D** = 20 + 45 + 12 = 77 Pa.
Using Equation 12.17, the flow rate is given by:

$$\Delta p = r \dot{V}^2$$
$$77 = 34.75 \dot{V}^2$$
$$\dot{V} = 1.49 \text{ m}^3/\text{s}.$$

∴ the flow rate through the heater increases by 0.29 m³/s, or by 24%, with the damper fully shut.

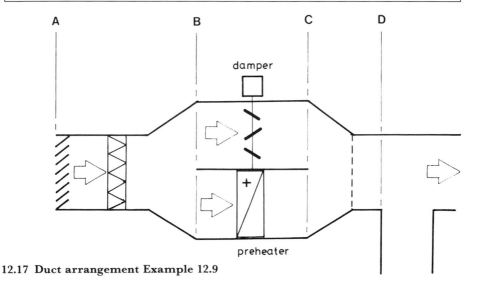

12.17 Duct arrangement Example 12.9

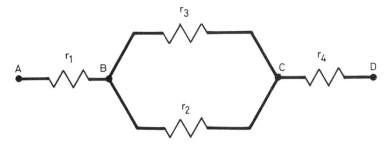

12.18 Resistances in Example 12.9

Example 12.10 Air is supplied to two outlets which discharge to atmospheric pressure. In the diagram of the system (Figure 12.19) the pressure at the nodal point **B** is 20 Pa when the flow rates are:

$$\text{outlet } 1 - 0.3 \text{ m}^3/\text{s}$$
$$\text{outlet } 2 - 0.5 \text{ m}^3/\text{s}$$

If the total flow rate in **AB** is increased to $1.2 \text{ m}^3/\text{s}$ determine the percentage change in the flow rates from the two outlets.

Solution

Using Equation 12.16 the resistance are:

$$\text{branch 1: } r_1 = \frac{20}{0.3^2} = 222$$

$$\text{branch 2: } r_2 = \frac{20}{0.5^2} = 80$$

The two branches are parallel resistances, both flows discharging to atmospheric pressure, which is the datum or zero pressure. Using Equation 12.18:

$$\frac{1}{\sqrt{r_t}} = \frac{1}{\sqrt{r_1}} + \frac{1}{\sqrt{r_2}}$$

$$\frac{1}{\sqrt{r_t}} = \frac{1}{\sqrt{222}} + \frac{1}{\sqrt{80}}$$

$$r_t = 31.2.$$

The increased pressure at **B** at the increased flow rate of $1.2 \text{ m}^3/\text{s}$ is given by Equation 12.16:

$$\Delta p = r\dot{V}^2$$

$$= 31.2 \times 1.2^2 = 44.9 \text{ Pa}.$$

The flow rate from outlet 1 is given by:

$$V_1 = \sqrt{\frac{44.9}{222}} = 0.45 \text{ m}^3/\text{s}.$$

The flow rate from outlet 2 is given by:

$$V_2 = \sqrt{\frac{44.9}{80}} = 0.75 \text{ m}^3/\text{s}.$$

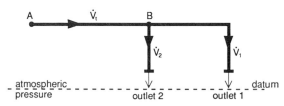

12.19 Duct system: Example 12.10

Comparing the new flow rates with the original values in Example 12.10 it is seen that both increase by the same percentage. That is, when the total flow rate supplying a pair of outlets is increased (or decreased) the flow rates in the outlets change in the same proportion relative to one another. This is an important principle which is used to develop the on-site proportional balancing procedures in Chapter 15.

Pressure loss coefficient and resistance

The relationship between the velocity pressure coefficient and resistance of a fitting, or a length of duct, is obtained as follows: it is assumed that the fitting has a characteristic area, A, from which the mean velocity, $\bar{v}$, is obtained:

Using resistance, the pressure drop is found from Equation 12.16:

$$\Delta p = r \dot{V}^2.$$

Using loss coefficient, the pressure loss is found from Equation 12.12:

$$\Delta p = K p_v = K 0.5 \rho v^2.$$

Equating these two relationships:

$$r \dot{V}^2 = K 0.5 \rho \bar{v}^2$$

$$r \dot{V}^2 = K 0.5 \rho \left(\frac{V}{A}\right)^2$$

$$\therefore r = K 0.5 \rho \left(\frac{1}{A}\right)^2. \tag{12.19}$$

Equation 12.19 shows that resistance is directly proportional to air density.

Example 12.11   A 350 mm diameter bend has a pressure loss coefficient of 0.3. Determine its resistance for an air density of 1.1 kg/m³.

Solution

Cross section area of bend $= \dfrac{\pi 0.35^2}{4} = 0.0962 \text{ m}^2$.

Using Equation 12.19:

$$r = K0.5\rho \left(\frac{1}{A}\right)^2 = 0.3 \times 0.5 \times 1.1 \left(\frac{1}{0.0962}\right)^2 = 17.8.$$

MEASUREMENT OF FLOW RATE

In the building services industry there are two basic methods for measuring the flow rate in pipes and ducts:

– by using a pressure drop device with a known flow characteristic;
– by obtaining the mean velocity and multiplying by the duct cross sectional area.

For commercial and industrial installations pressure difference devices are used extensively in water and steam pipe distribution networks. Because of the significant additional fan power requirement, large duct cross sections and inadequate straight lengths of duct, pressure difference devices are rarely used for air systems. In the latter case therefore, it is more usual to measure the mean velocity. The principal methods of obtaining the flow rates are described below, using data from current ISO and British Standards.

Pressure difference devices

Pressure difference devices include orifice plates and conical inlets and these are illustrated in Figure 12.20. Other devices include nozzles and venturi meters.

A single flow equation is used for these devices, ie:

$$\dot{V} = \alpha\varepsilon A_0 \sqrt{2\Delta p'/\rho} \tag{12.20}$$

For orifice plates, the values of the flow coefficient, α, depends on the orifice to duct diameter ratio, β, and also to some extent on the duct REYNOLDS number, Re_D, and the position of the pressure tappings. Table 12.1 gives typical values of flow coefficients for REYNOLDS number of 7×10^4 with tappings placed one duct diameter *upstream* and half a duct diameter *downstream* of the plate. These may be used to determine the flow with reasonable accuracy but if a more precise calculation is required then equations, tables or graphs, given in the Standards, may be used. The expansibility factor, ε, may be taken as unity for the majority of cases dealt with by air conditioning engineers.

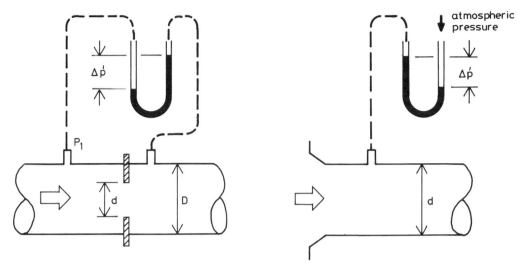

12.20 Pressure difference devices for the measurement of flow rate
(a) orifice plate (b) conical inlet

β	α	β	α	β	α
0.20	0.598	0.40	0.610	0.60	0.656
0.22	0.598	0.42	0.613	0.62	0.664
0.24	0.599	0.44	0.616	0.64	0.673
0.26	0.600	0.46	0.619	0.66	0.683
0.28	0.601	0.48	0.623	0.68	0.694
0.30	0.602	0.50	0.627	0.70	0.706
0.32	0.603	0.52	0.631	0.71	0.713
0.34	0.605	0.54	0.636	0.72	0.720
0.36	0.606	0.56	0.642	0.73	0.728
0.38	0.608	0.58	0.649	0.74	0.737

Table 12.1 Flow coefficients, a, for orifice plate with
D and D/2 tappings at $Re_D = 7 \times 10^4$

Pressure distribution through an orifice plate

For an orifice plate, mounted in a duct, the air contracts to pass through the orifice and then expands to fill the downstream duct. This expansion of the fluid means that there will be a regain in static pressure and the *net* pressure drop will be less than the pressure difference measured to determine the flow rate. The duct length for full pressure recovery is approximately five equivalent duct diameters. These relative pressure changes are illustrated in Figure 12.21.

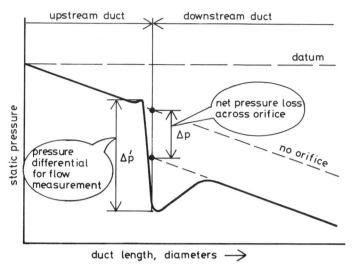

upstream duct | downstream duct

datum

static pressure

net pressure loss across orifice

Δp

pressure differential for flow measurement

Δṕ

no orifice

duct length, diameters ⟶

12.21 Pressure distribution in the vicinity of an orifice plate

The *net* pressure drop across an orifice plate is related to the measured differential pressure, $\Delta p'$, by the following expression:

$$\Delta p = \frac{(1 - \beta^2)}{(1 + \beta^2)} \Delta p'. \tag{12.21}$$

Conical inlet

With a conical inlet, as shown in Figure 12.20b, the differential pressure, $\Delta p'$, required for use in Equation 12.20 is the duct static pressure measured at tappings placed half a duct diameter from the inlet cone. To some extent the compound flow coefficient, $\alpha\varepsilon$, in Equation 12.20 depends on the REYNOLDS number, but may be taken as 0.96 when $Re_d \geqslant 3 \times 10^5$.

MEASUREMENT OF MEAN VELOCITY

To obtain the mean velocity it is usual to traverse the duct with a suitable instrument. The recommended methods for circular and rectangular ducts are detailed below and are independent of the instrument used to measure the point velocities, though it is of course important that the head of the instrument can be accommodated at the positions close to the duct wall.

In the traverses described, the mean velocity, $\bar{v}$, is obtained from the arithmetic average of the total of the individual velocities.

Traverse of a circular duct

For a circular duct, the standard traverse is known as the *log-linear traverse;*

this has 4, 6, 8 or 10-points per traverse line. The positions for locating the instrument head are given in Table 12.2 and the 8-point traverse illustrated in Figure 12.22. It is usual to make at least two traverses across the duct at 90° to one another and the engineer makes a judgement about the number of measuring points. If two 8-point traverse are selected, the mean velocity is obtained from the arithmetic mean of 16 readings of velocity. BS848[3] recommends three traverses at 60° to one another with 6 points per traverse line; however, because of access limitations to air ducts in buildings, this will often be impractical.

points per traverse line	distance from wall in duct diameters					
4	0.043	0.290	0.710	0.957		
6	0.032	0.135	0.321	0.679	0.865	0.968
8	0.021	0.117	0.184	0.345		
	0.655	0.816	0.883	0.979		
10	0.019	0.076	0.153	0.217	0.361	
	0.639	0.783	0.847	0.924	0.981	

Table 12.2 **Location of measuring points for log-linear traverse of a circular duct**

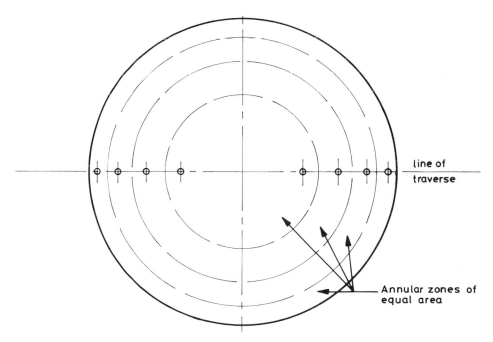

12.22 **Log-linear traverse of a circular duct (8 points per traverse line)**

Traverse of a rectangular duct

The recommended traverse of a rectangular duct is known as a *log-Tchebycheff traverse*. A number (m) of straight traverse lines are selected parallel to the smaller side of the rectangle and on each of them a number (n) of measuring points are located. The positions of the traverse lines and measuring positions are determined in accordance with the Table 12.3.

points per traverse line	proportional distance of measuring positions from inside wall of duct
5	0.074 0.288 0.5 0.712 0.926
6	0.061 0.235 0.437 0.563 0.765 0.939
7	0.053 0.203 0.366 0.5 0.634 0.797 0.947

Table 12.3 Positions of measuring points and traverse lines for a rectangular duct by the log-Tchebycheff rule

Again, choice of the number of points is left to the engineer, with a minimum of 25 and a maximum of 49. A 30-point traverse is illustrated in Figure 12.23.

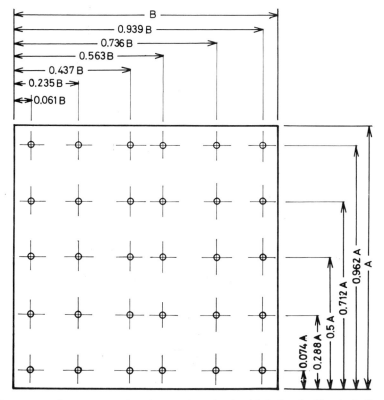

12.23 Traverse of a rectangular duct using the log-Tchebycheff rule (5 lines of traverse with 6 points per traverse line)

INSTRUMENTS FOR MEASURING VELOCITY

The two most reliable instruments for measuring point velocity are the Pitot-static tube and the vane anemometer. The application of these instruments to the on-site balancing procedures is described in Chapter 15.

Pitot-static tube

The principles behind the Pitot-static tube have already been outlined on page 216. The point velocity in an air stream is obtained from:

$$v = \sqrt{\frac{2Pv}{C\rho}} \qquad (12.22)$$

where C = calibration coefficient for the tube.

For suitably designed instruments the value of the tube coefficient, C, may be taken as unity but precise values are published in the various Standards (eg Reference[4]). In determining the flow rate from a Pitot-static tube *traverse*, additional corrections for effects such as tube blockage and turbulence, are applied to the mean velocity. These corrections, amounting to a total of approximately 2% in terms of the mean velocity, are usually reserved for laboratory work.

Because of its simplicity and ease of use, the Pitot-static tube, connected to a suitable manometer, is used extensively in testing ventilation systems. Normally it is not suitable for use below duct velocities of about 3.5 m/s, because of the probable inaccuracy in measuring the velocity pressure. Alignment of the probe relative to the flow direction affects accuracy, but this amounts only to about 1% error in the recorded velocity pressure for up to 10° for both yaw and pitch. Most tubes incorporate a short external rod to ensure the probe is aligned to the axis of the duct.

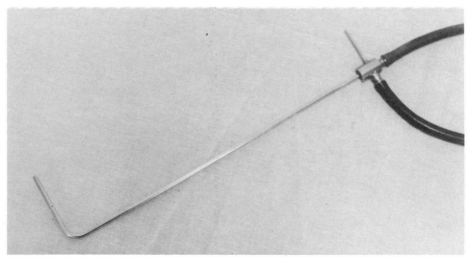

12.24 Pitot-static tube

Vane anemometers

These instruments consist of a number of light vanes supported on radial arms rotating on a common spindle. The differences in the type of instrument are in the method by which the air speed is obtained from the number of revolutions of the vane wheel. There are two main groups of the instrument, the mechanical type which requires timing over short periods, typically one minute, and the electronic type which gives a direct reading in metres per second. The main advantage of these instruments compared with a Pitot-static tube is that velocities can be measured down to about 0.5 m/s, but the speed range will depend on its manufacture, low, medium and high speed models being available up to a maximum speed of about 30 m/s.

12.25 **Vane anemometers (a) mechanical (Biram)**

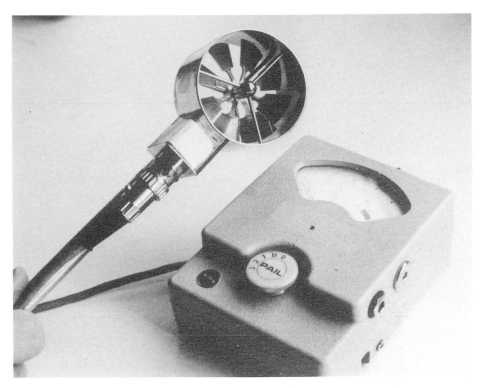

(b) **electronic**

A vane anemometer requires a calibration chart and this should be brought up-to-date periodically to ensure continued accuracy in the measurement of air velocity. The calibration is provided, the correction velocity, v_{corr}, added to the indicated velocity, v_i, to give the true velocity v_a. A calibrated instrument should have an accuracy of about $\pm 1\%$ over the speed range for which it was designed and with the instrument mounted in a steady uniform air stream. *Yaw* error is less than 1% up to $12°$ of yaw. Variations due to changes in air density are not significant for normal on-site measurements. With in-duct measurements there will be a blockage effect, causing an over-estimation of the velocity. Provided the instrument-to-duct diameter ratio is such as to accommodate the instrument at the traverse positions this effect should not be more than 3%. The recommended procedure for measuring the flow rate at the face of a grille is set out in Chapter 15, Table 15.4.

SYMBOLS

A	duct or pipe cross section area	Re_d	Reynolds number related to orifice diameter
A_o	orifice plate area	R_s	static pressure regain coefficient
b	duct breadth		
C	calibration coefficient for Pitot-static tube	R'	static pressure regain coefficient for maximum pressure recovery
D	diameter		
D_e	equivalent diameter of rectangular duct	r	resistance
		r_t	total resistance
d	diameter of orifice or conical inlet	t_a	air temperature
		$\dot{V}$	volume flow rate
f	friction factor	v	duct or pipe velocity
K	pressure loss coefficient	$\bar{v}$	mean duct or pipe velocity
K_b	bend pressure loss coefficient	W	duct width
		α	flow coefficient
K_e	expansion pressure loss coefficient	β	diameter ratio of orifice plate, d/D
K_f	straight duct pressure loss coefficient	Δp	pressure drop
		Δp_f	pressure loss due to friction
k	roughness of duct wall	Δp_{rs}	static pressure regain
L	duct length	$\Delta p'$	differential pressure
$\dot{m}$	mass flow rate	ε	expansibility (expansion) factor
p_{at}	atmospheric pressure		
p_s	static pressure	ρ	air density
p_t	total pressure	ν	kinematic viscosity
p_v	velocity pressure		
Re	Reynolds number		

Subscripts

Re_D	Reynolds number related to duct diameter	1, 2 ..	relates to specific duct or pipe section

13 Ducted Air Systems

Ducted air systems can be grouped into two categories with the following general characteristics:

Low velocity systems have air flows of velocities less than 10 m/s; total pressure drop in the ductwork 500 Pa and plant pressure drops up to 500 Pa;

High velocity systems use air flows in the velocity range 10 to 40 m/s; pressures on the fan discharge range from 500 to 2500 Pa and on the fan suction up to 500 Pa.

The data presented in this chapter allow the ductwork to be sized and the pressure losses to be determined. Duct sizing procedures are explained and illustrated, and special requirements described.

The design of ductwork distribution systems commences by locating and selecting the room outlet grilles and planning the distribution runs to these outlets in a logical manner, consistent with the building and system design. The layout should attempt to keep the total length of duct to a minimum and the index run as short as is practical. For systems in which the flow rates can be considered to be constant, each section of duct within the network must be sized to carry the sum of the individual outlet flow rates which it serves. For dual-duct and VAV systems, account should be taken of the load diversity on the main supply and extract ducts.

When sizing the ducts the engineer should aim at economic sizes, ie minimizing the total cost by a comparison of the capital cost of the ductwork and the fan energy cost, whilst at the same time ensuring that air velocities are such that any generated air noise is kept to acceptable levels. With high velocity systems air noise is not usually a problem, since the terminal units will incorporate acoustic attenuating linings.

As the ducts are sized, the pressure losses in the system are calculated, so that the fan pressure requirement can be determined. These pressure losses in air systems may be grouped under three headings:

– friction losses in straight ducts;
– dynamic losses in fittings;
– losses across plant items.

The fan pressure requirement will be the sum of all the losses from these sources along the *index run*.

PRESSURE LOSSES DUE TO FRICTION

Friction chart for circular ducts

Pressure losses due to friction may be obtained from Equation 12.8, page 222. However, it is more usual to make use of a *duct friction chart*, sometimes referred to as a *duct sizing chart*, as in Figure 13.1. This chart has been obtained from the friction equation for the following standard conditions:

- barometric pressure ... 1013.25 mbar;
- air temperature ... 20°C;
- surface roughness ... 0.15 mm.

The parameters on which the chart are based have the following units:

- flow rate ... m^3/s;
- pressure drop per unit length of duct ... Pa/m;
- duct size ... m;
- mean duct velocity ... m/s.

When using the chart, interpolation between a pair of lines will often be required, in which case exact values of the parameters may not be obtained. Chart values are sufficiently accurate for most practical calculations but if a precise value is required, this can be obtained by calculation.

Although ductwork can be made to any size, not all the duct sizes given on the friction chart are *standard* sizes. The ISO standard circular duct sizes are given in Table 13.1.

diameter (mm)	diameter (mm)	diameter (mm)
63	200	630
71	224	710
80	250	800
90	280	900
100	315	1000
112	355	1120
125	400	1250
140	450	
160	500	
180	560	

**Table 13.1 ISO standard sizes
of circular ducts**

Example 13.1   An 8 m length of duct is required to supply 0.8 m^3/s of air. Size the duct so that the mean velocity will be 6 m/s and determine the pressure loss due to friction.

Solution

Refer to Figure 13.2, a sketch of the friction chart. The intersection of flow

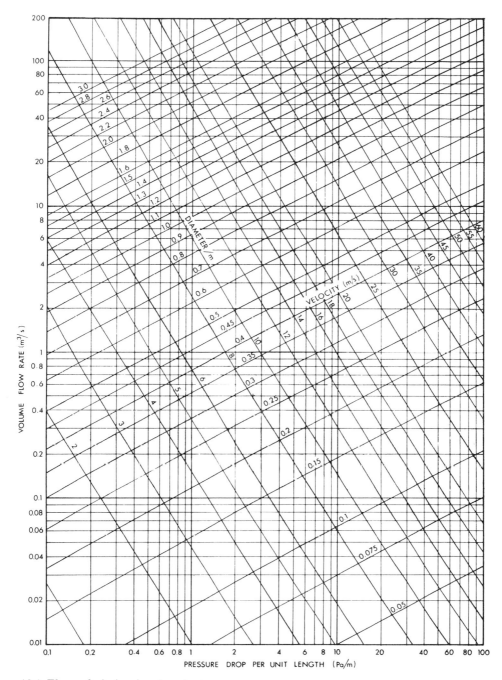

13.1 Flow of air in circular ducts
(Reproduced from Section C4 of the *CIBSE Guide*, by permission of the Chartered Institute of Building Services Engineers)

rate line 0.8 m³/s and duct velocity line 6 m/s requires interpolation between duct diameter lines 0.4 and 0.45 m. This gives:

$$\text{duct size} = 0.41 \text{ m}$$
$$\text{Pressure drop due to friction:} = 0.95 \text{ Pa/m}$$
$$\therefore \text{ total pressure loss due to friction} = 0.95 \times 8 = 7.6 \text{ Pa.}$$

In practice the duct will be sized to the nearest standard diameter size. In this case, if the ISO standard size of 400 mm is used for the flow rate of 0.8 m³/s, the duct velocity would be 6.3 m/s at a pressure drop per metre of duct length of 1.0 Pa.

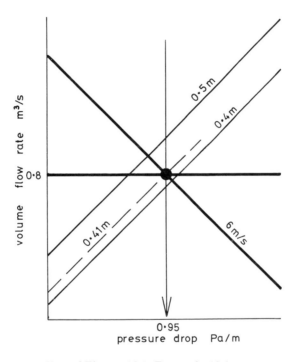

13.2 **Use of Figure 13.1: Example 13.1**

Example 13.2 An extract duct is required to handle 0.4 m³/s of air. Size the duct for a pressure drop of 3 Pa per unit length.

Solution

Refer to the sketch of the friction chart in Figure 13.3.
 Intersection of flow rate line 0.4 m³/s and pressure drop line of 3 Pa gives:

$$\text{duct size} = 0.25 \text{ m}$$
$$\text{duct velocity} = 8.15 \text{ m/s (by calculation)}.$$

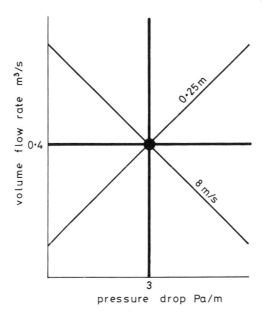

13.3 Use of Figure 13.1: Example 13.2

Rectangular duct sizes

Friction charts are available for rectangular ducts but it is more usual to use a conversion table in conjunction with the circular duct friction chart. Equivalent sizes of circular ducts are required so that there will be *equal friction pressure loss*, with *equal surface roughness*, for the *same flow rate*. This will mean that the mean velocity in the rectangular duct will differ from that in the circular duct. The equivalent circular duct size for this requirement is given by:

$$D = 1.265 \left(\frac{(bW)^3}{b+W} \right)^{0.2} \tag{13.1}$$

Table 13.2 gives the rectangular sizes of equivalent circular ducts. Smaller sizes are read above the stepped line, larger sizes below. Recommended standard sizes of rectangular ducts are given in Table 13.3.

The aspect ratio, AR, of a rectangular duct is defined by the ratio of the dimension of the horizontal side, b, to the vertical side, W, ie:

$$AR = \frac{b}{W}. \tag{13.2}$$

Dimen. of side, b	100	125	150	175	200	225	250	300	350	400	450	500	550	600	650	700	750	800	850	900	Dimen. of side, b
100	110	123	134	145	154	163	171	185	199	211	222	232	242	251	260	268	276	284	291	298	100
125	347	138	151	162	173	183	192	209	225	239	251	263	275	285	295	305	314	323	331	339	125
150	385	394	165	178	190	202	212	231	248	264	278	291	304	316	327	338	348	358	368	377	150
175	421	430	440	193	206	218	230	251	269	287	303	317	331	344	357	369	380	391	401	411	175
200	454	464	474	484	220	233	246	269	289	308	325	341	356	371	384	397	409	421	433	444	200
225	485	496	507	517	527	248	261	286	308	328	346	364	380	395	410	424	437	450	462	474	225
250	515	527	538	549	560	570	275	301	325	346	366	385	402	419	434	449	463	477	490	503	250
300	570	583	596	608	620	632	643	330	357	381	403	424	443	462	479	496	512	527	542	556	300
350	620	635	649	662	676	689	701	714	385	412	436	459	481	501	520	539	556	573	589	605	350
400	667	683	698	713	727	741	755	768	794	441	467	492	515	537	558	578	597	616	633	650	400
450	710	727	744	760	776	791	806	820	848	874	496	522	547	571	594	615	636	655	674	693	450
500	751	770	787	804	821	837	853	869	898	927	954	551	577	603	627	650	672	693	713	733	500
550	790	810	828	847	864	882	898	915	946	976	1005	1033	606	633	658	682	706	728	749	770	550
600	827	848	867	887	905	924	941	959	992	1024	1054	1084	1112	661	688	713	738	761	784	806	600
650	862	884	905	925	945	964	982	1001	1036	1069	1101	1132	1162	1191	716	743	769	793	817	840	650
700	896	919	940	962	982	1002	1022	1041	1077	1113	1146	1179	1210	1240	1269	771	798	824	849	873	700
750	928	952	975	997	1018	1039	1059	1079	1118	1154	1190	1223	1256	1287	1318	1347	826	853	879	904	750
800	959	984	1008	1031	1053	1075	1096	1117	1157	1195	1231	1267	1301	1333	1365	1396	1426	881	908	934	800
850	989	1015	1039	1063	1086	1109	1131	1152	1194	1234	1272	1308	1344	1378	1411	1443	1474	1504	936	963	850
900	1018	1044	1070	1095	1119	1142	1165	1187	1230	1271	1311	1349	1385	1421	1455	1488	1520	1551	1582	991	900
950	1046	1073	1100	1125	1150	1174	1198	1221	1265	1308	1349	1388	1426	1462	1498	1532	1565	1597	1629	1659	950
1000		1101	1128	1155	1180	1205	1230	1254	1299	1343	1385	1426	1465	1503	1539	1575	1609	1642	1675	1706	1000
1050			1156	1184	1210	1236	1261	1285	1332	1378	1421	1463	1503	1542	1580	1616	1652	1686	1719	1752	1050
1100				1211	1239	1265	1291	1316	1365	1411	1456	1499	1540	1580	1619	1657	1693	1729	1763	1797	1100
1150					1266	1294	1320	1346	1396	1444	1490	1534	1577	1618	1658	1696	1734	1770	1806	1840	1150
1200						1322	1349	1375	1427	1476	1523	1568	1612	1654	1695	1735	1773	1811	1847	1882	1200
1250							1377	1404	1456	1507	1555	1602	1646	1690	1732	1772	1812	1850	1888	1924	1250
1300								1432	1486	1537	1587	1634	1680	1725	1768	1809	1850	1889	1927	1965	1300
1400									1542	1596	1648	1697	1745	1792	1837	1881	1923	1964	2004	2043	1400
1500										1652	1706	1758	1808	1857	1904	1949	1993	2036	2078	2119	1500
1600											1762	1816	1868	1919	1968	2015	2061	2106	2149	2192	1600
1700												1872	1926	1979	2029	2078	2126	2173	2218	2262	1700
1800													1982	2036	2089	2140	2189	2237	2284	2330	1800
1900														2092	2147	2199	2250	2300	2348	2396	1900
2000															2203	2257	2310	2361	2411	2459	2000
2100																2313	2367	2420	2471	2521	2100
2200																	2423	2477	2530	2582	2200
2300																		2533	2587	2640	2300
2400																			2643	2697	2400
2500																				2753	2500
	950	1000	1050	1100	1150	1200	1250	1300	1400	1500	1600	1700	1800	1900	2000	2100	2200	2300	2400	2500	

$$d = 1.265 \left[\frac{(ab)^3}{a+b}\right]^{0.2}$$

Table 13.2 Equivalent diameters of rectangular ducts for equal volume flow rate, pressure drop and surface roughness
(Reproduced from Section C4 of the *CIBSE Guide*, by the permission of the Chartered Institute of Building Services Engineers)

long side (mm)	short side (mm)										
	100	150	200	250	300	400	500	600	800	1000	1200
150	134	165									
200	154	190	220								
250	171	212	246	275							
300	185	231	269	301	330						
400	211	264	308	346	381						
500		291	341	385	424	492	551				
600		316	371	419	462	537	603	661			
800			421	477	527	616	693	761	881		
1000				527	583	682	770	848	984	1101	
1200					632	741	837	923	1075	1205	1321
1400						794	898	992	1156	1299	1427
1600						843	954	1054	1231	1385	1523
1800							1005	1112	1301	1465	1612
2000							1053	1166	1365	1539	1695

Table 13.3 Recommended standard sizes of rectangular ducts with equivalent diameter of circular ducts

Example 13.3 Determine the recommended standard size of a rectangular duct whose maximum width is 250 mm for the circular duct in Example 13.2. Compare the mean velocity in this duct with that of the circular duct.

Solution

From the solution to Example 13.2, duct diameter = 250 mm.
  Refer to the sketch of Table 13.2 in Figure 13.4. Taking the nearest recommended size:
  Dimensions of side b = 200 mm.
  Mean duct velocity:

$$v = \frac{\dot{V}}{A} = \frac{0.4}{0.250 \times 0.2} = 8.0 \text{ m/s}.$$

This compares with 8.41 m/s for the equivalent circular duct diameter of 0.246 m and 8.15 for the 0.25 m diameter duct.

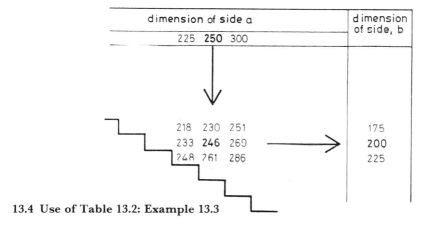

13.4 Use of Table 13.2: Example 13.3

Example 13.4 A rectangular duct with dimensions 500 mm 1200 mm is to supply 4.0 m³/s of air at 20°C to a large hall. Determine the pressure drop due to friction if the duct is 30 metres in length.

Solution

Refer to the sketch of Table 13.2 in Figure 13.5:

duct diameter = 837 mm, or 0.84 m.

  From the friction chart, Figure 13.1:

0.84 m diameter duct with a flow rate of 4.0 m³/s
pressure drop per unit length = 0.55 Pa/m

∴ pressure drop in duct = 30 × 0.55 = 16.5 Pa.

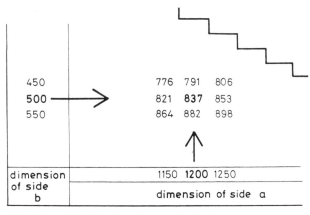

13.5 Use of Table 13.2: Example 13.4

Corrections for air at other densities

The pressure losses obtained from the friction chart may be corrected for other air densities by using the following expression:

$$\Delta p_2 = \Delta p_1 \left(\frac{\rho_2}{\rho_1} \right) \tag{13.3}$$

where Δp_1 = pressure loss at standard density, ρ_1, $(1.2 \ \mathrm{kg/m^3})$

Δp_2 = pressure loss at air density, ρ_2.

Corrections for ducts of other materials

The friction chart is based on an average roughness coefficient of 0.15 mm, assumed to be typical of commercial galvanized sheet steel ductwork. For ducts of other materials, with different roughness values, the pressure drops taken from the chart can be corrected in accordance with the factors given in Table 13.4.

pressure drop from chart	spiral wound	plastic	builders work			
			fair-faced brick or concrete		rough brickwork	
(Pa/m)			200 mm	1000 mm	200 mm	1000 mm
0.2	0.96	0.90	1.35	1.35	2.02	1.96
0.5	0.95	0.88	1.42	1.42	2.18	2.05
1.0	0.94	0.85	1.48	1.46	2.34	2.12
2.0	0.93	0.84	1.53	1.49	2.45	2.17
5.0	0.92	0.80	1.60	1.54	2.58	2.23
10.0	0.91	0.77	1.63	1.56	2.67	2.25

Table 13.4 Pressure drop correction factors for ducts of different materials (galvanized sheet steel = 1.0)

Example 13.5 Determine the pressure drop for the duct given in Example 13.4, if the duct is built of fair-faced concrete.

Solution

From Example 13.5 the pressure drop = 0.55 Pa/m.
   The equivalent duct diameter = 0.84 m.
   From Table 13.4, the correction factor for a fair-faced concrete duct = 1.42

$$\therefore \text{ pressure drop due to friction} = 30 \times 0.55 \times 1.42$$
$$= 23.4 \text{ Pa}.$$

PRESSURE LOSSES IN FITTINGS

The general principles of determining the pressure loss across duct fittings are given in Chapter 12. The specific loss coefficients for different types of fittings, drawn in the main from MILLER[2] and the *ASHRAE Handbook*,[3] are given below together with examples to illustrate calculation methods. There are some variations in the loss coefficients quoted by these sources; the values given in the following tables and graphs are considered to be those which best reflect the probable loss. In selecting these data, priority has been given to the fittings recommended in the HCVA *Specification for Sheet Metal ductwork*.[4]

It will be appreciated that where more than one velocity is associated with a fitting, the loss coefficient must be applied to the appropriate velocity pressure. For example, in the case of expansions and contractions, the velocity pressure in the smaller area is used to calculate the loss; for branch pieces, the velocity pressure in the duct carrying the total flow. (Other texts may use different reference velocities.)

In general, pressure losses in fittings apply to both supply and extract systems, the main exception being those for branch pieces.

Bends
Circular ducts   The loss coefficients for circular duct bends are given in Table 13.5.
   The following bend construction details are recommended:

– for ducts up to, and including, 300 mm:
   long radius pressed bends, $r/D = 1.5$;
– for ducts up to, and including, 400 mm:
   medium radius pressed bend, $r/D = 1.0$;
– for ducts above 400 mm:
   segmented bends as Table 13.5.

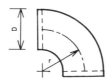

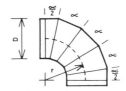

90° radiused pressed bend   90° segmented bend

bend angle	radius to duct diameter ratio r/D					segmented bends	
	–	medium radius	long radius	–	–	no. of segments	α
	0.75	1.0	1.5	2.0	2.5		
90°	0.40	0.24	0.18	0.16	0.15	5	22.5
60°	0.22	0.15	0.12	0.11	0.11	4	20
45°	0.14	0.10	0.09	0.09	0.09	3	15
30°	0.07	0.06	0.06	0.06	0.06	2	15

Table 13.5 Loss coefficients, K_b, for circular duct bends

Example 13.6 Determine the pressure loss across a medium radius, 90° bend for a 350 mm diameter circular duct. The air flow rate is 0.6 m³/s at standard air density.

Solution

The mean velocity is:

$$\bar{v} = \frac{\dot{V}}{A} = \frac{0.6}{\pi 0.35^2/4} = 6.24 \text{ m/s.}$$

The velocity pressure is:

$$p_v = 0.6 \, \bar{v}^2 = 0.6 \times 6.24^2 = 23.3 \text{ Pa.}$$

For a medium radius bend r/D = 1.0. From Table 13.5 the loss coefficient = 0.24.
From Equation 12.12, the pressure loss across bend is obtained:

$$\Delta p_b = K_b p_v = 0.24 \times 23.3 = 5.6 \text{ Pa.}$$

Rectangular ducts   Bends for rectangular ducts are described as *hard* or *easy*. A hard bend rotates in the plane of the longer side and an easy bend rotates in the plane of the shorter side. These are illustrated in Figure 13.6. The loss coefficients are given in Tables 13.6 and 13.7.

The following bend construction details are recommended:

– for ducts up to 300 mm in duct width:
    short radius bend, *throat* radius = 100 mm;

– for ducts over 300 mm in duct width:
 short radius bend with splitters as specified in Table 13.7;
– for all duct widths:
 medium radius bends, r/W = 1.0
 long radius bends, r/W = 1.5.

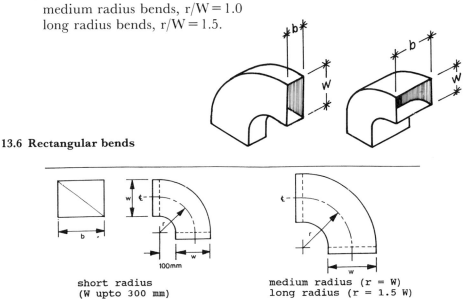

13.6 Rectangular bends

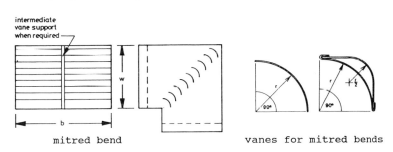

short radius
(W upto 300 mm)

medium radius (r = W)
long radius (r = 1.5 W)

intermediate
vane support
when required

mitred bend

vanes for mitred bends

bend description	$\frac{r}{W}$	aspect ratio, b/W					
		hard bends		–	easy bends		
		0.25	0.5	1.0	2.0	3.0	4.0
short radius	0.8	0.57	0.50	0.44	0.4	0.4	0.4
medium radius	1.0	0.27	0.25	0.21	0.18	0.18	0.19
long radius	1.5	0.22	0.20	0.17	0.14	0.14	0.15

Table 13.6 Loss coefficients, K_b, for 90° rectangular duct bends

Notes:
 (i) dimensions apply to other turning angles
 (ii) multipliers for bend angles less than 90°: 60°–0.8 45°–0.6 30°–0.3
 (iii) 90° mitred bends, no turning vanes, $K_b = 1.1$, 90° mitred bends with turning vanes, $K_b = 0.15$

Example 13.7 Determine the pressure loss across a long radius, 90° bend for a rectangular duct width 200 mm and height 400 mm. The air flow rate is 0.6 m³/s at standard air density.

Solution

The mean velocity is:

$$\bar{v} = \frac{0.6}{0.4 \times 0.2} = 7.5 \text{ m/s}.$$

The velocity pressure is:

$$p_v = 0.6\,\bar{v}^2 = 0.6 \times 7.5^2 = 33.7 \text{ Pa}.$$

For a long radius bend $r/W = 1.5$.
The aspect ratio $b/W = 200/400 = 0.5$.
From Table 13.6 the loss coefficient $= 0.20$.
∴ pressure loss across bend is obtained:

$$\Delta p = K_b p_v = 0.20 \times 33.7 = 6.7 \text{ Pa}.$$

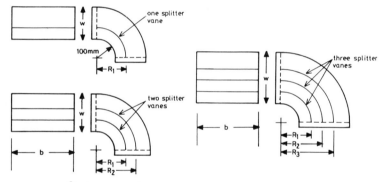

90° short radius bends (W over 300 mm)

$$R_1 = 100/C_r, \quad R_2 = 100/C_r^2, \quad R_3 = 100/C_r^3$$

where C_r = radius curve ratio for splitter vane

no. of splitters	duct width W	radius curve ratio C_r	aspect ratio, b/W					
			hard bends		–	easy bends		
			0.25	0.5	1.0	2.0	3.0	4.0
1	300	0.50	0.13	0.09	0.08	0.07	0.08	0.08
	400	0.45	0.18	0.13	0.11	0.11	0.12	0.13
	500	0.41	0.22	0.16	0.14	0.15	0.16	0.17
2	500	0.55	0.09	0.07	0.06	0.06	0.06	0.06
	700	0.50	0.12	0.09	0.08	0.08	0.09	0.10
	1000	0.45	0.17	0.13	0.11	0.13	0.15	0.16
3	1000	0.55	0.07	0.05	0.06	0.06	0.07	0.07
	2000	0.47	0.11	0.10	0.12	0.14	0.16	0.18

The loss coefficient for a *mitred bend* is approximately 1.1. By including turning vanes (see Table 13.6) the bend loss will be reduced to that of the lowest value of an equivalent arc bend. The vanes, single or double skinned, usually have an inner radius, r, of 50 mm and the recommended number of vanes, n, is given by the relationship:

$$n = \frac{1.5W}{r}. \qquad (13.4)$$

Corrections for REYNOLDS number
Pressure loss coefficients have some dependence on REYNOLDS number, although the data available to allow for its effect is limited. The data for bends is included here to illustrate how a basic loss coefficient is modified by a REYNOLDS number correction factor, C_{Re}, applied to the basic loss coefficient, K_b. The variations of C_{Re} are given in Figure 13.9.

$$\Delta p = C_{Re} K_b p_v \qquad (13.5)$$

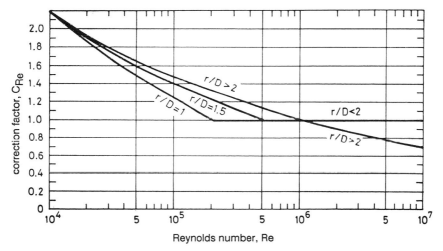

13.7 Reynolds number correction factors for bends
(Redrawn from Figure 9.4, D S Miller, *Internal Flow Systems*, 2nd edition 1990, BHRA, Cranfield, UK, by permission)

◄ **Table 13.7 Loss coefficients, K_b, for 90° rectangular duct bends with splitters**

    Notes:
        (i) **dimensions apply to other turning angles**
        (ii) **multipliers for bend angles less than 90°:**
                      60° − 0.8   45° − 0.6
        (iii) **splitters not applicable to bends less than 45°**

Example 13.8 For the bend in Example 13.7, determine the REYNOLDS number correction factor to be applied to the basic loss coefficient.

Solution

For the bend given, hydraulic diameter is given by Equation 12.7:

$$D_e = \frac{2A}{(b+W)} = \frac{2(0.2 \times 0.4)}{(0.2+0.4)} = 0.267.$$

REYNOLDS number:

$$R_e = \frac{D\bar{v}}{v} = \frac{0.267 \times 7.5}{1.5 \times 10^{-5}} = 1.33 \times 10^5.$$

From Figure 13.9, REYNOLDS number correction factor for $r/D = 1.5$:

$$C_{Re} = 1.3$$

∴ pressure loss across bend using Equation 13.5:

$$\Delta p = C_{Re} K_b p_v$$
$$= 1.3 \times 0.24 \times 23.3 = 7.3 \text{ Pa}.$$

Correction for interaction between bends

When bends follow each other there is an interaction affecting the total pressure loss of the system and this applies to ductwork in which the intervening (spacer) length is less than 30 diameters. To arrive at the net loss of a pair of bends, a correction factor, C_{b-b}, is applied to the sum of the individual loss coefficients. Then the pressure loss is given by:

$$\Delta p = C_{b-b}(K_{b1} + K_{b2})p_v. \tag{13.6}$$

A typical set of interaction effect factors are given in Table 13.8 for circular duct bends with an r/D ratio of 1.0. The bend configurations referred to in this table are defined by the sketches. The factors assume that the outlet discharge duct from the bend arrangement is more than 30 diameters in length. This data can also be used for rectangular duct bends but where the bends incorporate turning vanes there is *no* interaction effect.

Example 13.9 Two 90° medium radius, 400 mm diameter bends are separated by 1.6 m of straight duct and are arranged with a combination angle of 180°. Determine the net pressure loss due to the bends if the flow rate is 1.2 m³/s at standard air density.

Solution

Mean duct velocity:

$$\bar{v} = \frac{\dot{V}}{A} = \frac{1.2}{\pi 0.4^2/4} = 9.55 \text{ m/s}.$$

Velocity pressure:

$$p_v = 0.6 \times 9.55^2 = 54.7 \text{ Pa}.$$

From Table 13.8: $K_{b1} = K_{b2} = 0.24$.
The length of the spacer in duct diameters is given by:

$$L_s = 2/0.5 = 4.$$

From Table 13.8: $C_{b-b} = 0.71$.
Using Equation 13.6:

$$\Delta p = C_{b-b}(K_{b1} + K_{b2})p_v$$
$$= 0.71(0.24 + 0.24)54.7 = 18.6 \text{ Pa}.$$

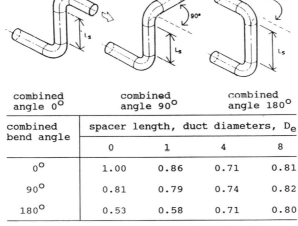

combined angle 0°	combined angle 90°	combined angle 180°	
combined bend angle	spacer length, duct diameters, D_e		
	0	1	4	8
0°	1.00	0.86	0.71	0.81
90°	0.81	0.79	0.74	0.82
180°	0.53	0.58	0.71	0.80

Table 13.8 Typical bend interaction effect factors, C_{b-b}

Expansions

Ducts which expand in cross sectional area are variously termed expansions, expanders, diffusers and enlargements.

Expansions are classified in two groups:

– ducted discharge;
– no discharge duct (free discharge).

Pressure loss coefficients for these two groups of fittings are given in Tables 13.9 and 13.10. The coefficients are for both circular and rectangular ducts and are applied to the inlet velocity pressure.

	area ratio A_2/A_1	velocity ratio v_2/v_1	half-angle, θ 5°	10°	15°	22.5°
	1.2	0.83	0.06	0.06	0.06	0.06
	1.5	0.67	0.06	0.08	0.09	0.11
	2.0	0.50	0.06	0.12	0.17	0.21
	2.5	0.40	0.08	0.17	0.20	0.30
	3.0	0.33	0.10	0.20	0.30	0.40
	3.5	0.29	0.20	0.35	0.45	0.50
	4.0	0.25	0.27	0.42	0.50	-

Table 13.9 Loss coefficient, K_e, for circular and rectangular expansions with discharge duct

Note: maximum angles: circular duct $\quad\theta = 15°$

rectangular duct $\quad\theta = 22.5°$

description	area ratio A_2/A_1	velocity ratio $\bar{v}_2/\bar{v}_1$	half-angle, θ 5°	10°	15°	20°
circular, concentric	1.5	0.67	0.57	0.62	0.70	-
	2.0	0.50	0.44	0.57	0.70	-
	2.5	0.40	0.37	0.55	0.65	-
	3.0	0.33	0.32	0.50	0.65	-
	3.5	0.29	0.30	0.55	0.70	-
	4.0	0.25	0.30	0.58	0.70	-
rectangular, concentric	1.5	0.67	0.52	0.60	0.70	0.75
	2.0	0.50	0.37	0.55	0.65	0.75
	2.5	0.40	0.30	0.48	0.65	0.75
	3.0	0.33	0.30	0.45	0.65	0.80
	3.5	0.29	0.30	0.45	0.65	0.80
	4.0	0.25	0.30	0.45	0.65	0.80
rectangular, concentric with splitters	2.0	0.50	-			
	3.0	0.33	- }	0.45	0.45	0.45
	4.0	0.25	-			

Table 13.10 Loss coefficients, K_e, for circular and rectangular duct expansions with free discharge

Note: coefficients include the discharge velocity pressure from the expansion

It is recommended that the slope of an expansion should not exceed 15° on any side for circular ducts and 22.5° for rectangular ducts.

For rectangular ducts, splitters (vanes) may be used to minimize the pressure loss. For vaned expanders with a free discharge, a reasonable performance will be achieved with a half angle:

θ up to 20° – two vanes
θ up to 25° – three vanes

The recommended angle, α, of the expansion vanes, from the axis is given by:

$$\alpha = \frac{\theta}{2.9}.$$

Expansion with a short outlet pipe
When an outlet duct is added to an expansion, there is a reduction in the loss coefficient related to the length of the outlet duct. The loss coefficient, K_{es}, is obtained from the following relationship:

$$K_{es} = 1 - C_e(1 - K_e) \tag{13.7}$$

where K_e is obtained from Table 13.9
where C_e is obtained from Table 13.11.

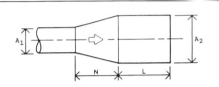

L/D$_2$	area ratio A$_2$/A$_1$	velocity ratio v$_2$/v$_1$	N/D$_1$		
			1	2	3
1	1.5	0.67	1.04	1.00	1.00
	2.0	0.50	1.15	1.08	1.05
	2.5	0.40	1.25	1.13	1.08
	3.0	0.33	1.30	1.17	1.08
2	1.5	0.67	1.07	1.03	1.00
	2.0	0.50	1.28	1.13	1.08
	2.5	0.40	1.50	1.24	1.16
	3.0	0.33	1.60	1.32	1.15
4	1.5	0.67	1.10	1.00	1.00
	2.0	0.50	1.45	1.20	1.08
	2.5	0.40	-	1.40	1.22
	3.0	0.33	-	1.53	1.24

Table 13.11 **Loss factors, C_e, for short outlet pipes following an expansion**
Note: **the factors are applied to the loss coefficients, K_e, in Table 13.9, using Equation 13.7**

Example 13.10 An expansion in a circular duct with inlet diameter 0.2 m, discharge duct diameter 0.3 m and angle of expansion on both sides of 10°. If the air flow rate is 0.2 m³/s at an air density of 1.15 kg/m³, determine the pressure loss:

(a) with long discharge duct;
(b) with no discharge duct;
(c) with discharge duct 0.6 m in length.

Solution

Inlet duct area $A_1 = \pi 0.2^2/4 = 0.0314\ m^2$.
Outlet duct area $A_2 = \pi 0.3^2/4 = 0.0707\ m^2$.
  The velocity at the inlet is:

$$\bar{v}_1 = \frac{\dot{V}}{A_1} = \frac{0.2}{0.0314} = 6.37\ m/s.$$

The inlet velocity pressure is:

$$p_{v1} = 0.5\ \rho\bar{v}_1{}^2 = 0.5 \times 1.15 \times 6.37^2 = 23.3\ Pa.$$

The area ratio $= A_2/A_1 = 0.0707/0.0314 = 2.25$.

(a) From Table 13.9, the loss coefficient $K_e = 0.145$
     pressure loss across expansion with *long outlet* duct is:

$$\Delta p = K_e p_v = 0.145 \times 23.3 = 3.4\ Pa$$

(b) From Table 13.10, the loss coefficient $K_e = 0.56$
     pressure loss across expansion with *free discharge* is:

$$\Delta p = K_e p_v = 0.56 \times 23.3 = 13.0\ Pa$$

(c)
$$N = \frac{0.05}{\tan 10°} = 0.284\ m$$

$$\therefore N/D_1 = 0.284/0.2 = 1.42$$
$$L/D_2 = 0.6/0.3 = 2$$

From Table 13.11, the loss factor $C_e = 1.26$ (by interpolation).
Using Equation 13.7:
$$K_{es} = 1 - C_e(1 - K_e)$$
$$= 1 - 1.26(1 - 0.56) = 0.45$$

pressure loss across expansion with *short outlet* duct is:

$$\Delta p - K_{es} p_v = 0.45 \times 23.3 = 10.5\ Pa.$$

Contractions

Ducts which reduce in cross sectional area are variously termed contractions and reducers.

  Pressure loss coefficients for abrupt contractions, given in Table 13.12a, are applied to the velocity pressure in the outlet duct.

	area ratio A_2/A_1	K_c
	0.2	0.55
	0.3	0.48
	0.4	0.45
	0.5	0.38
	0.6	0.30
	0.7	0.20
	0.8	0.08

Table 13.12 Loss coefficients, K_e, for contractions
         (a) abrupt contraction

The recommended maximum slopes of gradual contraction pieces are given in Table 13.12b. There is a single loss coefficient, K_c, of 0.05.

contraction	maximum values of angle θ	
	rectangular	circular
concentric	22.5°	15°
	$K_c=0.05$	
eccentric	22.5°	30°
	$K_c=0.05$	

Table 13.12 Loss coefficients, K_c, for contractions
(b) gradual contractions

Example 13.11 Determine the pressure loss across an abrupt contraction with an inlet area of 0.4 m² and an outlet area of 0.15 m². The air flow rate is 0.9 m³/s at standard air density of 1.2 kg/m³.

Solution

The mean velocity at the outlet is:

$$\bar{v}_2 = \frac{\dot{V}_2}{A} = \frac{0.9}{0.15} = 6.0 \text{ m/s}.$$

The velocity pressure is:

$$p_{v2} = 0.5 \, \rho\bar{v}^2 = 0.5 \times 1.2 \times 6.0^2 = 21.6 \text{ Pa}.$$

The area ratio is:

$$\frac{A_2}{A_1} = \frac{0.4}{0.15} = 2.7.$$

From Table 13.12a the pressure loss factor for an abrupt contraction, $K_c = 0.50$

∴ pressure loss across contraction is:

$$\Delta p = K_c p_{v2} = 0.50 \times 21.6 = 10.8 \text{ Pa}.$$

Branches

The alternative arrangements for making a branch connection are shown in Figures 13.8 and 13.9. These apply to both *supply* and *extract* systems. It is recommended that a branch should be taken off the main duct and not off a taper. Therefore, when a change of section is required to accommodate the next section of ductwork, expansion or contraction may be necessary as illustrated in Figure 13.10.

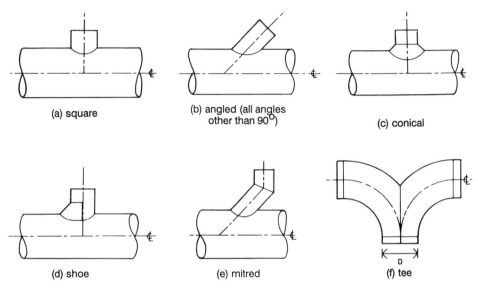

(a) square

(b) angled (all angles other than 90°)

(c) conical

(d) shoe

(e) mitred

(f) tee

13.8 Branches for circular ducts
(connections are made off the duct or as a separate fitting)

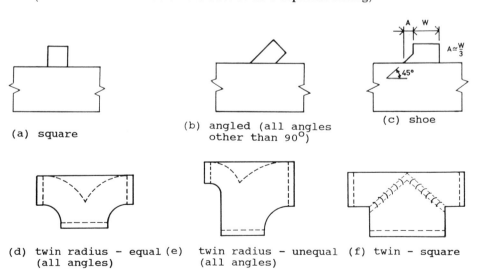

(a) square

(b) angled (all angles other than 90°)

(c) shoe

(d) twin radius – equal (all angles)

(e) twin radius – unequal (all angles)

(f) twin – square

Continued

(g) branch (for all turning
    angles; splitters in
    accordance with Table 13.7)

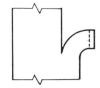

13.9 Branches for rectangular ducts

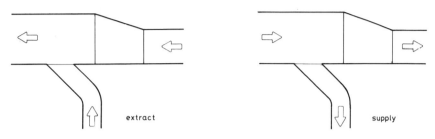

13.10 Typical arrangement of supply and extract branches

For *supply* ducts the loss coefficients K_{31} and K_{32} refer to the velocity pressure in the duct with the *combined* flow, and this duct is designated as *duct 3*.

K_{31} refers to the loss to branch 1

K_{32} refers to the loss to branch 2, (usually the main duct)

The loss coefficients, K_{31}, for 45° and 90° square branch connections for circular ducts are given in Figure 13.11. These figures give the contours of constant loss coefficients and are typical of data given by MILLER.[3]

For conical and shoe branch connections, correction factors, C_{31}, in Table 13.13 should be applied to the square branch coefficient, K_{31}, ie:

$$\Delta p = C_{31} \, K_{31} \, p_{v3} \qquad (13.8)$$

duct area ratio A_1/A_3	flow rate ratio $\dot{V}_1/\dot{V}_3$						
	0.2	0.3	0.4	0.5	0.6	0.7	0.8
0.4	0.95	0.88	0.80	0.75	0.67	0.64	0.60
0.6	0.97	0.92	0.87	0.82	0.77	0.72	0.68
0.8	0.98	0.95	0.90	0.84	0.81	0.77	0.73

Table 13.13 Supply branch (conical or shoe) correction factor, C_{31}, for square
   branches
   Note: factors are applied to the loss coefficients, K_{31}, in
      Figure 13.11, using Equation 13.8

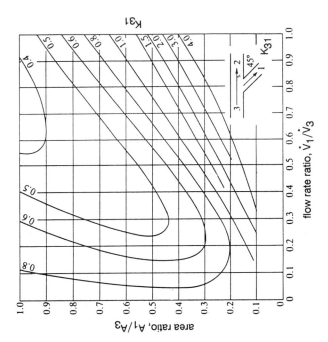

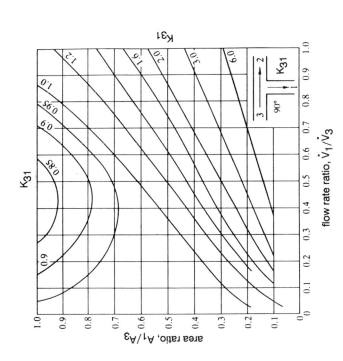

13.11 Loss coefficients for supply branches with sharp edges
(Based on Figures 13.19 and 13.21, D S Miller, *Internal Flow Systems*, 2nd edition
1990, BHRA, Cranfield, UK, by permission)

Loss coefficients, K_{32}, for the pressure loss in the main are given in Table 13.14; these are independent of any chamfers or branch angles.

	flow rate ratio $\dot{V}_2/\dot{V}_3$					
	0.2	0.3	0.4	0.5	0.6	0.7
K_{32}	0.22	0.15	0.09	0.03	–	–

Table 13.14 Loss coefficients, K_{32}, for supply branch, flow to main
Note: **coefficients are independent of type of offtake at branch 1; duct area ratio $A_2/A_3 = 1$**

Loss coefficients, K_{31}, for twin radius tee pieces with equal area ducts are given in Table 13.15.
For *rectangular* ducts the loss coefficients should be increased by 10% over the circular duct loss.

	flow rate ratio $\dot{V}_1/\dot{V}_3$						
	0.2	0.3	0.4	0.5	0.6	0.7	0.8
K_{31}	0.48	0.41	0.35	0.30	0.27	0.23	0.2

Table 13.15 Loss coefficients, K_{31}, for supply 90° twin radius tee, equal area ducts

Example 13.12 Determine the pressure losses for a 90° branch shown in Figure 13.12 at a standard air density of 1.2 kg/m³:

(a) for a branch with sharp edges;
(b) for a branch with a shoe connection.

Solution

The relevant data is set out in the following table:

duct	air flow rate	duct diameter	duct area	velocity	velocity pressure
	$\dot{V}$ (m³/s)	D (m)	A (m²)	v (m/s)	p_v (Pa)
1	0.1	0.16	0.02	–	–
2	0.3	0.25	0.049	–	–
3	0.4	0.25	0.049	8.15	39.8

Loss to branch 1

(a) The area ratio $A_1/A_3 = 0.02/0.049 = 0.41$
     The flow rate ratio $\dot{V}_1/\dot{V}_3 = 0.1/0.4 = 0.25$
     From Figure 13.11 the loss coefficient $K_{31} = 1.0$
     $\therefore$ the pressure loss $= K_{31}\ p_{v3} = 1.0 \times 39.8 = 39.8$ Pa;

(b) From Table 13.13, $C_{31} = 0.91$
     Pressure loss using Equation 13.6:
     $$\Delta p = C_{31}\ K_{31}\ p_{v3}$$
     $$= 0.91 \times 1.0 \times 39.8 = 36.2 \text{ Pa.}$$

Loss to branch 2

The flow rate ratio $\dot{V}_2/\dot{V}_3 = 0.3/0.4 = 0.75$.
    From Table 13.14 the loss coefficient $K_{32} = 0$.
    There is therefore negligible pressure loss through the branch.

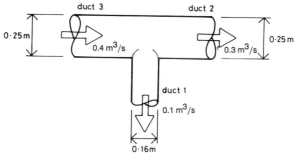

13.12 Supply branch: Example 13.12

For *extract* ducts, the loss coefficients K_{13} and K_{23} refer to the velocity pressure in the duct with the *combined* flow, and this duct is designated as *duct 3*. As with the supply duct branch pieces, the loss coefficient refers to the velocity pressure in the duct with the combined flow:

K_{13} refers to the loss from branch 1 to duct 3;
K_{23} refers to the loss from branch 2 to duct 3.

The loss coefficients, K_{13} and K_{23}, for 45° and 90° circular duct branch connections are given in Figure 13.13. This figure gives the contours of constant loss coefficients and is typical of the data given by MILLER. Loss coefficients for a combining tee are given in Table 13.16.

	flow rate ratio $\dot{V}_1/\dot{V}_3$						
	0.2	0.3	0.4	0.5	0.6	0.7	0.8
K_{13}	−	0.05	0.12	0.18	0.21	0.23	0.27

Table 13.16 Loss coefficients, K_{13}, for extract 90° twin radius tee, equal area ducts

Due to the *suction* effect caused by the flow in the main, the loss factor can take a negative value. This means that when the total pressure drop is calculated for the branch duct, the loss at the branch is *subtracted* from the total of the other losses in the branch.

For conical or shoe connections, the loss coefficients are *reduced* by 10%. For *rectangular* ducts the loss coefficients should be *increased* by 10%.

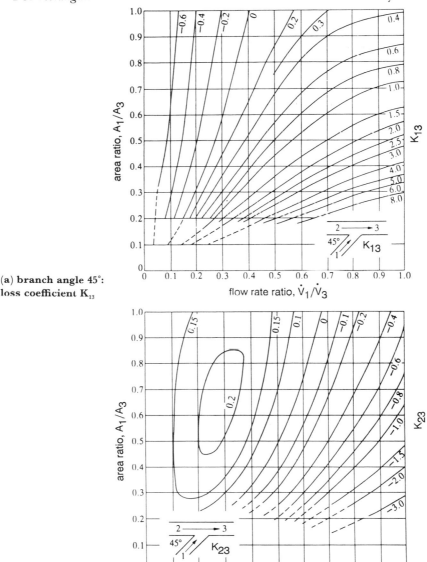

(a) branch angle 45°: loss coefficient K_{13}

(b) branch angle 45°$_{13}$: loss coefficient K_{23}

13.13 Loss coefficients for extract branches with sharp edges

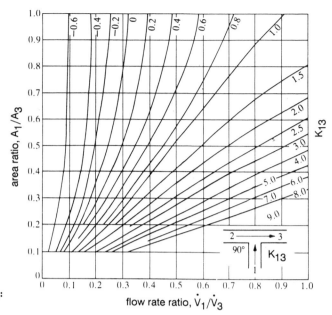

(c) branch angle 90°₂₃:
loss coefficient K₁₃

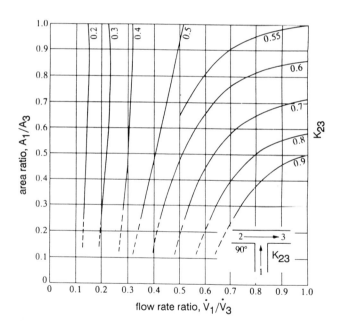

(d) branch angle 90°:
loss coefficient K₂₃

13.13 Loss coefficients for extract branches with sharp edges
(Based on Figures 13.6, 13.7, 13.10 and 13.11, D S Miller, *Internal Flow Systems*, 2nd
edition 1990, BHRA, Cranfield, UK, by permission)

Example 13.13 Determine the pressure losses for a 45° suction branch piece shown in Figure 13.14 with standard air density of $1.2 \, \text{kg/m}^3$.

duct	air flow rate	duct area	velocity	velocity pressure
	$\dot{V}$ (m^3/s)	A (m^2)	v (m/s)	p_V (Pa)
1	0.1	0.02	5.0	–
2	0.4	0.04	10.0	–
3	0.5	0.04	12.5	93.8

Solution

Flow rate ratio: $\dot{V}_1/\dot{V}_3 = 0.1/0.5 = 0.2$.
Area ratio: $A_1/A_3 = 0.2/0.4 = 0.5$.

From Figure 13.13a, the loss coefficient for the suction branch to the main is:

$$K_{13} = -0.2.$$

∴ pressure *gain* across branch is:

$$\Delta p = K_{13} \, p_{v3} = 0.2 \times 94 = 18.8 \, \text{Pa}.$$

From Figure 13.13b, the loss coefficient in main is:

$$K_{23} = 0.2.$$

∴ pressure *loss* in main is:

$$\Delta p = K_{23} \, p_{v3}$$
$$= 0.2 \times 94 = 18.8 \, \text{Pa}.$$

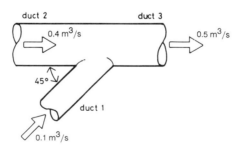

13.14 Extract branch: Example 13.13

Duct entries

Pressure loss coefficients for duct entries and intake louvres are given in Tables 13.17 and 13.18 overleaf.

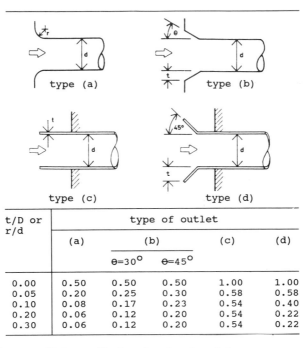

t/D or r/d	type of outlet				
	(a)	(b)		(c)	(d)
		$\theta=30°$	$\theta=45°$		
0.00	0.50	0.50	0.50	1.00	1.00
0.05	0.20	0.25	0.30	0.58	0.58
0.10	0.08	0.17	0.23	0.54	0.40
0.20	0.06	0.12	0.20	0.54	0.22
0.30	0.06	0.12	0.20	0.54	0.22

Table 13.17 Loss coefficients, K_i, for circular duct inlets
Note: **for rectangular ducts, increase corresponding circular duct coefficient by 10%**

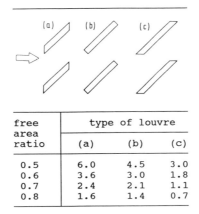

free area ratio	type of louvre		
	(a)	(b)	(c)
0.5	6.0	4.5	3.0
0.6	3.6	3.0	1.8
0.7	2.4	2.1	1.1
0.8	1.6	1.4	0.7

Table 13.18 Loss coefficients for 45° intake louvres

Example 13.14   Determine the pressure loss across the intake louvre with dimensions 400 mm and 300 mm. The air flow rate is 0.6 m³/s at standard air density.

Solution

The mean velocity is:

$$\bar{v} = \frac{\dot{V}}{A} = \frac{0.4}{0.4 \times 0.3} = 3.3 \text{ m/s.}$$

The velocity pressure is:

$$p_v = 0.6 \; \bar{v}^2 = 0.6 \times 3.3^2 = 6.5 \text{ Pa.}$$

From Table 13.18, and assuming a type c louvre with a free area ratio of 0.6, the loss coefficient = 1.8.

∴ pressure loss across intake louvre is:

$$\Delta p = K_i \; p_v = 1.8 \times 6.5 = 11.8 \text{ Pa.}$$

Duct exits

Pressure loss coefficients for duct exits are given in Table 13.19. These are losses for discharges into rooms, plenums and large open spaces and are in addition to any grille covering the outlet. The air discharges to datum pressure which is most often at atmospheric pressure.

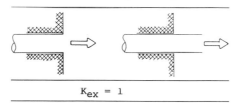

$$K_{ex} = 1$$

Table 13.19 Loss coefficient, K_{ex}, exhaust discharge to a plenum or to atmosphere

Example 13.15 Determine the pressure loss at a circular, plain 300 mm diameter supply discharge to a room. The flow rate is 0.5 m³/s at standard air density.

Solution

The mean velocity is:

$$\bar{v} = \frac{\dot{V}}{A} = \frac{0.5}{\pi 0.3^2 / 4} = 7.07 \text{ m/s.}$$

The velocity pressure is:

$$p_v = 0.6 \; \bar{v}^2 = 0.6 \times 7.07^2 = 30 \text{ Pa.}$$

From Table 13.18, the pressure loss factor = 1.0.

∴ pressure loss at the outlet is:

$$\Delta p = K \; p_v = 1.0 \times 30 = 30 \text{ Pa.}$$

DUCT SIZING PROCEDURES

Reference to the friction chart, and as indicated in Examples 13.1 and 13.2, suggests the two most commonly used methods of obtaining the size of a duct to carry a given flow rate, ie either by setting a duct velocity or by setting a pressure drop per unit length. These two methods are usually referred to as:

     – sizing by velocity;
     – sizing by pressure drop per unit length.

The duct sizing procedure using the second of these methods is demonstrated by the following example. (The sizing procedure would be similar for the second method.) The example also illustrates a number of important points which are discussed in the solution. Selection of suitable velocities and pressure drops for low velocity systems are given in Table 13.20.

item	type of building		
	residential hospitals hotels	commercial schools theatres	industrial
outdoor air intake	2.0 – 3.0	2.0 – 3.0	2.0 – 3.0
fan inlet	3.0 – 4.0	3.5 – 4.5	4.5 – 5.5
fan outlet	5.0 – 8.0	6.5 – 10.0	8.0 – 12.0
main duct	3.5 – 4.5	5.0 – 6.5	6.0 – 9.0
branch duct	2.5 – 3.5	3.0 – 4.5	4.0 – 5.0
outlet duct	2.0 – 2.5	3.0 – 3.5	3.5 – 4.5

(a) air velocities in m/s

application	pressure drop per unit length of duct
	Pa/m
concert halls, broadcasting studios, recording studios	0.4
hospitals, libraries, theatres	0.65
general offices, department stores	0.8
factories and workshops	1.0 – 1.5

(b) pressure drops per unit length of duct

Table 13.20 Recommended air velocities and pressure drops for sizing low velocity ductwork systems

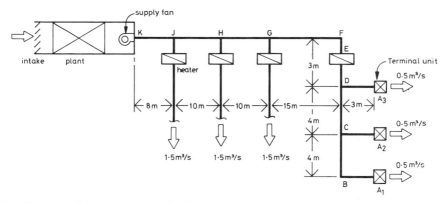

13.15 Ductwork layout: Example 13.16

Example 13.16 A diagram of a system of rectangular ductwork as shown in Figure 13.15. Each terminal unit supplies 0.5 m³/s of air to the air conditioned spaces. The ductwork is to be sized on a pressure drop of 1.0 Pa/m length of duct; loss coefficients for fittings to be obtained from the preceding tables. Determine the total system pressure loss.

Plant pressure drops:

terminal units	20 Pa
heaters	45 Pa
plant (to fan suction)	250 Pa

Solution

It is often convenient to calculate the losses in a supply system *against the flow*, in order to generate sub-totals. This enables the index run to be identified, at the same time providing information for balancing the branches ducts. (With an extract system the sizing procedure will be in the same direction as the flow.) The majority of the calculations are completed in Table 13.21 and relevant notes are as follows:

- The pressure losses of the terminal unit, heater battery and central plant are obtained from manufacturer's information.
- Where the final discharge from the system is in the form of a jet the velocity pressure from the discharge outlet is a system pressure loss and should be included in the sum of the total losses with which the fan has to deal. In systems supplying air for comfort air conditioning this will be a relatively small loss, as the outlet velocity will be about 2 m/s and the velocity pressure if omitted from the total system pressure loss would not significantly affect the system/fan performance. However it is good practice always to include it, so that when it is relatively large compared with the total system loss (for example in an industrial exhaust system with high discharge velocity) it may not be omitted.

- Rectangular duct sizes from Table 13.3 have been used, such that the equivalent circular duct diameter gives a friction pressure loss of *approximately* 1.0 Pa/m of duct length. Duct lengths are measured along the axis of the duct; the fittings are assumed to have zero length (in accordance with the definition of loss coefficients).

Bend B: short radius bend, aspect ratio, $b/W = 0.4/0.25 = 1.6$ from Table 13.6, $K_b = 0.42$.

Bend F: medium radius bend, aspect ratio, $b/W = 0.4/0.6 = 0.67$ from Table 13.6, $K_b = 0.24$.

Branches: take branch as Figure 13.8(g)

flow to main – branch C, flow to main
$K_{32} = 0.03$, followed by a contraction (Table 13.12)
$K_c = 0.05$; total loss coefficient $= 0.08$;

flow to branch – branch C taken as a medium radius bend.
- The pressure drop sub-total from branch C to outlet A_1 is 34.5 Pa.
- The total pressure drop in the system is 404.7 Pa. As the fan has not yet been sized, this total does not include any loss in the fan discharge expansion. The fan duty is therefore 6 m^3/s at 405 Pa and the fan selected on *fan static pressure* as described in Chapter 14, page 298–9.
- REYNOLDS number and interaction effects between fittings have not been included in these calculations.

Index run and system balance

Referring to Example 13.16 and Table 13.21, from terminal A_2 the duct A_2C has been sized for the same friction pressure drop as the duct supplying terminal A_2. The pressure loss up to the point of offtake has a sub-total of 27.4 Pa, which is 7.1 Pa *less* than duct A_1BC. These two branch ducts are therefore out of *balance*. The index run is therefore taken as A_1 through to the fan discharge. The out-of-balance pressure drop in this branch (and other branches similarly) can be dealt with in one of three ways:

- resize the duct at a higher velocity so that the duct and fittings together absorb pressure drop of 34.5 Pa;
- include a fitting such as a perforated plate, sized to give a net pressure loss of 7.1 Pa;
- include a damper which can be adjusted on site, to provide the out-of-balance pressure drop.

Though a balance at the design stage is theoretically possible with either of the first two methods, the calculations are based on published pressure loss coefficients. It is most unlikely that this can be achieved because the duct friction loss is based on an assumed roughness coefficient and the loss coefficients are experimental values obtained, for the most part with fully developed, turbulent flow conditions under laboratory conditions. The

item fitting, straight duct plant	air flow rate $\dot{V}$ (m³/s)	duct size rectangular b×W (m m)	duct size circular D (m)	velocity $\bar{v}$ (m/s)	fitting velocity pressure pv (Pa)	fitting loss coeff. K	straight duct length l (m)	straight duct pressure loss/m Δp_f (Pa)	pressure loss (Pa) item K pv or lΔp_f	pressure loss (Pa) sub-total	pressure loss (Pa) accumulated total
terminal A₁	0.5			5.0	-	-	-	-	20.0		
duct A₁BC	0.5	0.40×0.25	0.346	5.0	-	-	7.0	0.9	6.3		
bend B	0.5				15.0	0.42	-	-	6.3		
branch C	1.0/0.5			6.25	23.4	0.08	-	-	1.9	34.5	
duct CD	1.0	0.40×0.40	0.441	6.25	-	-	4.0	1.1	4.4		
branch D	1.5/1.0			6.25	23.4	0.05	-	-	1.2	5.6	40.1
duct DEFG	1.5	0.40×0.60	0.537	6.25	-	-	18.0	0.8	14.4		
heater E	1.5			6.25	-	-	-	-	45.0		
bend F	1.5			6.25	23.4	0.24	-	-	5.6		
branch G	3.0/1.5			8.3	41.3	0.05	-	-	2.1	67.1	107.2
duct GH	3.0	0.60×0.60	0.661	8.3	-	-	10.0	1.2	12.0		
branch H	4.5/3.0			9.4	54.0	0.05	-	-	2.7	14.7	121.9
duct HJ	4.5	0.80×0.60	0.848	9.4	-	-	10.0	1.1	11.0		
branch J	6.0/4.5			9.4	54.0	0.05	-	-	2.7	13.7	135.6
duct JK	6.0	0.80×0.80	0.381	9.4	-	-	8.0	1.0	8.0		
fan discharge	6.0			***	****	****	-	-	****	8.0	143.6
plant	6.0								250.0		
intake grille	6.0	2.00×1.20	-	2.5	3.75	3.0	-	-	11.1	261.1	404.7
terminal A₂	0.5			-	-	-	-	-	20.0		
duct A₂C	0.5	0.40×0.25	0.346	5.0	-	-	3.0	0.9	2.7		
branch C	1.0/0.5			6.25	23.4	0.2	-	-	4.7	27.4	

**** pressure drop not available until fan has been selected

Table 13.21 Duct sizing calculations for Example 13.18

difference between these and the actual losses can be substantial, so that in practice a balance cannot be achieved by design calculations. The third solution of placing dampers in the network should therefore be adopted; their location has to be based on the principles of proportional balancing procedures described in Chapter 15.

LAYOUT CONSIDERATIONS

Ductwork distribution systems have to be integrated into the building design and coordinated with the other services. There are a number of ways in which these ducts can be arranged within the building. Often the simplest and cheapest method is to accommodate them above a false ceiling in a central corridor, with connections at high level facing the external windows. In this case, access to services is relatively easy. Other types of distribution include the following, individually or in combination:

- vertical risers, eg toilet extract;
- horizontal perimeter ducts to serve window sill units; the duct runs may be under the window or in a bulkhead at ceiling level on the floor below;
- ducts above false ceilings, supplying ceiling diffusers; the space above the ceiling can be used for plenum extract, a common approach when extract air lighting fittings are installed;
- air handling units above the false ceiling;
- ducts under a false floor;
- inter-floor plant space.

Balancing requirements

The ductwork distribution systems should be designed in such a way as to assist the balancing procedures and included facilities for measuring and regulating the air flow rates. These requirements are described in Chapter 15.

Connections to plant items

As with the fan suction and discharge connections (see page 299), duct connections between plant items can have an adverse effect on the performance of the system unless they are made correctly. The fresh air intake grille should be positioned away from the exhaust outlets to minimize the risk of bringing in humid and/or contaminated air. To avoid heat gains from the plant room itself, the intake should be connected, either by sheet metal or by builders' work, to the first item of plant in the system.

Connections to plant items which produce asymmetrical velocity profiles at the upstream face of the equipment may lead to increased pressure drop, increased noise generation, lower efficiency and moisture carry-over. They should therefore, be designed to ensure an even flow over such items as filters, heating and cooling coils, humidifiers, dampers and silencers.

Smoke and fire precautions

The ductwork system, including any linings, must be designed and installed to minimize the risk of smoke and fire spreading through the building via the ductwork itself. The materials used must therefore be non-combustible or, at the very least, difficult to ignite, possess a good rating against the surface spread of flame and not generate smoke or toxic fumes in the presence of fire or heat. Special attention should be given to the insulating materials and adhesives, flexible connections, filters and acoustic linings.

When a duct passes through a fire resisting builders work enclosure, any holes through which the duct passes should be as small as possible and when installed sealed with fire-stopping material. Similar treatment should be applied to ducts which pass through floors and partitions designed to prevent the spread of fire and smoke.

To reduce the risk of fire and smoke spreading through the trunking, fire dampers must be installed where the ductwork passes through a fire partition. Detectors for smoke and fire warning signals are often included in the ductwork. It is important to follow local authority and fire brigade regulations in all these aspects of design.

SYMBOLS

A	duct or pipe cross section area	K_e	pressure loss coefficient, expansion
b	duct breadth	K_{es}	pressure loss coefficient, expansion with short outlet pipe
C_{b-b}	bend interaction effect factor		
C_e	loss factor for expansions with short outlet pipe	K_{ex}	pressure loss coefficient, exhaust discharge
C_{Re}	Reynolds' number correction factor	K_i	pressure loss coefficient, intake
C_{31}	branch correction factor	K_{31} }	pressure loss coefficients,
C_r	curve ratio for guide vanes in rectangular ducts	K_{32} }	supply branch
		K_{13} }	pressure loss coefficients,
D	diameter	K_{23} }	extract branch
D_e	equivalent diameter of rectangular duct	L	duct length
		N	length of expansion
K	pressure loss coefficient	n	number of vanes in mitred bend
K_b	pressure loss coefficient, bend		
		P_{at}	atmospheric pressure
K_c	pressure loss coefficient, contraction	p_v	velocity pressure
		Re	REYNOLDS' number

r   radius
t   air temperature
$\dot{V}$ volume flow rate
$\bar{v}$ mean duct or pipe velocity
W   duct width
Δp pressure drop
Δp_f pressure loss due to friction
ρ air density
α angle of expansion vanes,
            from axis
θ angle of an expansion

Subscripts
1, 2 .. relates to specific duct or
            pipe section

Abbreviations
AR   aspect ratio of rectangular
            duct $(= b/W)$

14 Fans

Fan selection is necessarily important to the overall performance of air conditioning systems. The required duty is obtained by an analysis of the flow rates and pressure losses in the ductwork system as described in previous chapters. The choice of an appropriate unit will then depend on the application, together with considerations such as plant arrangement, space available for the installation, corrosive and hazardous gases. Since the annual energy cost for fan operation represents a relatively high proportion of the total energy costs of an air conditioned building, it is also important to consider the fan's overall efficiency.

CHARACTERISTICS

By running a fan at constant speed and measuring the pressure developed and the power input at different flow rates, the manufacturer obtains data on the fan's performance. These measurements, together with the computed efficiency, when graphed or tabulated, are known as the *fan's characteristics*. The tests are carried out according to recognised national and international standards.

FAN TYPES

The following are the five main categories of fans used for a variety of applications in air conditioning systems and equipment:

- centrifugal;
- axial;
- mixed flow;
- propeller;
- cross flow.

These fans are described below, together with graphs of their typical characteristic curves of:

- pressure versus flow rate;
- efficiency versus flow rate;
- power versus flow rate.

Not every part of a fan's characteristic will give satisfactory operation in the system. Therefore, on each of the fan curves an indication is given of the *normal working range*.

Centrifugal

Centrifugal fans may be single or double inlet and are either belt driven or direct driven by an electric motor. A single inlet fan is illustrated in Figure 14.1.

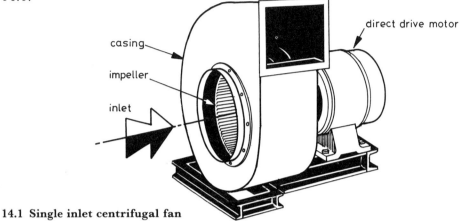

casing

direct drive motor

impeller

inlet

14.1 Single inlet centrifugal fan

Centrifugal fans are classified by their impeller blade shape, ie:

– backward curved;
– radial;
– forward curved.

The blade shapes and characteristic curves for these fan types are illustrated in Figures 14.2, 14.3 and 14.4.

The *backward curved* impeller with 6–16 blades, has a power characteristic which rises to a maximum at the middle of its flow rate range and then falls at its highest flow rates. This is known as a non-overloading power characteristic. These fans have efficiencies of up to 90% where aerofoil section blades are used. They are most commonly used when continuous high efficiency operation is required.

The *radial* or *paddle bladed* impeller, with 6–16 blades, has a continuously rising, or overloading, power characteristic. With flat blades, this fan type has an efficiency up to 60%; with blades slightly curved at the heel up to 75%. It is most often used where corrosion and blade wear may be a problem, as in industrial exhaust systems, the blades being self cleaning and often replaceable.

The *forward curved* fan has a 40–60 blade impeller with an over-loading power characteristic and an efficiency up to 75%. It is compact in size and because of this is the most commonly used centrifugal fan for AC and mechanical ventilation systems.

The overall design of centrifugal fans makes a single inlet fan a suitable choice when the plant installation requires a right-angled turn. Double inlet fans are usually installed in plenum chambers and are often used in this way in packaged AC units.

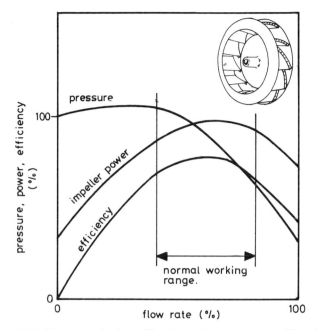

14.2 Characteristics of backward curved centrifugal fan

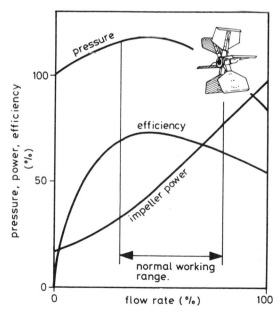

14.3 Characteristics of radial centrifugal fan

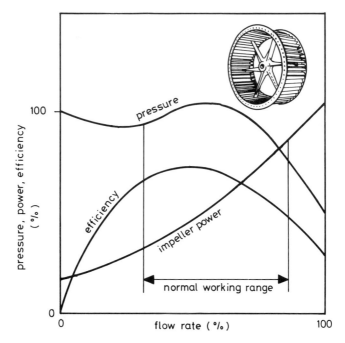

14.4 Characteristics of forward curved centrifugal fan

Axial flow

The axial flow fan has an impeller with 6–12 aerofoil section blades with a non-overloading power characteristic; efficiencies of over 85% have been obtained. To achieve these high efficiencies, guide vanes are used either to give pre-rotation on the upstream side or to remove air rotation from the downstream side of the fan. An alternative method of improving efficiency is to have two stages rotating in opposite directions (contra-rotating). The second impeller then acts in a similar manner to the upstream guide vanes. The fan is most often directly driven by a synchronous speed motor inside the fan casing but where the installation requires, eg in hot, dirty or corrosive air streams, the fan can be belt driven by an external motor.

Variable pitch axial flow fans are used for VAV air conditioning systems. The pitch angle of the blades are varied through a controller, eg a pneumatic actuator, with increasing blade angle for increasing flow rates, and vice versa for reducing flow rates. The characteristics for a typical fan are given in Figure 14.6.

The axial fan is compact and can easily be accommodated into the system. It can be placed at low or high level in the plant room, in false ceilings and in either vertical or horizontal ducts. The fan's configuration makes it suitable for installations in which the intake and discharge connections are opposite one another.

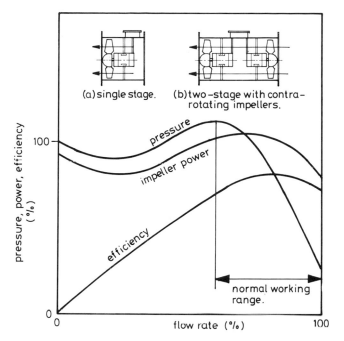

14.5 **Characteristics of axial flow fan**

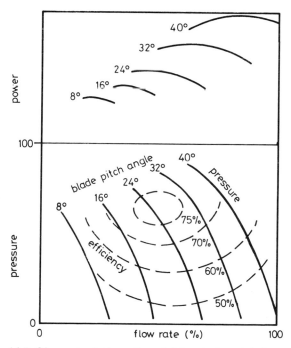

14.6 **Characteristics of variable pitch axial flow fan**

Mixed flow

Mixed flow fans are intermediate between the centrifugal and axial fan types in pressure development and compactness. The air path through the impeller has a radial as well as an axial component. These fans are normally constructed with the intake and discharge on the same axis, though they can be fitted with a radial discharge. Efficiencies are up to 80% with a non-overloading characteristic.

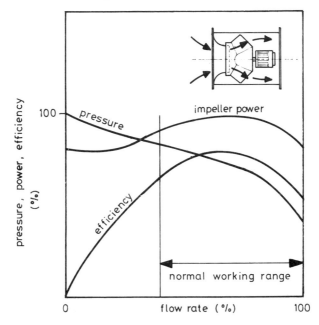

14.7 Characteristics of mixed flow fan

Propeller

The propeller fan is a comparatively simple form of fan with a sheet metal, 3–6 bladed impeller directly driven by a motor mounted in the air stream. Efficiencies of up to 75% can be obtained but pressure development is low and overloading can occur if the fan is installed in a system with too high a resistance. Performance can be improved by using aerofoil blades which give efficiencies up to 80%. The fan is useful for low pressure drop systems, eg through-the-wall kitchen extract. It is also used extensively for unit equipment such as refrigeration air cooled condensers and for cooling towers.

Cross flow

The cross flow fan has a multi-bladed, cylinder-like impeller with blade shape similar to the forward curved centrifugal fan. The ends of the impeller are blanked off and, because of the casing shape, air enters along one of the cylindrical surfaces and discharges from the other. The maximum efficiency is about 40%. The principal application is in small domestic electric heaters.

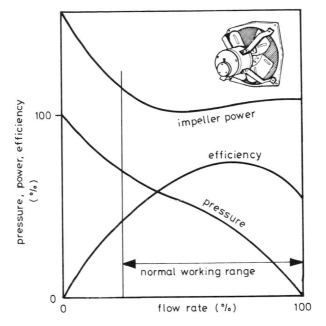

14.8 **Characteristics of propeller fan**

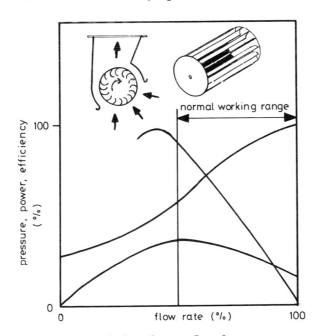

14.9 **Characteristics of cross flow fan**

FAN LAWS

It is not practical to test every fan and therefore the performance of geometrically similar fans is obtained by applying the fan laws. These laws apply to a particular point on a fan characteristic and cannot be used to predict other points. The fan volume flow rate and pressure are related to the fan speed, diameter and air density according to the following equations:

Fan flow rate is given by:

$$\dot{V} = C_{Fv} \, n \, D^3. \tag{14.1}$$

Fan pressure is given by:

$$p_F = C_{Fp} \, n^2 \, D^2 \, \rho \tag{14.2}$$

where C_{FV} = constant of proportionality for fan flow rate
C_{Fp} = constant of proportionality for fan pressure.

Air power is given by:

$$P_a = \dot{V} \, p_F. \tag{14.3}$$

Combining Equations 14.1 and 14.2

$$P_a = C_{FP} \, n^3 \, D^5 \, \rho \tag{14.4}$$

where C_{FP} = constant of proportionality for fan power.

Fan efficiency, η, is given by:

$$\eta = \frac{P_a}{P_i} \tag{14.5}$$

where P_i = impeller power.

Notes:
(i) The constants of proportionality C_{FV}, C_{Fp} and C_{FP} are specific for the type of fan and the point of operation.
(ii) The *overall* fan efficiency must be related to the motor power input, which takes account of motor and drive efficiencies.

Example 14.1 A centrifugal fan with a 0.6 m diameter impeller, running at 13 rev/s, delivers 2 m³/s of air with an air density of 1.2 kg/m³, at a total pressure of 500 Pa. Determine the duty of a geometrically similar fan with a 0.8 m diameter impeller running at 20 rev/s if the air density is the same for both fans. Determine the air power of this second fan.

Solution

To obtain constant of proportionality for flow rate, C_{FV}, use Equation 14.1:

$$\dot{V} = C_{FV} \; n \; D^3$$
$$2 = C_{FV} \; 13 \times 0.6^3$$

$$\therefore \; C_{FV} = \frac{2}{13 \times 0.6^3} = 0.712.$$

The new fan flow rate is given by Equation 14.1:

$$\dot{V} = 0.712 \times 20 \times 0.8^3 = 7.29 \; \text{m}^3/\text{s}.$$

To obtain constant of proportionality for pressure, C_{Fp}, use Equation 14.2:

$$p_F = C_{Fp} \; n^2 \; D^2 \; \rho$$
$$500 = C_{Fp} \; 13^2 \times 0.6^2 \times 1.2$$

$$\therefore \; C_{Fp} = \frac{500}{13^2 \times 0.6^2 \times 1.2} = 6.85.$$

The new fan pressure is given by Equation 14.2:

$$p_F = 6.85 \times 20^2 \times 0.8^2 \times 1.2 = 2104 \; \text{Pa}.$$

Air power is given by Equation 14.3:

$$P_a = \dot{V} \; p_F = \frac{7.29 \times 2104}{1000} = 15.3 \; \text{kW}.$$

Example 14.2 For the centrifugal fan with a 0.8 m diameter impeller, in Example 14.1, determine the new duty and air power if the air density reduces to 1.05 kg/m^3.

Solution

The air volume flow rate is unaffected by air density,
$\therefore$ flow rate $= 7.29$ m^3/s.

From Example 14.1, the constant of proportionality for pressure: $C_{Fp} = 6.85$.

The new fan pressure is given by Equation 14.2:

$$p_F = C_{Fp} \; n^2 \; D^2 \; \rho$$
$$= 6.85 \times 20^2 \times 0.8^2 \times 1.05 = 1841 \; \text{Pa}.$$

New air power is given by Equation 14.3:

$$P_a = \dot{V} \; p_F = \frac{7.29 \times 1841}{1000} = 14.6 \; \text{kW}.$$

Operating point

In an air flow system consisting of ducts, fittings and plant items, there will be a total system pressure loss, Δp_t, against the total flow rate, $\dot{V}$. This determines the system characteristic which is assumed to obey the square law;

$$\Delta P = r_t \dot{V}^2 \tag{14.6}$$

where r_t, is known as the *system total resistance*.

A fan is selected for the design duty and the system and fan characteristics intersect, as shown in Figure 14.10, to give the fan and system operating point. The volume flow rate and pressure at the operating point may be obtained either graphically or by solving simultaneously the system and fan characteristic equations.

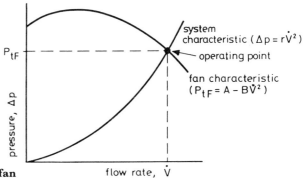

14.10 Operating point of a fan

For most fans, the fan volume/pressure characteristic in the normal working range can be defined by a quadratic equation of the form:

$$p_F = A' - B' \dot{V}^2 \tag{14.7}$$

where A' and B' are fan characteristic constants determined from a curve fit of manufacturer's data.

The values of A' and B' in Equation 14.7 apply to a particular fan at a given speed and air density.

Example 14.3   Determine the operating point for a fan which is connected to a system with a design duty of 2 m³/s at a total pressure drop of 440 Pa. The fan characteristic constants for Equation 14.7 are $A' = 600$ and $B' = 20$.

Solution
The system resistance is obtained using Equation 14.6:

$$\Delta P = r_t \dot{V}^2$$
$$440 = r_t \, 2^2$$
$$r_t = 440/4 = 110.$$

∴ the system characteristic is given by:

$$\Delta P = 110 \ \dot{V}^2.$$

The fan characteristic is given by Equation 14.7:

$$p_F = A - B \ \dot{V}^2$$
$$p_F = 600 - 20 \ \dot{V}^2.$$

At the operating point, $\Delta p = p_F$

$$\therefore \ 100 \ \dot{V}^2 = 600 - 20 \ \dot{V}^2$$

$$\dot{V} = \sqrt{\frac{600}{110 + 20}} = 2.15 \ \text{m}^3/\text{s}.$$

The fan pressure is then given by:

$$p_F = 600 - 20 \ \dot{V}^2 = 600 - 20 \times 2.15^2 = 508 \ \text{Pa}.$$

The operating point of the fan and system is therefore:

$$2.15 \ \text{m}^3/\text{s at a pressure of } 508 \ \text{Pa}.$$

ADJUSTMENT OF THE TOTAL FLOW RATE

If the air flow rate is not at the design level when the system has been balanced to the design flow rate, adjustment of the total flow rate in constant flow systems may be achieved by one of the following methods:

- throttling damper
- inlet guide vanes (centrifugal fans) } for reducing flow rates

- change of fan speed
- change of pitch angle (axial fans) } for increasing or reducing flow rates.

Throttling damper

The simplest method of reducing the total flow rate is to restrict the flow with a throttling damper. This has the effect of increasing the system resistance, the operating point *retreating* along the fan characteristic curve. This is illustrated in Figure 14.11.

Example 14.4 Determine the pressure drop across a main throttling damper to reduce the flow rate in Example 14.3 to the design duty.

Solution

Using the fan characteristic equation, at the design flow rate of 2 m³/s the fan pressure is given by:

$$p_F = 600 - 20 \times 2^2 = 520 \ \text{Pa}.$$

The system pressure drop at the design flow rate = 440 Pa
∴ the pressure drop across the damper = 520 − 440 = 80 Pa.

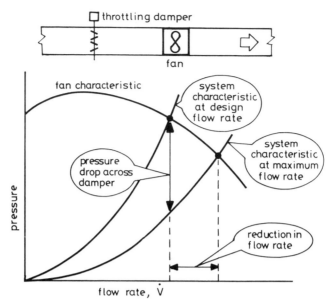

14.11 Reduction in flow rate with a throttling damper

Speed control

Since fan power is proportional to the *cube* of fan speed, reducing the speed of the fan is the most energy efficient method of reducing the flow rate. For a belt driven centrifugal fan this may be achieved either by changing the pulley size or by incorporating continuously variable pulleys in the fan drive.

Example 14.5   Determine the new fan speed to reduce the flow rate in Example 14.4 to the design duty if the original speed was 13 rev/s. Compare the air power requirement with that of the throttling damper in Example 14.3.

Solution

From Example 14.3 the actual flow = 2.15 m³/s
                   the design duty = 2 m³/s.

   Referring to Equation 14.1, the flow rate is proportional to speed. Therefore the reduced speed is given by:

$$n = \frac{2}{2.15} \times 13 = 12.1 \text{ rev/s.}$$

Referring to Equation 14.4, the air power is proportional to (speed).[3] Therefore if the fan is now made to run at 12.1 rev/s the percentage reduction in fan *air* power is given by:

$$\left[1 - \left(\frac{12.1}{13} \right)^3 \right] \times 100 = 19.6\%.$$

SERIES AND PARALLEL OPERATION

Series operation   Two fans, of equal duty, which are connected in series have a combined flow/pressure characteristic which doubles the pressure at the same flow rate compared with a single fan.

This combined characteristic is illustrated in Figure 14.12.

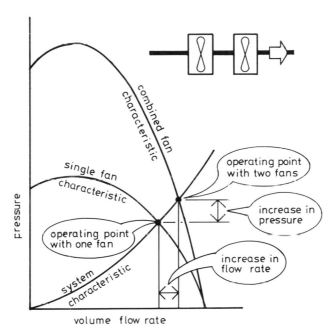

14.12 **Characteristic of two fans in series**

Parallel operation   Two fans, of equal duty, which are connected in series have a combined flow/pressure characteristic which doubles the flow rate at the same pressure, compared with a single fan. This combined characteristic is illustrated in Figure 14.13.

However, these fan arrangements will not produce in the system *double the pressure* or *double the flow rate*; as with a single fan the operating point for the combined fan characteristics will depend on the characteristic of the system in which the fans are installed.

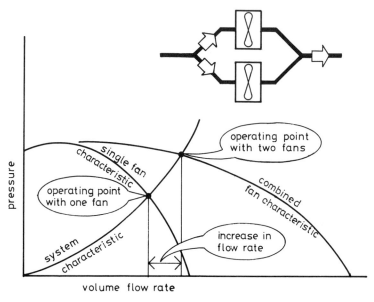

14.13 **Characteristic of two fans in parallel**

Example 14.6   Two identical fans, with characteristic constants $A' = 500$, $B' = 10$, are connected to a system having a resistance of 80. Determine the combined duty of the fans:

(a) in series;
(b) in parallel

Solution

(a) *series operation*
The single fan characteristic is given by:

$$p_F = A' - B' \, \dot{V}^2$$
$$= 500 - 10 \, \dot{V}^2$$

∴ the characteristic of two fans in series is given by:

$$p_F = 2(500 - 10 \, \dot{V}^2).$$

The system characteristic is given by:

$$\Delta p = 80 \, \dot{V}^2$$

At the operating point

$$\Delta p = p_F$$
$$80 \, \dot{V}^2 = 2(500 - 10 \, \dot{V}^2)$$

$$\therefore \dot{V} = \sqrt{\frac{1000}{(80 + 20)}} = 3.16 \text{ m}^3/\text{s};$$

(b) *parallel operation*

The single fan characteristic is given by Equation 14.7:

$$p_F = 500 - 10 \ \dot{V}^2$$

$$\therefore \ \dot{V} = \sqrt{\frac{500 - p_F}{10}}$$

∴ the characteristic of two fans in series is given by:

$$\dot{V} = 2\sqrt{\frac{500 - p_F}{10}}.$$

From the system characteristic the flow rate is also given by:

$$\dot{V} = \sqrt{\frac{\Delta p}{80}}.$$

At the operating point $\Delta p = p_F = p$

$$\therefore 2\sqrt{\frac{500 - p}{10}} = \sqrt{\frac{p}{80}}$$

$$\therefore p = 485 \ \text{Pa}$$

$$\therefore \ \dot{V} = \sqrt{\frac{485}{80}} = 2.46 \ \text{m}^3/\text{s}.$$

CONTROL FOR A VAV SYSTEM – CENTRIFUGAL FANS

Inlet guide vanes

This method has been a popular form of control where centrifugal fans are used. The guide vane assembly may either be bolted onto the fan inlet or built into the fan suction eye. The guide vanes must be selected for the correct handing of the fan so that the vanes direct air onto the impeller blades. The relative power requirements are normally presented against a percentage of the design flow rate as shown in Figure 14.14.

Speed control

For economy of operation, speed control of a centrifugal fan is probably the most satisfactory, since fan power is proportional to the cube of fan speed. However, at the present initial costs of suitable control equipment, it makes this the most expensive of the different methods of flow rate control.

Variation of fan speed may be achieved in a number of ways. One method is by using an A.C. motor whose speed is controlled through a variable frequency static inverter. The efficiency of this equipment remains high over the range of system load variations and the equipment has good reliability. Other methods of speed control include variable pulley drive, fluid coupling and D.C. motor.

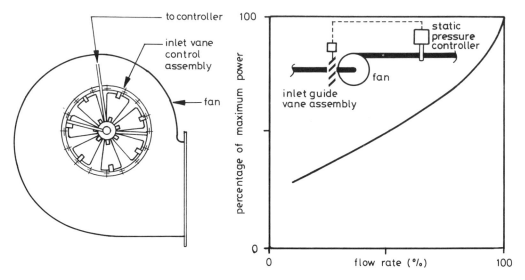

14.14 Flow rate control of a centrifugal fan using inlet guide vanes

SELECTING THE FAN FOR THE SYSTEM

Fan static pressure is defined by:

$$p_{sF} = p_{tF} - p_{\bar{v}F} \qquad (14.8)$$

The fan velocity pressure, $p_{\bar{v}F}$, is calculated using the mean velocity at the fan discharge, $\bar{v}_F$.

The fan static pressure should not be confused with the actual static pressure rise across the fan, (as illustrated in Figure 14.18). It is defined by Equation 14.8 for two reasons:

- for fan selection when the losses at fan suction and discharge are not known;
- to allow for the case when the fan velocity pressure is not available to deal with any of system losses.

Initially, when calculating the total pressure loss through the system, the losses at the fan suction and discharge connections cannot be calculated, since the fan at this stage will not have been selected. The total system pressure loss may therefore exclude losses at these connections and the fan is selected on fan static pressure. When installed the fan must operate on the total pressure developed and therefore, by definition, one fan velocity pressure will be available to deal with the losses at the fan connections. Alternatively, the losses at the fan suction and discharge connections can now be calculated, based on the first fan selected, added to the original total and the fan *re-selected* on fan total pressure. This is rarely done however as the pressure losses at the fan connections, especially in the discharge, are likely to be considerably larger than those obtained from *text-book* losses, where the loss coefficients are for fully developed, turbulent pipe flow conditions.

One further important point must be made. Even with one fan velocity pressure available to deal with the losses at suction and discharge connections, excessive losses can occur if a centrifugal fan is not *handed* correctly. This would also be the case if the suction and discharge connections are not designed according to good practice. It should be noted in passing that when a fan/system fails to deliver the design duty, the cause can often be traced to an unsatisfactory duct arrangement close to the fan. Some investigations have been made into what is termed the fan-system effect in order that any additional losses may be accounted for. These effects are similar to the interference effects between fittings described in Chapter 12. In detailing the connections the aims are:

– to produce a smooth, even, flow into the fan suction eye;
– to ensure the fan is handed correctly in relation to the ductwork on its discharge;
– to make any expansion of the discharge duct take place gradually.

Well designed connections will also ensure minimum air noise generation from these fittings. Some typical poor and improved duct arrangements are shown in Figure 14.15.

FAN NOISE

Fans generate noise in a variety of ways; for example broad band noise from turbulent flow over the blades and discrete tones due to the interaction between the wakes leaving the impeller blade and the vanes. This noise radiates directly into the upstream and downstream ducts. Invariably a fan will generate least noise at a duty corresponding to maximum efficiency so that it is important to match the fan correctly to the required system design point. Where a number of fans are available for the same duty/efficiency point the fan with the lower tip speed will generate least noise.

The general shape of the sound power level spectrum of centrifugal and axial flow fans is given in Figure 14.16. Subjectively centrifugal fans appear to be less noisy to the ear than axials; this is because they generate more low frequency noise. However, it is sometimes cheaper to install a higher speed, smaller diameter axial fan with an attenuator, than a slower speed centrifugal.

The physical size of the fan will be related to both the duct sizing and siting of the fan to give adequate space for maintenance, repairs, cleaning and possible replacement, but the fan should be as large as possible to keep discharge velocities as low as possible with a recommended maximum of 10 to 16 m/s.

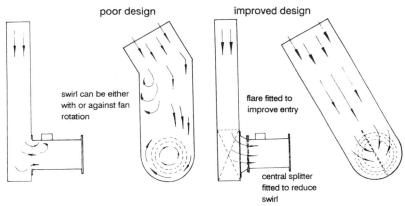

poor design   improved design

swirl can be either with or against fan rotation

flare fitted to improve entry

central splitter fitted to reduce swirl

(a) effect of lateral bends in fan inlet ducting
(axial fan shown - applies equally to centrifugal fans)

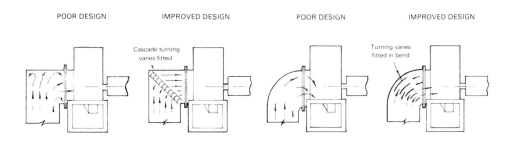

POOR DESIGN   IMPROVED DESIGN   POOR DESIGN   IMPROVED DESIGN

Cascade turning vanes fitted

Turning vanes fitted in bend

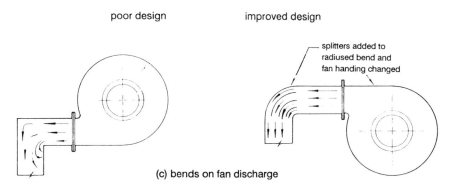

poor design   improved design

splitters added to radiused bend and fan handing changed

(c) bends on fan discharge

14.15 Examples of poor and improved fan suction and discharge connections
(Based on Figures 9.5, 9.6 and 9.10 from *Fan Application Guide*, 1981, the Fan
Manufacturers Association, by permission)

With CAV systems the flow rate may vary to some extent due to increased system resistance from such plant items as clean/dirty filters. It is therefore good practice to select the fan with an operating point to the right of the maximum efficiency point. With a VAV fan greater care will be needed to select a fan, together with its method of flow rate control, which will meet the acoustic requirements of the system.

Where the building is continuously occupied, and at night lower noise levels are desirable, eg in hospital wards, two fans in parallel (or a two-stage axial fan) selected for the maximum duty, provide the opportunity to reduce noise levels by operating only one of the fans.

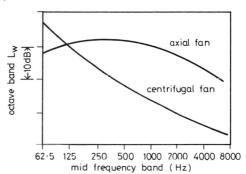

14.16 Typical sound power level characteristics of centrifugal and axial flow fans

Flexible connections and anti-vibration

Flexible connections should be used where the fan connects directly to the ductwork on the suction and discharge. Anti-vibration mounts should be used to isolate the fan from the building structure. With indirect drives the fan and motor must be mounted on a common base with the base isolated from the structure. Care must be taken that no bridging straps connect such connections and anti-vibration mounts.

TESTING FANS FOR GENERAL PURPOSES

British Standard BS848:1980; Part 1[2] for testing fans includes the important provision for site tests of fans. Four types of fan installation are recognized:

- Type A ...   free inlet, free outlet
- Type B ...   free inlet, ducted outlet (blowing fan)
- Type C ...   ducted inlet, free outlet (extract fan)
- Type D ...   ducted inlet, ducted outlet (booster fan)

Duct connections modify fan performance. Therefore a fan adaptable for more than one type of installation may have more than one standardized performance characteristic.

Site installations rarely conform to standard test duct arrangements, and pressures obtained from a site test are strictly applicable to the site installation. To ensure satisfactory site measurements the system should be designed to include the necessary test lengths and the recommendations of the British Standard are shown in Figure 14.17.

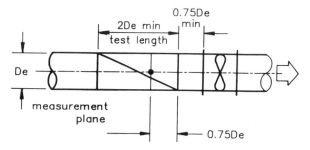

(i) test length on inlet side of fan

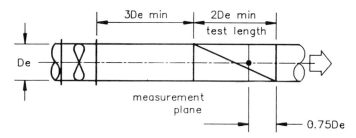

(ii) test length on discharge side of fan

(a) flow measurement planes

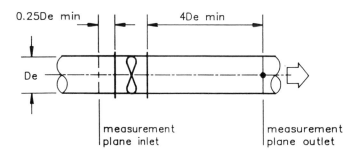

(b) pressure measurement planes

14.17 Recommended test lengths and measurement planes for fan testing

The *test length* is a straight length of duct of uniform cross section free from obstructions. On the discharge side of the fan the test duct has a minimum length of three duct diameters following the fan discharge or discharge expansion. An air straightener is included in the discharge test duct.

Pressure measurements should be made as close to the fan as possible, to ensure minimum error from the pressure loss due to friction. However, if the test length is sufficiently close to the fan, measurements should be made in the flow measurement plane.

Pressure for a booster fan

The example of a fan with a ducted inlet and ducted outlet is described to illustrate the site measurements required to determine fan pressure.

The pressure distribution is shown in Figure 14.18. In the context of fan performance the following should be noted:

CD – fan total pressure
CG – fan static pressure
DG – fan velocity pressure
EF – system discharge velocity pressure.

The fan velocity pressure is available to provide part of the systems pressure loss requirements. Therefore the fan performance is given in terms of fan total pressure.

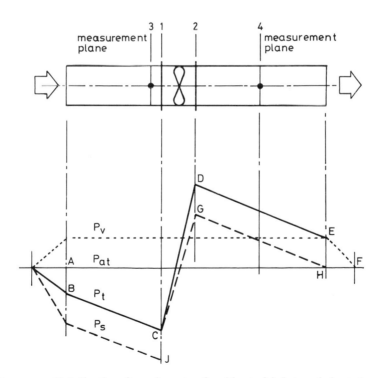

14.18 Pressure distribution for a booster fan (ducted inlet and ducted outlet)

Fan total pressure is given by:

$$P_{tF} = P_{t1} - P_{t2}$$
$$= (P_{t4} + P_{f24}) - (P_{t3} - P_{f31}).$$

Since the duct cross-section 3 is on the suction side of the fan, p_{t3} is negative; therefore ignoring signs:

$$P_{tF} = P_{t4} + P_{f24} + P_{t3} + P_{f31}. \qquad (14.9)$$

Where
p_{t4} = total pressure at the outlet measurement plane
p_{t3} = total pressure at the inlet measurement plane
p_{f31} = pressure loss due to friction between inlet measurement plane and fan inlet
p_{f24} = pressure loss due to friction between outlet measurement plane and fan outlet.

Measurement required:

– total pressures at measurement planes 3 and 4, ie Pitot (total pressure) tube readings.

Example 14.7   Determine the fan total pressure for the ducted air system shown in Figure 14.19 with the following test data:

total pressures relative to atmospheric pressure:
measurement plane 3:	100 Pa
measurement plane 4:	250 Pa
measured flow rate:	1.2 m³/s
air density:	1.15 kg/m³.

Solution

Fan cross section area $= \pi 0.4^2/4 = 0.126 \text{ m}^2$
  Mean velocity at fan suction and discharge $= 1.2/0.126 = 9.55 \text{ m/s}$
  Velocity pressure $= 0.5 \times 1.15 \times 9.55^2 = 54.4 \text{ Pa}$.
  Loss coefficients for contraction and expansion obtained from data in Chapter 13.
  Pressure loss in expansion $= K_e \, p_v = 0.2 \times 54.7 = 11 \text{ Pa}$.
  Pressure loss in contraction $= K_c \, p_v = 0.05 \times 54.7 = 2.6 \text{ Pa}$.
  Friction pressure loss in 0.8 m diameter straight duct is negligible.

  Friction pressure loss in 0.5 m diameter straight duct (from friction chart, Figure 13.1) and correcting for air density:

$$\Delta p_f = l \Delta p = 3 \times 0.8 \times 1.15/1.2 = 2.3 \text{ Pa}.$$

Fan total pressure is given by Equation 14.9:

$$P_{tF} = P_{t4} + P_{t3} + P_{f31} + P_{f24}$$
$$= 250 + 100 + 2.6 + (11 + 2.3) = 365.9 \text{ Pa}.$$

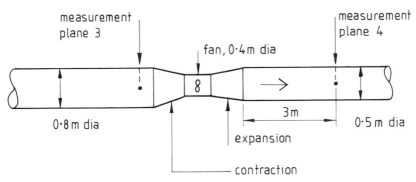

measurement plane 3

measurement plane 4

fan, 0·4m dia

8

→

3m

0·8 m dia

0·5 m dia

expansion

contraction

14.19 Fan and duct arrangement: Example 14.7

Comparison with published characteristics

Site measurements of fan performance often do not correspond to the catalogue data. The fan manufacturer's published characteristics are obtained under *ideal* (laboratory) conditions and therefore almost any system will modify its stated performance. In addition to any obvious duct friction losses the effect of the duct configuration close to the fan must be considered. All site readings must be corrected to standard air conditions and fan running speed obtained at the same time as the flow rate and pressure measurements.

SYMBOLS

A	duct or pipe cross section area	P_{at}	atmospheric pressure
A' }	constants of fan	P_s	static pressure
B' }	characteristics	P_t	total pressure
		P_v	velocity pressure
C_{FV}	constant of proportionality for fan flow rate	P_{sF}	fan static pressure
		P_{tF}	fan total pressure
C_{Fp}	constant of proportionality for fan pressure	P_{vF}	fan velocity pressure
		r_t	total system resistance
C_{FP}	constant of proportionality for fan power	t_a	air temperature
		$\dot{V}$	volume flow rate
D	diameter	$\bar{v}_F$	mean velocity at fan discharge
K_c	pressure loss coefficient, contraction	Δp_f	pressure loss due to friction
K_e	pressure loss coefficient, expansion	η	fan efficiency
		ρ	air density
l	duct length		
n	fan speed	**Subscripts**	
P_a	fan air power	1, 2 .. relates to specific duct or pipe section	
P_i	fan impeller power		

15 Balancing of Fluid Flow Systems

As discussed in Chapter 13, design engineers should aim to balance the fluid flow circuits at the design stage of a project so that, when the systems are brought into operation, each outlet or unit will operate at the design flow rate, within specified tolerances. This is a theoretical ideal since it is probable that pressure losses of the installed system will differ from the calculated losses and it will be found from initial measurements that this balance has not been achieved. It is therefore often necessary to incorporate devices for regulating the flow in the ductwork and pipework systems, together with facilities for measuring the flow rates, to allow on-site balancing. Even with systems which are considered to be self-balancing it may still be necessary to include flow measuring devices and other facilities to verify design performance or, alternatively, to allow investigations into the reasons for an apparent failure of a system to deliver its rated output. Whichever philosphy is adopted, the necessary facilities must be included at the design stage of the project. The understanding of on-site balancing procedures is necessary so that the design engineer can plan the distribution runs of pipes and ducts and locate the regulating devices in the correct positions to facilitate those procedures.

On-site balancing of a fluid flow system can be defined as the adjustment of the flow rates to correspond to those specified in the design. The reasons for measuring and balancing are:

- for air systems, to achieve:
  (i) the design ventilation rate
  (ii) the air movement pattern within the space or building;
- for water systems, to ensure that heat transfer equipment is supplied with design flow rates within specified tolerances;
- to provide a basis for assessing the ability of the system to maintain design conditions of temperature and humidity;
- to allow accurate testing of plant components;
- to ensure efficiency of operation: an unbalanced system is likely to work at a lower overall efficiency than a balanced one.

THEORY OF PROPORTIONAL BALANCING

Two methods of balancing which have been used with success are given below. Both methods are known as *proportional* methods, the absolute (design) flow rate, within the required tolerances, being obtained only at the end of the balancing procedure. The methods are based on the following theory.

Consider a duct or pipe **WX** supplying two branches 1 and 2, with volume flow rates $\dot{V}_1$ and $\dot{V}_2$, as shown in Figure 15.1. For a pipe system there will be a return **YZ**, while for an air system it will be usual for $\dot{V}_1$ and $\dot{V}_2$ to discharge to a common datum, close to atmospheric pressure.

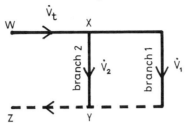

15.1 Flow network: proportional balancing theory

Using the concept of system resistance for fully developed turbulent pipe flow (Equation 12.16):

$$\Delta p = r\dot{V}^2.$$

Branches 1 and 2 have resistances r_1 and r_2. The pressure drop from **X** to **Y** is the same for both branches; therefore:

$$\Delta p_1 = \Delta p_2$$
$$r_1 \dot{V}_2 = r_2 \dot{V}_2^{\,2}$$

$$\therefore \frac{\dot{V}_1}{\dot{V}_2} = \sqrt{\frac{r_2}{r_1}} - \text{constant.}$$

Assuming that the friction factor and velocity pressure coefficients do remain constant, and with no further damper or valve adjustment, the ratio of the flow rates, $\dot{V}_1/\dot{V}_2$, will therefore remain the same irrespective of any change in the total flow rate, $\dot{V}_t$, in **WX**. In terms of a balancing procedure, this means that the outlets on any branch which are furthest away from the fan/pump must first be adjusted so that the ratio of the *measured volume flow rate*, $\dot{V}_M$, to the *design volume flow rate*, $\dot{V}_D$, are equal. This ratio, in balancing parlance, is known as the **R** ratio, ie:

$$R = \frac{\dot{V}_m}{\dot{V}_d}. \tag{15.1}$$

When the **R** ratios are equal, the individual flow rates from the outlets being compared are in the same proportion to one another. Then, when the total flow rate is adjusted to the design flow rate, the balanced flows will be maintained, with all outlets delivering their design flow rates.

To illustrate these principles consider the three outlets on one branch in the simple ducted air system in Figure 15.2.

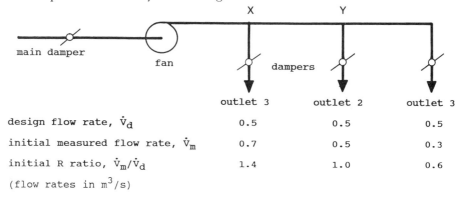

	X	Y	
	outlet 3	outlet 2	outlet 3
design flow rate, $\dot{V}_d$	0.5	0.5	0.5
initial measured flow rate, $\dot{V}_m$	0.7	0.5	0.3
initial R ratio, $\dot{V}_m/\dot{V}_d$	1.4	1.0	0.6

(flow rates in m^3/s)

15.2 Simple system to illustrate proportional balancing procedure

Each outlet has the same design flow rate but when the fan is operating with all the dampers fully open, the actual flow rates do not correspond to the design flow rates, so the system requires balancing. Outlet 1 is obviously the *index* outlet since it has the lowest measured flow rate compared with the design and its damper must remain fully open. It should also be noted that the **R** ratio of 0.6 at outlet 1 is the lowest of the three outlets. This **R** ratio indicates that the outlet is delivering 40% less air than design. When the damper in outlet 2 is adjusted to give the same measured flows in both outlets 1 and 2, these two will be in proportional balance, though not necessarily delivering their design flow rates. Note that the flow rate in outlet 1 will change as the damper for outlet 2 is adjusted since this will increase the resistance of branch 2. Therefore each time damper 2 is altered a comparison of both flow rates must be made. Moving towards the fan, the damper for outlet 3 must now be adjusted to bring its flow rate into proportional balance with the other two but, in doing this, the flow rate in **XY** will increase, thus changing the flow rates from outlets 1 and 2, though not upsetting their *balance*. Again it is important to note that the reduced flow in outlet 3 must be compared with the increased flow in outlet 1 until their **R** ratios correspond. When this proportional balancing is complete, the total flow delivered by the fan is adjusted to give the design flows from all three outlets. Once more, the change in total flow will not upset the balance of the three outlets, and each will deliver the design flow rates within the tolerance of the balance and within the accuracy of the measurement technique and instrumentation employed.

Two methods of proportional balancing, which are described in greater detail below, refer to the *index* outlet (in some literature this is termed *the least favoured outlet*). This is the outlet having the smallest **R** value in any group of outlets served by a branch duct before the balancing starts, and when all dampers/valves are fully open. Ideally the damper/valve on the index branch will be fully open after the balancing procedure is complete. With the HARRISON and GIBBARD method this is not always easy to achieve when the end outlet is not the *index*, whereas the MA method easily overcomes this difficulty.

Preliminary checks and adjustments

- The system must be capable of being balanced, eg the system must include dampers/valves in suitable positions, and it must be possible to measure the flow rates with an accuracy consistent with the required tolerance of balance;
- The information required includes drawings of plant and ductwork/ pipework layouts. A simple line diagram of the system with all outlets labelled will be of considerable help. Such a diagram should contain all the design flow rates;
- Ascertain that, for air systems the plant section and ductwork are airtight, that the fan or pump is running correctly and that there are no blockages. Wet systems must be thoroughly flushed through, refilled and vented;
- All outlet devices and fittings should be in place and all dampers/valves must be fully open on the branch being balanced. It is also necessary to have any associated systems, such as air extract, running.

Having taken all these precautions, and adjusted the fan/pump to pass approximately the total design volume flow, everything is ready for the balancing procedure to commence.

The detailed procedures and examples given below are for ducted air supply systems but the same principles also apply to extract air systems and water systems. The on-site balancing process is performed within a stated *tolerance of balance* (see pages 316 and 332), though for illustration purposes in the given examples approximately equal values of R are quoted at the end of each stage in the procedure.

HARRISON AND GIBBARD'S METHOD

The method proposed by HARRISON and GIBBARD[1] has been adopted for the CIBSE Commissioning Codes for air and water distribution systems (respectively, Series AD[2] and W[3]). In many ways this method is the most attractive procedure for air systems because it requires no special equipment or fittings. Calibrated dampers are not required (unlike the MA method) and the method is generally appropriate to systems that are already designed and installed. A suitable anemometer is required for measurements at the outlets, with in-duct measurement facilities being preferred in the main branches.

The procedure listed below is illustrated in Table 15.1.

(a) Find the *index* outlet by inspection or measurement.

(b) If the *index* is not the end outlet, reduce the flow in the end outlet to make it so.

(c) Measure the flow rates in outlets 1 and 2.

(d) Calculate the **R** ratios for outlets 1 and 2, ie:

$$R = \frac{\text{measured volume flow rate}}{\text{design volume flow rate}} = \frac{\dot{V}_m}{\dot{V}_d}.$$

(e) Adjust damper 2 to give the same value of **R** for both outlet 1 and 2. Following each adjustment measure the flow rate of the index outlet

	outlet no.	5	4	3	2	1
	design flow rate, $\dot{V}_d$	0.11	0.14	0.17	0.15	0.10
outlet 1 is found to be the index from initial measurements						
measured flow rates at outlets 1 and 2: $\dot{V}_m$ (m^3/s)					0.18	0.09
$R = \dot{V}_m/\dot{V}_d$					1.20	0.90
adjust damper 2 to give, after a few trials, equal **R** ratios					0.98	0.98
measure flow rate at outlet 3: $\dot{V}_m$ (m^3/s)				0.23	–	–
$R = \dot{V}_m/\dot{V}_d$				1.35	–	0.98
adjust damper 3 to give, after a few trials, equal **R** ratios				1.05	(1.05)	1.05
measure flow rate at outlet 4: $\dot{V}_m$ (m^3/s)			0.21	–	–	–
$R = \dot{V}_m/\dot{V}_d$			1.50	–	–	1.05
adjust damper 4 to give, after a few trials, equal **R** ratios			1.10	(1.10)	(1.10)	1.10
measure flow rate at outlet 5: $\dot{V}_m$ (m^3/s)		0.19	–	–	–	–
$R = \dot{V}_m/\dot{V}_d$		1.73	–	–	–	1.10
adjust damper 5 to give, after a few trials, equal **R** ratios		1.13	(1.13)	(1.13)	(1.13)	1.13

Table 15.1 Example of HARRISON and GIBBARD method of proportional balancing

Notes:

(i) after adjusting damper 5, all outlets are delivering 13% more than their respective design flow rate. If this was the only branch of the ducted air system, the total flow rate would finally be reduced to deliver the design flow rate of 0.67 m³/s.

(ii) the bracketed figures indicate flow rates which are not measured but which remain in proportional balance.

and recalculate values of **R**. It is unlikely that equality will be achieved with the first adjustment and a number of trials may have to be made.

(f) Measure flow rate at outlet 3 and calculate **R**.

(g) Adjust damper 3 until the value of **R** is the same for both outlets 1 and 3. (See note (e).)

(h) Repeat steps (f) and (g) for as many more outlets as are supplied by the branch duct.

Multi-branch systems

A more complex system is now considered with a number of branches each supplying a group of outlets, as shown in Figure 15.3.

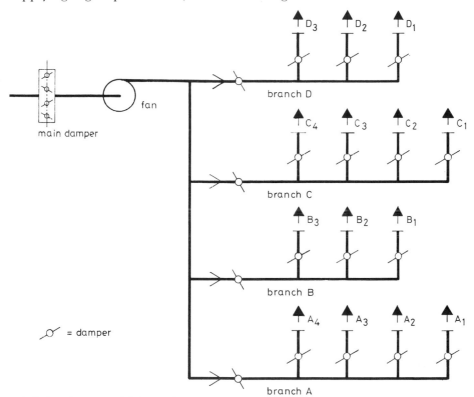

15.3 Balancing a multi-branch network

In this case each group of outlets on individual branches is brought into proportional balance independently of the other branches. Each branch will then have a different **R** ratio, eg the **R** values may be $R_a = 0.7$, $R_b = 0.9$, $R_c = 1.2$, $R_d = 1.4$. The requirement is now to balance these four branches using the procedure already described. In this example branch B is balanced to branch A (the index), then branch C to A and finally branch D to A. Only when this has been completed will the total flow in the system be adjusted to the design flow rate. This example illustrates the necessity for the designer to include dampers in the branches before the first outlets, ie in branches B, C and D.

These procedures can be followed through with any complexity of circuit, provided the principles are understood both by the system designer and by the commissioning engineer.

Effect of branch balancing dampers

Balancing procedures rely on the static pressure gradients remaining uniform along the duct when flow rates are either reduced or increased. When a damper is in a partially closed position, the pressure distribution along the duct, downstream of the damper, has the same general characteristics as that of an orifice plate – that is, there will be static depression followed by pressure recovery. If the first offtake on a supply branch duct is too close to the balancing damper, then the proportional balance of the downstream outlets will be affected when that damper is closed. To avoid this problem the first offtake on a branch duct should be a *minimum* of two to three duct diameters downstream of the branch damper.

Last outlet not the index

Consider the branch of a system with the pre-balanced flow rates shown in Figure 15.4. Here the index outlet is identified as No. 4 which is *closest to the fan* in the sequence in which the ducts are taken off the main duct. The reasons for this outlet being the index are either that it has the longest duct run, (the figure is diagrammatic), or the duct and fittings are relatively small diameter compared with other branch ducts. If the balancing proceeds from outlets 1 and 2, without first identifying the index as outlet 4 then, when the commissioning engineer reaches that outlet, balance could not be achieved without back-tracking, something which good procedures seek to avoid. Hence, step (b) of the procedure is to identify the index outlet by measurement. Then, if the end branch (either 1 or 2) is not the index, its damper is adjusted to make it an *artificial* index. Experience is required in making a suitable assessment on the amount of reduction – the CIBSE Commissioning Code suggests making it equal to the actual index. The expectation is, that when the operator reaches outlet 4, the damper should be fully open. If the original judgement of closing damper 1 was incorrect, too much closure should mean that damper 4 should either have to be closed or the procedure would have to be repeated. At the end of balancing the actual index outlet should be fully open. If it is not, there may be penalty on fan energy consumption. There also may be unwanted noise generated at the other dampers since they would be unnecessarily closed.

Two design points are illustrated by this. First the end branch should have a damper (even if the designer has taken it to be the index outlet). Secondly, systems should be designed to facilitate the balancing procedures. In this particular example, the design engineer should try to design the system in such a way as to make the last outlet the index so that the problem does not arise, although this is a simple layout to illustrate one particular aspect of design. There are, of course, other practical and theoretical considerations to be taken into account when designing the circuits. The implications of commissioning and balancing are only part of the overall design considerations and must be reconciled with other requirements to obtain an *optimum* design.

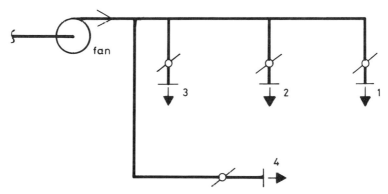

15.4 Balancing a branch in which the end outlet is not the index outlet

MA'S METHOD

The method of balancing developed by MA[4] makes use of dampers which have a calibration relating damper position to reduction in flow rate. Since the reduction in flow rate in any outlet must be related to the index outlet, this calibration has to be in terms of a proportional flow rate ratio, ie a ratio, F, is calculated as:

$$F = \frac{\text{R ratio obtained in the index outlet}}{\text{R ratio obtained in outlet to be regulated}} = \frac{R_i}{R_o}.$$

For circular ducts a suitable, commercially available, damper is the iris shutter type for which the calibration curve in Figure 15.5 was obtained over a wide range of damper sizes (100 to 350 mm diameter) and air velocities (3 to 7 m/s). For rectangular ducts an opposed blade damper would be suitable, provided an appropriate calibration were available.

Taking any one branch with a group of outlets, the procedure is to set all dampers, in one operation, relative to the index. Because of the complexities of the pressure changes within the system, the initial settings are unlikely to give a balance within required tolerances. The measurements have therefore usually to be repeated, and finer adjustments made, to give a more precise balance. In other words it is an iterative procedure which continues until balance has been achieved to the specified tolerances. However, in practice, only one or two repetitions will be required.

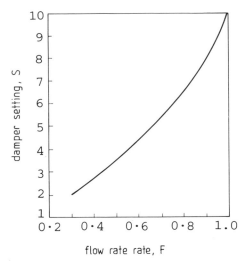

15.5 Calibrated iris damper characteristic
(for MA method of proportional balancing)

After the initial preliminary checks, the balancing procedure listed below is illustrated in Table 15.2.

Stage 1

(a) Measure air flow rate to each outlet.

(b) Calculate the R ratio for all outlets.

$$R = \frac{\text{measured volume flow rate}}{\text{design volume flow rate}} = \frac{\dot{V}_m}{\dot{V}_d}.$$

(c) Divide lowest $R(R_o)$ value (the *index* outlet R_i ratio) by the other ratios. This gives the proportion by which rate of airflow must be reduced, ie:

$$F = \frac{R_i}{R_o}.$$

(d) Enter the F ratios calculated in (c), individually into calibration curve to find damper setting, S. (Figure 15.5).

(e) Adjust damper to setting S.

Stage 2 (first iteration)

(f) Measure air flow rates to each outlet and calculate new R values.

(g) Obtain new values for R_i divided by values of R_o ($=F'$). Note that the R_i value is for the *original* index outlet.

(h) Multiply ratios, F', by damper settings obtained in (d):

$$S' = F' \times S$$

(j) Adjust damper to new setting S'.

Stage 3 (second iteration, if required)
   (k) Measure air flow rates and, if necessary, proceed to a third adjust-
        ment following steps (f) to (j).

Stage 4
   (l) Adjust the main balancing damper or the fan speed to obtain the
        design air flow rate.

		4	3	2	1
outlet no.		4	3	2	1
design flow rate, $\dot{V}_d$		0.35	0.25	0.20	0.30
stage 1: initial measurements					
measure flow rates, $\dot{V}_m$ (m^3/s)		0.40	0.27	0.16	0.20
$R = \dot{V}_m/\dot{V}_d$		1.14	1.08	0.80	0.67
$F = R_i/R_o$		0.59	0.62	0.83	1.00
damper setting, S (Fig. 15.5)		4.4	4.7	6.8	10.0
adjust dampers to setting S					
stage 2: first iteration					
measure flow rates, $\dot{V}_m$ (m^3/s)		0.31	0.20	0.18	0.25
$R = \dot{V}_m/\dot{V}_d$		0.89	0.80	0.90	0.83
$F' = R_i/R_o$		0.93	1.04	0.92	1.00
dampor setting, $S' = S \times F'$		4.1	4.9	6.3	10.0
adjust dampers to setting S'					
stage 3: final measurements					
measure flow rates, $\dot{V}_m$ (m^3/s)		0.30	0.21	0.17	0.26
$R = \dot{V}_m/\dot{V}_d$		0.86	0.84	0.85	0.87

Table 15.2 Example of MA method of balancing, using calibrated dampers

   Notes:
   (i) outlet 1 is the index and its damper remains open throughout
   (ii) after stage 3, each outlet is delivering approximately 86% of the
        design flow rate; if this was the only branch of the system, the
        flow rate would be increased to deliver the design flow rate

Where systems have more than one branch, each branch is treated
indepedently and afterwards proportionally balanced to each other.
Finally, the fan total volume flow is adjusted to give design flow rates from

each outlet in a similar manner to that described on page 311–2 for the system shown in Figure 15.3.

One of the advantages of the MA method of balance is that it is not dependant on the end branch being the index outlet. If the procedure is applied correctly, the damper on the index will remain fully open.

TOLERANCE OF AIR FLOW RATE BALANCE

It is unnecessary and impractical to obtain identical **R** values for all outlets or branches within a group, nor is it necessary to regulate the total flow to an accurate design total air flow rate. Realistic tolerances should be established so that balance can be achieved with economy of effort. The tolerances suggested in the CIBSE Code[2] for air distribution systems are given in Table 15.3.

Both the methods described above are capable of achieving balanced systems within these recommended tolerances.

damper location	low velocity supply and extract branch or sub-branch with terminals serving		supply to induction units
	single space	more than one space	
terminals % of proportional flow at index terminal	+ 20 − 0	+ 15 − 0	+ 5 − 5
branches and sub-branches % of proportional flow at index sub-branch	+ 10 − 0	+ 10 − 0	+ 5 − 0
fan % of design flow	+ 10 to − 0.0		

Table 15.3 Tolerances on air system balance

INDUCTION, DUAL-DUCT AND VAV SYSTEMS

The same principles of proportional balancing can be applied to high velocity systems and the methods previously described can be used. In fact, their application is sometimes made simpler because the pressure loss in an induction unit (or mixing box) is high in comparison with the loss in the main distributing ducts, and an alteration to one unit will have only a small influence on adjacent units.

With *induction* systems the balancing procedure is carried out in the same way as before, except that this time, instead of measuring the air flow rate, the pressures of the induction units are measured and compared with the design pressures. These ratios are then adjusted to the same value before finally correcting the fan volume to the design flow.

With *dual-duct* systems, constant volume regulators are fitted to the mixing boxes which in theory should make the system self-balancing. However, some adjustments will usually be necessary and an appreciation of the principles of proportional balancing is essential for any further regulation. This may be required not only for the mixing boxes but also for the set of outlets supplied by each box.

Variable air volume (VAV) systems are also, in principle, self balancing, though care is required in setting up the system if the full potential or energy economy is to be realised. No specific Code for this system has yet been published, but a suggested procedure has been outlined by HOLMES[5].

MEASUREMENT OF AIR FLOW RATES AND BALANCING

Measurement at the outlets

In many ducted air systems, a vane anemometer is the most convenient instrument with which to make measurements at the outlets. In the CIBSE Commissioning Code for air distribution, the recommended method for measuring the mean velocity, $\bar{v}_a$, at the face of a grille is set out in Table 15.4. The air volume flow rate is then calculated as:

$$\dot{V} = \bar{v}_a A_g \tag{15.2}$$

where A_g = the gross area of the grille.

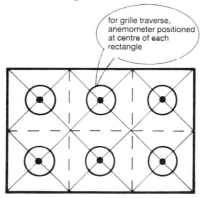

for grille traverse, anemometer positioned at centre of each rectangle

15.6 Anemometer traverse positions at grille face

Provided the outlets are similar, it is usually unnecessary to consider the effect that the *free area* has on this calculation, when the outlets are being proportionally balanced against one another. This is because only the **R** ratios are being compared and any *grille effect factor*, C_g, is considered to be constant. When an accurate flow rate is required from the measurement, then the measured flow rate must be multiplied by the grille effect factor

---

The grille face is divided into, say 6, equal areas, as in Figure 15.6

The mean face velocity is obtained as follows:-

Mechanical instrument

(1) Check direction of flow through instrument.

(2) Check the instrument off position and zero.

(3) Position the instrument at the centre of the top left hand rectangle with the back of the instrument touching the front of the grille.

(4) Start the instrument, observing the second hand of clock or watch. (Operators hand should be removed as quickly as possible.)

(5) After 10 seconds have elapsed, move the instrument to the centre of the next rectangle without stopping the rotation. Repeat this until all rectangles have been covered. At the end of the 10 second period of the sixth position stop the anemometer.

(6) Read the instrument to obtain v_i.Correct v_i according to the calibration, to obtain the mean air velocity, $\bar{v}_a$

Electronic instrument

(1) Check direction of flow through instrument.

(2) Position the anemometer in the centre of each rectangle in turn, with the back of the instrument touching the front of grille. For each rectangle record the indicated velocity, v_i. If there is needle fluctuation, try to judge the mean position.

(3) Calculate the arithmetic mean of all 6 values of v_i, to obtain v_i.

(4) Correct v_i according to instrument calibraton to give v_a

(Note that an average correction factor is sufficiently accurate for site measurements)

Table 15.4 Obtaining the mean grille face velocity with a vane anemometer

which is either supplied by the manufacturer or obtained from a laboratory calibration. The flow rate is then obtained from:

$$\dot{V} = C_g \bar{v}_a A_g \qquad (15.3)$$

From laboratory tests[6], C_g is of the order of 0.85 for supply grilles and 0.90 for extract grilles, where the grille sizes are above approximately 0.1 m². Grille free areas should *not* be used as grille effect factors.

With many grilles, the balancing damper is often incorporated as a composite unit, behind the blades of the grille. To balance the air quantities it will be necessary partially to close the damper to provide the *out-of-balance* resistance in relation to the index outlet. When a damper is partially closed the higher velocity jets strike the anemometer vanes causing the instrument to indicate an average velocity higher than would be indicated with the damper fully open, at the same flow rate. The grille effect factor then varies considerably and if not taken into account, accuracy of flow rate measurement becomes much greater than the required tolerance against which the system is being balanced. Variation of the grille effect factor, C_g, with the damper position is given in Figure 15.7.

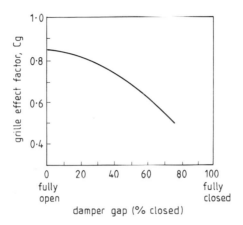

15.7 **Grille effect factor, C_g, varying with damper position (damper immediately behind grille face)**

Alternatives should be considered to overcome this problem of comparative accuracy between flow measurements at supply air outlets which incorporate dampers behind the grille face. These include:

- *use an anemometer mounted in a hood.* The advantage of this method is that the velocity only need be recorded and the **R** ratio calculated as $v_a/\dot{V}_d$. One disadvantage is that the additional resistance imposed by the hood will introduce an additional uncertainty into the accuracy of balance. Also a number of different sizes of hood might be required for different sizes of outlet;
- *measure the flow rate upstream of the damper.* For this, access to the ductwork must be provided at the design stage of the project;
- *separate the damper from the grille with a length of duct.* Again, this is a solution which must be incorporated into the original design. It has the additional advantage of allowing acoustic treatment of the duct where air noise generated by the damper is likely to cause annoyance;
- *use an extension duct.* Here, an instrument traverse is made away from the grille face. This is not often practical since several extension pieces may be required.

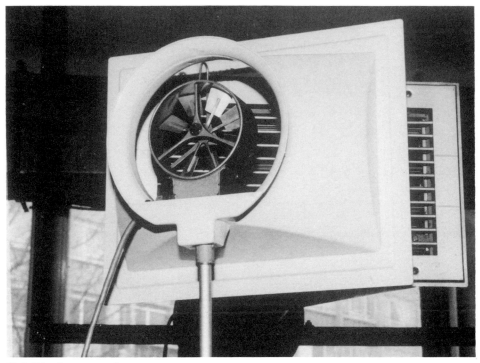

15.8 Hood mounted anemometer

Anemometers mounted in a hood are recommended when measurements are made at supply diffusers. Measurements in extract systems are unaffected by a damper immediately behind the grille.

In-duct measurements

Balancing in a multi-branch system, as illustrated in Figure 15.3, can be achieved by using the index outlets on each branch, as representative individual flow rates. However, it is more accurate and often more convenient to make in-duct measurements close to the dampers to be adjusted. The traverse methods of obtaining the flow rate in a duct have been described in Chapter 12. For air velocities above 3.5 m/s it will be usual to use Pitot-static tubes, and for velocities below 3.5 m/s either a small diameter vane anemometer or a hot-wire instrument will be used.

Flow rate using a centre constant

If in-duct measurements are made to bring a group of branches into proportional balance, repeated traverses become time consuming. Single point measurements to obtain a flow rate are particularly useful when repeated measurements have to be made in the same duct, as in proportional balancing procedures. But unless the duct flow is fully developed with a symmetrical velocity profile (and outside the laboratory it rarely is), it is not possible to rely on published centre constants since the installed duct lengths are relatively short between fittings and the velocity profiles will be asymmetrical.

To obtain a centre constant, C_c, requires making a preliminary traverse of the duct to determine the ratio of the mean to axial velocities, ie:

$$C_c = \frac{\bar{v}}{v_c}.$$ (15.4)

The centre constant is then multiplied by the centre-point measurement of velocity to determine the flow rate, ie:

$$\dot{V} = C_c v_c A_d.$$ (15.5)

This method of obtaining the flow rate would not be suitable if the velocity distribution in the duct changed due to varying inlet conditions, eg from a balancing damper upstream of the measuring station. (For this reason it is important to have the measuring position upstream of the damper.) It is also limited by the accuracy of the centre constant and the accuracy of the subsequent centre point velocity measurement.

BALANCING WATER SYSTEMS

As with ducted air systems, it should be the aim of the design engineer to balance water circuits at the design stage of a project. Where this is not possible it will be necessary to include devices for regulating and measuring the flow rates in the piping circuits to allow on-site balancing.

Most plant items require isolating valves for maintenance. Taditionally the valve in the return pipe has also had a regulating function. Current balancing techniques use both valves for flow measurement.

As yet there are no standard terms for water flow balancing equipment. The abbreviations given below have been adopted from various publications.

MEASUREMENT OF WATER FLOW RATE AND BALANCING

The flow rate through an orifice plate is given by Equation 12.20. Although the flow coefficient, α, in this equation depends to some extent on REYNOLDS' number, these variations may be discounted for site balancing procedures. The expansibility factor may be taken as constant, at $\varepsilon = 1$. If the density of water is also assumed to be constant then, for a fixed orifice to pipe diameter ratio, Equation 12.20 can be written as:

$$\dot{V} = K_{vs}\sqrt{\Delta p'}.$$ (15.6)

Thus the characteristic of a fixed orifice flow measuring devise may be expressed by a single constant, K_{vs}. The proportional balancing procedures compare the ratios of measured flow rate to design flow rate, R. From Equation 15.3 it follows that:

$$R = \frac{\dot{V}_m}{\dot{V}_d} = \sqrt{\frac{\Delta p'_m}{\Delta p'_d}}$$

(15.7)

Therefore, when fixed orifice measuring devices are used for measuring the flow rates, the pressure difference ratio only is required for balancing purposes. The valve flow rate constant, K_{vs}, is required only to determine the design pressure difference, $\Delta p'_d$. The CIBSE Commissioning Code W[3] recommends that hot water systems are balanced cold, which for HTHW systems is important for safety reasons. If measurements of hot water flow rates are made, the correction to Equation 15.6 due to water density variations will be $\pm\sqrt{1000/\rho}$. For water at 100°C this correction amounts to approximately $\pm 2\%$. Proportional balancing assumes that the circuit resistances obey the *square law*, ie that both the friction factors and the fitting pressure loss coefficients remain constant. In fact both depend, to some extent, on the REYNOLDS' number and this means that the system balance will alter when the total flow rate and the water temperature change. Consequently the tolerance against which the system is balanced must be such that it takes account of the possible variations in REYNOLDS' number. These points are discussed and quantified by CAMPBELL and NEWSON.[7]

The engineer will be aware that the pressure difference for measuring the flow rate is not the same as the net pressure drop. This point is illustrated in Figure 12.20. The downstream pressure tapping sits in the low pressure region, and there is a pressure recovery after the device, before the net pressure loss is obtained. It is important to remember this when referring to tables and charts published by manufacturers who sometimes use terms incorrectly. At least one manufacturer uses the term *signal* to describe the measured pressure difference and the adoption of this word may lead to less confusion in the presentation of information.

Flow measurement valve (FMV)

Flow measurement valves may be an orifice valve (OV) shown in Figure 15.9a or a flow measuring device (FMD) such as an orifice plate which is close-coupled to a gate valve (FOGV) shown in Figure 15.9b. As well as giving the flow rate constants, K_{vs}, as in Table 15.5, manufacturers also present the pressure difference/flow rate characteristics as log-log graphs as shown in Figure 15.10.

Double regulating valve

A double regulating valve (DRV), shown in Figure 15.11, is used for regulation and isolation. Once a branch has been regulated (brought into proportional balance) the valve can be locked off to prevent further opening. Then, if the valve has to be closed to isolate the circuit, the former when reopened will return to its original setting. Most of these valves

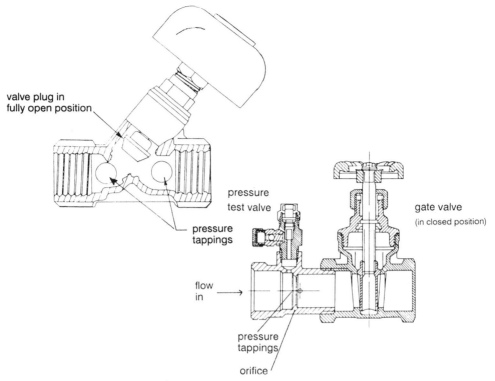

valve plug in
fully open position

pressure
test valve

gate valve
(in closed position)

pressure
tappings

flow
in

pressure
tappings

orifice

15.9 Flow measurement valves

    (a) orifice valve (Courtesy Holmes Valves Ltd)

    (b) fixed orifice gate valve (Courtesy Crane Fluid Systems)

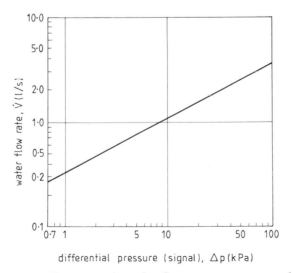

differential pressure (signal), Δp (kPa)

15.10 Flow rate chart for flow measurement valve

nominal valve size (mm)	K_{vs}
15	0.085
20	0.204
25	0.334
32	0.71
40	0.95
50	1.60
65	2.40

Table 15.5 Typical values of flow rate coefficient, K_{vs}, for fixed orifice valves
Note: the constants in this table are for flow rates measured in litres/s and for pressures in kPa; where flow rates are given in m³/hour and pressures in bars (MPa), the constants in the table are multiplied by 36

include a vernier scale on the stem to allow accurate setting. Hand wheels can be removed to reduce the risk of unauthorized tampering.

A DRV will usually be placed in the return pipe and used in conjunction with a FMD installed in the flow pipe. When used in this way there is no need for a DRV to include pressure tappings as in the variable orifice double regulating valve described below. A DRV is sometimes close-coupled with an orifice plate to form a fixed orifice double regulating valve (FODRV) which is illustrated in Figure 15.12. This arrangement is sometimes referred to as a *single valve commissioning set*, its rationale being that the engineer can more readily observe the change in pressure difference as the valve is adjusted.

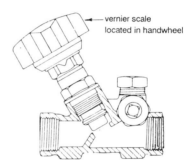

15.11 Double regulating valve (Courtesy Crane Fluid Systems)

vernier scale located in handwheel

Variable orifice double regulating valve

A variable orifice double regulating valve (VODRV) is a DRV with pressure tappings. The valve has the functions of the DRV plus the ability to measure the flow rate. The flow rate/pressure difference characteristics are usually presented graphically on a log-log graph as in Figure 15.13, but these can also be obtained by using Equation 15.3, with constants, K_{vs}, determined for the whole range of valve settings. For reasonable accuracy of measurement a vernier scale for setting the valve is essential. Even so the accuracy of measurement, as given by one manufacturer, deteriorates from approximately $\pm 5\%$ when fully open to $\pm 20\%$ with the valve 80% closed.

Where VODRVs are used for regulating and measuring the flow rates it will be necessary to determine the flow rate for the **R** ratio, rather than using the differential pressures, since K_{vs} varies with valve position.

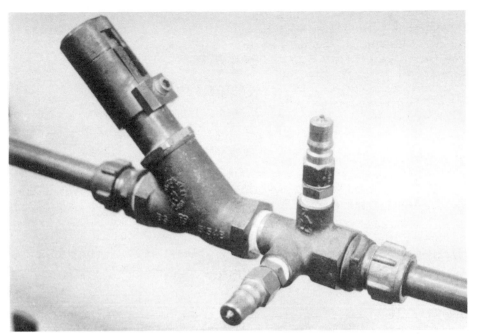

15.12 Fixed orifice close coupled to a double regulating valve (single valve commissioning set)

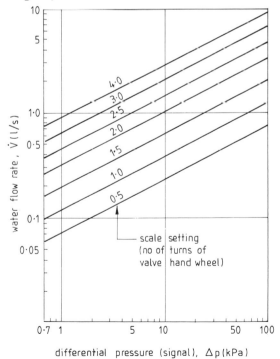

15.13 Flow rate chart for variable orifice double regulating valve

15.14 Venturi for measuring flow rate

Nozzles, orifice plates and venturi meters

Orifice plates, nozzles and venturis are marketed as either *flow measuring devices* or *metering stations*. Orifice plates are often close-coupled to a gate valve, or to a DRV, thus forming a composite unit as previously described. When close-coupled, the static pressure distribution through the unit may be affected by the valve and the flow coefficient, α, will then differ from those published in various Standards, eg Reference[8]. In these cases the pressure difference/flow rate characteristics given by the manufacturer should be used, not the published values.

Pressure tappings and pressure test valves

Pressure tappings (PT) allow a manometer to be connected to the flow measuring device. These devices are self-sealing and usually suitable for cold and LTHW systems. The manometer tubes must also incorporate matching self-sealing connectors; these may be push-on units for quick connect/disconnect to the system but more usually a threaded fitting is used with a cap to protect it from dust.

A pressure test valve (PTV) is a self-sealing ball unit which also incorporates a needle valve operated by a standard aircock key. This device is therefore safer to use where the pipe fluid is at high pressure or temperature and allows the pressure tapping to be isolated for cleaning the ball seat. Where measurements are to be made on a live high temperature hot water system, copper bleed tubes should be taken from the PTV and they should terminate at the pipe with a needle valve. Both items are illustrated in Figure 15.15.

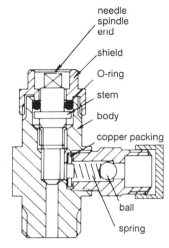

15.15 Pressure test valve and tapping
(Courtesy Crane Fluid Systems)

Selection of valves and flow measuring devices

Balancing valves and flow measuring devices will impose additional system resistances on the circuit. Therefore selection should be made to ensure that these are kept to a minimum, while ensuring sufficient pressure difference to allow accurate measurement of the flow rate. The selection should be considered at the same time as sizing the pipe since it is desirable that both should be the same size to avoid using reducers and/or expanders.

The *net* pressure loss across a valve or an FMD is obtained from the coefficient supplied by the manufacturer.

A representative piping circuit is shown in Figure 15.16 with an associated table giving alternative valve combinations, selected from the different devices described. The procedure for balancing the circuit will be similar to that described for the air systems. To achieve this, the control valves are initially fully open to load. When design flow rates have been obtained within the required tolerances, each of the bypass valves will then be regulated in turn, with the control valve closed to load, using the same flow measuring device used for the circuit balancing.

Installation

The flow measuring device, including flow measurement valves, should be positioned upstream of any associated valve and installed with a reason-

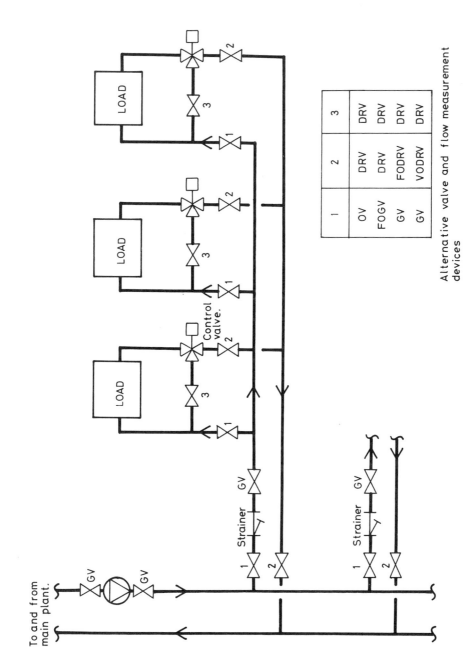

LOAD

LOAD

LOAD

Control valve.

Strainer GV

Strainer GV

To and from main plant.

GV

GV

	1	2	3
	OV	DRV	DRV
	FOGV	DRV	DRV
	GV	FODRV	DRV
	GV	VODRV	DRV

Alternative valve and flow measurement devices

15.16 Representative pipework circuit

able length of straight pipe upstream and downstream of the device. Where a measuring accuracy of 5% is sufficient these lengths should be a minimum of ten and three equivalent pipe diameters respectively. Standard lengths recommended in the various Standards should be adhered to if a more stringent measuring accuracy is required.

When coupling FMDs to the pipe, burrs should be removed from pipe fittings in the vicinity of the device. This will reduce the risk of creating a flow disturbance in the fluid entering the device and ensure that accuracy is maintained. Tappings should be placed either on the top or on the side of the device (or pipe) to reduce the risk of dirt accumulating in the pressure leads.

Manometers

The three types of manometer used for pressure measurements are:

– liquid;
– electronic digital;
– diaphragm.

All manometers will require an isolating and equalizing (bypass) valve manifold as illustrated in Figure 15.17. This is used to allow the instrument to be zeroed and purged of air.

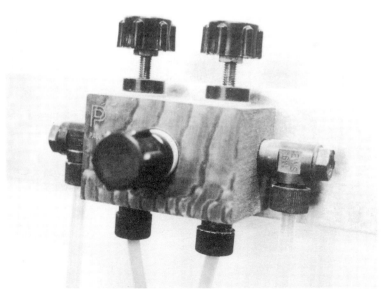

15.17 Valve block for zeroing and air purging manometers

An electronic digital manometer (Figure 15.18) may have interchange-able transducers of different ranges, each of which connects to the same display unit. It will be necessary to zero the instrument for each range and each time it is used. It is desirable that the instrument is recalibrated on a regular basis. Instruments are now available with direct read out in flow rate, programmable for different flow measurement devices: that is for the instrument to accept the appropriate value of the flow coefficient, K_{vs}.

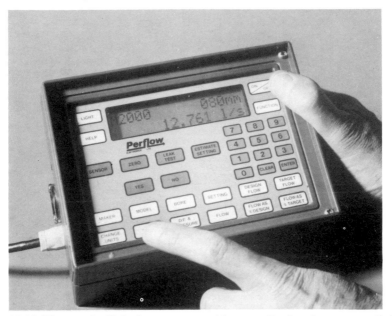

15.18 Electronic digital manometer (Courtesy Perflow Instruments Ltd)

Liquid manometers (Figure 15.19) use either water-over-mercury or water-over-fluorocarbon, with ranges typically 0 to 60 kPa and 0 to 4.6 kPa respectively. These instruments do not have to be recalibrated. For proportional balancing it is good practice to have a pair of compatible manometers. One of these will be positioned at the index (least favoured or reference branch) while the other is used as a *wander* for the units which are being proportionally balanced to the index. Tubes from the manometer must terminate in self-sealing connectors which match the pressure tap-pings of the flow measuring device. Tubing must be suitable for the temperature and pressure of the pipe fluid.

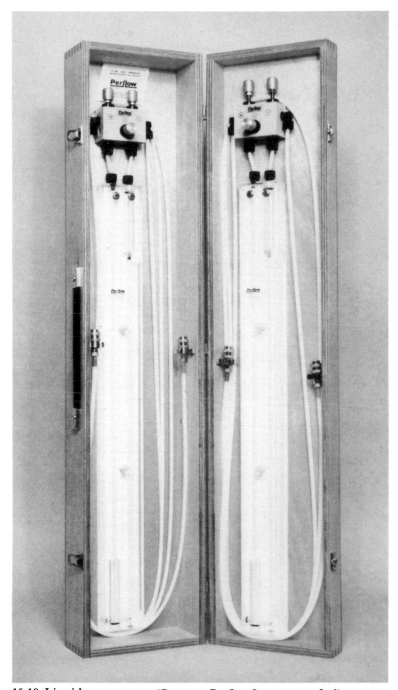

15.19 Liquid manometer (Courtesy Perflow Instruments Ltd)

TOLERANCE OF WATER FLOW RATE BALANCE

It is not necessary to balance flow rates within a group of loads precisely. The allowable tolerances suggested in the BSRIA Application Guide 1/79[9] are given in Table 15.5. It should be noted that, to guarantee a tolerance of balance, the accuracy of measuring flow rates may be considerably less than the tolerance of balance required by the specification.

flow measuring station location	type of system		cooling
	heating		
	design water temp. drop (K)		
	10 to 20	20 to 30	
at terminal unit as % of proportional flow at index unit	+ 20 - 0	+ 15 - 0	+ 15 - 0
major coils, heat exchangers, branches as % of proportional flow at index branch	+ 10 - 0		+ 10 - 0
pumps, boilers and refrigeration plant % of design flow	+ 10 - 0		+ 10 - 0

Table 15.6 Recommended working tolerances of indicated flow rates for water system balancing

SYMBOLS

A_d cross section area of duct
K_{vs} characteristic constant associated with flow measuring device, defined by Equation 15.6
C_c centre constant of velocity traverse
C_g grille effect factor, applied to grille face velocity
F proportional balancing ratio, MA method
R proportional balancing ratio
r resistance
S setting of a calibrated damper

$\dot{V}$ volume flow rate
v point velocity
$\bar{v}$ mean velocity
v_c centre point velocity
$\Delta p'$ pressure difference
ρ density

Subscripts
d   design
m   measured
a   vane anemometer
i   index outlet
o   outlet other than index

16 Control Dampers and Valves

Air conditioning systems are installed with a maximum capacity determined from indoor and outdoor design conditions and design loads but it is only on comparatively rare occasions that these conditions require full plant capacity. Therefore, controls are applied to maintain appropriate indoor conditions with efficient plant capacity reduction whilst also ensuring safe and consistent operation.

OVERVIEW OF CONTROL SYSTEMS

There is an abundance of published work describing control systems design both on a theoretical and practical basis. However, dampers and valves form only one part of the complete control arrangement and before going on to discuss them in detail, a brief description of the overall area of controls is necessary.

A control system is designed to provide a variety of functions which include:

- sustaining indoor conditions;
- providing optimum running conditions for both efficient system performance and minimum maintenance;
- monitoring of space and system status;
- simplifying end-user control;
- ensuring safe plant operation.

Whatever its function, the control system comprises three basic components: *sensor, controller* and *correcting unit*. The correcting unit usually consists of two items, the actuator and the final control element, eg a damper motor and damper blades respectively. In some cases the controller and the actuator are a single unit, as in the case of a self-acting thermostatic radiator valve.

Sensing elements

The sensor is the component which feeds information to the controller relating to a variation in a property of the observed medium, which for air conditioning systems is usually water or air. The scope of sensors which are available is continually broadening but those more commonly used monitor temperature, humidity, pressure and air flow.

Temperature sensors are used for monitoring dry and wet-bulb temperatures. Dry-bulb sensors include the thermocouple, thermistor, bimetallic elements and fluid filled bellows; wet-bulb sensors consist of a dry-bulb sensor enclosed in a permanently wetted wick.

Relative humidity sensors, rely on either a wet-bulb temperature sensor or a semi-conductor based material. These may be used to control plant items directly or through the use of associated elements such as in the resetting of a thermostat which is sensing dew-point temperature. In combination with a dry-bulb sensor, a humidity sensor may also be used to determine air enthalpy. Enthalpy monitoring is now being more widely used to provide certain control strategies. Nevertheless, the accuracy of humidity sensors has been shown to be uncertain[1] and, if improperly selected, fitted or maintained, may cause incorrect control actions leading to unsatisfactory space conditions and system inefficiencies.

Pressure sensors are used for measuring and controlling a variety of system conditions. Examples include the maintenance of base static pressure in a VAV system and the minimum ventilation flow rate (see Chapter 7, page 134 ff). The differential pressure across an item of plant, such as a filter, provides information about its condition, eg its state of cleanliness.

Location of the sensing element

The position of a sensor in the system is important in ensuring that a representative condition is observed. Temperature and humidity sensors should be placed in well mixed airstreams – avoiding positions where there is a risk of stratification or where there is likely to be a significant temperature gradient. Temperature sensors are affected by radiant heat transfer and should therefore also be screened from high or low temperature surfaces.

The location of pressure sensors in a fluid flow system to provide a representative measure of total flow rate have also caused problems[2]. Static pressure sensors should be placed well downstream of modulating dampers, (a minimum of 4 to 6 duct diameters), and flow measuring sensors should be placed upstream of these devices.

The consequence of poor sensor performance may lead to excessive energy consumption,[3] poor plant performance or, in severe cases, failure of the system itself.

The controller

The function of the controller is to interpret the input signal from the sensor and provide appropriate output to the correcting unit. The action taken will depend on the design of the controller which, increasingly, is being based on microprocessors which allow complex actions to be taken. Direct digital control (DDC) is a widely used term for this form of control. The inputs and outputs to a DDC system are (contrary to the suggestion of the term *direct digital*) are likely to be from and to traditional analog or continuous control elements such as the thermocouple and damper motor.

The correcting unit

Whether considering pipework systems for liquid flow, or ductwork systems for air flow, there is invariably a need for regulation of flow rates through the employment of a final control element.

The correcting unit is most likely to comprise of two parts, the actuator and the final control element.

The actuator provides the motive power to the final control element in response to a signal from the controller and may take the form of an electrical or pneumatic device. The selection of the actuator type will depend on the type of final control element and on the control strategy for the overall system.

Dampers are used for regulating and controlling the flow rates in mechanical ventilation and air conditioning ductwork systems. The most common application is in the *static* balancing of air flow networks to the design requirements, as described in Chapter 15.

Dampers are often also included with the central plant to allow the system to be operated in an efficient manner, as in the examples of the air conditioning systems utilising recirculated air described in Chapter 6.

For water systems, valves are employed to provide corresponding facilities to those provided by dampers in air systems, ie double regulating valves for *static* balancing and modulating valves for controlling the flow rates to heat exchangers such as cooler and heater batteries.

What ever type of fluid is being regulated, there are certain parameters that apply to the control device, (the damper or valve), which are common. These are used to determine the appropriate size and type of fluid flow control device, (FCD), for each particular application.

AUTHORITY

The authority, N, indicates the ability of an FCD to influence the flow rate in a system. An oversized FCD would not have as much influence over the flow rate in the duct or pipe network as a smaller one. In pressure terms, the greater the pressure drop across the device when fully open, the greater the influence of the device on the flow rate through the system. The authority is defined by:

$$N = \frac{\Delta p_\varphi}{\Delta p_s + \Delta p_\varphi} \qquad (16.1)$$

where Δp_φ = pressure drop in the FCD in the fully open position

Δp_s = pressure drop through the remainder of the system in which the flow is being controlled, but excluding the FCD pressure drop.

As the device closes the pressure drop increases across the FCD, reducing the fluid flow rate. As suggested above, the authority must be significant if satisfactory control of fluid flow is required across the whole range of device aperture. If an FCD is oversized it will have a small authority and control becomes effectively on-off or two position in nature.

Example 16.1 A short length of ductwork has a design pressure drop, excluding the damper, of 30 Pa. To maintain reasonable control of air flow rate it is assumed that the damper has an authority of 0.1. What should the pressure drop be across the fully open control damper?

Solution

The system has a pressure drop of 30 Pa; rearranging Equation 16.1 the required pressure drop across the damper is given by:

$$\Delta p_\varphi = \frac{N\Delta p_s}{(1-N)}$$

$$= \frac{0.1 \times 30}{(1-0.1)} = 3.33 \text{ Pa.}$$

INHERENT CHARACTERISTIC

The inherent characteristic, γ, of a FCD is defined at an authority of $N=1$. It is the relationship between the stroke of the FCD (or angular position for damper blades) and the flow rate relative to the maximum flow in the fully open position, whilst maintaining constant pressure drop across the device. That is:

$$\gamma = \frac{\dot{V}_\theta}{\dot{V}_\varphi}$$

It is then shown that (see Appendix, page 354):

$$\gamma = \sqrt{\frac{K_\varphi}{K_\theta}}$$

(16.2)

INSTALLED CHARACTERISTIC

The installed characteristic, γ', is the relationship between the stroke of the FCD and the flow rate through the system relative to the flow rate with the FCD when fully open. The general case is where the total pressure available in the system varies with the flow rate. An example of this is where the control device is used to *throttle* the total flow rate of a pump or fan. If the prime mover characteristic is defined by Equation 14.7

(ie $p_F = A' - B'\dot{V}^2$), then it can be shown that the installed characteristic is given by:

$$\gamma' = \sqrt{\frac{r_\varphi/N + B'}{r_\varphi(1/N - 1 + 1/\gamma^2) + B'}}. \tag{16.3}$$

For a system with a constant pressure drop, $B' =$ zero. Therefore, the installed characteristic becomes:

$$\gamma' = \frac{\gamma}{\sqrt{(N + \gamma^2(1 - N))}}. \tag{16.4}$$

Thus for a given design (eg fan/system) and inherent characteristic, there is a set of installed characteristic curves for FCD, each member of which is determined by the authority, N. A typical set of curves is illustrated in Figure 16.1.

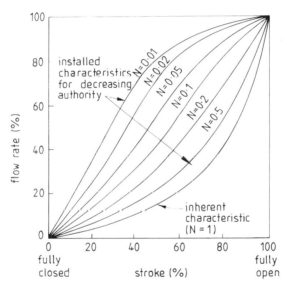

16.1 **Inherent and installed characteristics of a modulating fluid flow control device**

The derivation of the Equations 16.3 and 16.4 for inherent and installed characteristics of a flow control device is given in the appendix at the end of the chapter.

DAMPERS

If a system is to operate satisfactorily, the design and selection of a damper should be treated with appropriate importance relative to the other components within the system. Unfortunately this has often been neglected in the past, as few manufacturers publish the necessary data to allow a selection to be made and the damper is often sized to fit the ductwork which itself has been sized to accommodate other plant items. There have been a number of reports[4,5] which suggest that dampers are either not of a very high standard of manufacture and/or are not capable of adequately controlling the air flow rate, resulting in high system energy consumption.

Damper pressure loss coefficients

Control dampers in most common use for ducted air systems are normally made for rectangular ducts. These consist of a number of blades each with a central spindle, linked together to allow all the blades to move together varying the angle of inclination, θ, to the axis of the duct, as shown in Figure 16.2. With two or more blades, the spindles are linked to give either parallel or opposed motion, over 90 degrees angle from a start angle of φ degrees relative to the duct axis.

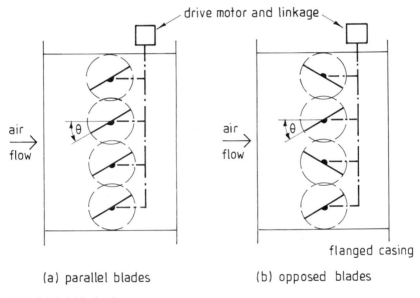

(a) parallel blades   (b) opposed blades

16.2 Multi-blade dampers

Commercial dampers have a variety of cross-section blade shapes including flat plates with crimped edges, aerofoil and wedge. The leading and trailing edges of the blades often incorporate material to ensure a reasonable degree of air tightness when in the closed position. For relatively small circular and rectangular ducts, a single leaf damper is sometimes referred to as a *butterfly damper*.

It has been shown experimentally with these devices that the loss coefficient, $K_{d\theta}$ is defined by the empirical relationship:

$$\log_e K_{d\theta} = a + b\theta \qquad\qquad (16.5)$$

generally in the blade angle ranges:

single and opposed blades $\qquad\qquad 10° \leqslant \theta \leqslant 60°$
parallel blades $\qquad\qquad\qquad\quad 10° \leqslant \theta \leqslant 70°$.

This relationship is illustrated in Figure 16.3. A number of dampers have been tested[6], and shown to obey Equation 16.5; this relationship was also confirmed by the analysis of data in other publications.[7,8,9] The value of the experimental constants a and b relate to parameters such as blade size and shape, projections on the blade, the relationship between the duct wall and blade configuration.

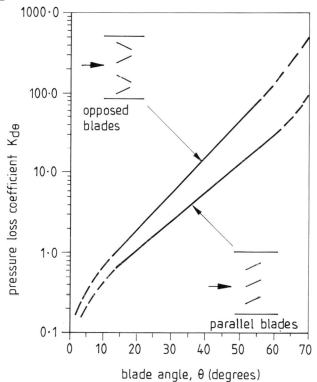

16.3 **Damper pressure loss coefficients**

Since the inherent characteristic for a damper may be obtained from pressure loss coefficients (Equation 16.2), it follows from Equation 16.5 that the inherent damper characteristic becomes an exponential decay, ie:

$$\gamma = \sqrt{\frac{K_{d\varphi}}{K_{d\theta}}} = \sqrt{\frac{e^{(a+b\varphi)}}{e^{(a+b\theta)}}}$$

$$= e^{(\varphi-\theta)b/2} \qquad\qquad\qquad\qquad\qquad (16.6)$$

For Equation 16.6 to apply for blade angles less than 10°, then the blade and/or duct wall projections should be sized for:

$$K_{d\varphi} = e^a \qquad\qquad (16.7)$$

where $\varphi =$ zero degrees

Typical values of a and b for crimped blade dampers are given in Table 16.1. Opposed blade dampers produce loss coefficients which are substantially greater than those of parallel blade dampers by virtue of the fundamental difference in the flow patterns resulting from the different orientation of the blades. This in turn leads to the former damper having a more extreme inherent characteristic than the latter.

type	number of blades	constants		damper sizing constant, G	
		a	b	$\theta = 0°$	$\theta = 10°$
single	1	−1.5	0.105	0.50	0.39
opposed	> 2	−1.5	0.105	0.50	0.39
parallel	2	−1.5	0.088	0.75	0.61
parallel	> 3	−1.5	0.080	0.91	0.77

Table 16.1 Typical values of empirical constants a and b (Equation 16.5) for crimped blade dampers, together with damper sizing constants, G

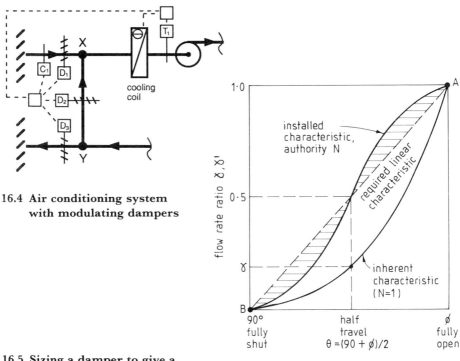

16.4 Air conditioning system with modulating dampers

16.5 Sizing a damper to give a near-linear characteristic

Damper size and selection

It is generally accepted that a linear installed characteristic is desirable for flow control in air handling systems. To obtain a near linear installed characteristic where the system pressure drop varies with flow rate, an iterative solution of Equation 16.3 will normally be required. However, the particular but typical case is where a *part of the system* (in which the damper controls the flow rate) operates at a constant pressure drop. Examples of this are the three dampers D_1, D_2, and D_3 in Figure 16.4 where the pressures at nodal points **X** and **Y** can be assumed to be constant for a CAV system.

Referring to Figure 16.5, if an installed characteristic is required which fits closest to the linear relationship **AB,** the authority can be calculated which minimises the shaded areas and a close approximation for this would be for $\gamma' = 0.5$ at a mid-range blade angle of $(90 + \varphi)/2$. Therefore, from Equation 16.4:

$$N = \frac{3\gamma^2}{1 - \gamma^2} \tag{16.8}$$

Then it can be shown that

$$\dot{V} = A_d G \sqrt{\Delta ps/\rho} \tag{16.9}$$

where:

$$G = \sqrt{\frac{2N}{K_{d\varphi}(1 - N)}} \tag{16.10}$$

= a constant for a particular damper design

The loss coefficient, $K_{d\varphi}$, is determined from Equation 16.5. The authority, N, is obtained from Equation 16.8 in which the inherent characteristic is defined by Equation 16.6 at the mid-range angle of $(90 + \varphi)/2$.

Here are examples of the use of these equations:

Example 16.2 For an opposed blade damper using empirical values of $a = -1.5$ and $b = 0.105$, determine the value of the damper sizing constant, G, for a start angle of zero degrees.

Solution

For a start angle of *zero* degrees, the damper should be designed for the loss coefficient $K_{d\varphi}$ to be given by Equation 16.7:

$$\begin{aligned} K_{d\varphi} &= e^a \\ &= e^{-1.5} = 0.223. \end{aligned}$$

At $\theta = 45°$ the value at $\gamma' = 0.5$, and taking $\varphi = 0$; using Equation 16.6:

$$\begin{aligned} \gamma &= e^{(\varphi - \theta)b/2} \\ &= e^{(0 - 45)b/2} \\ &= e^{(-45 \times 0.105/2)} \\ &= 0.0942. \end{aligned}$$

341

Using Equation 16.8:

$$N = \frac{3\gamma^2}{1-\gamma^2}$$

$$N = \frac{3 \times 0.0942^2}{1-0.0942^2} = 0.0269.$$

Using Equation 16.10, the damper sizing constant is therefore given by:

$$G = \sqrt{\frac{2N}{K_{d\varphi}(1-N)}}$$

$$= \sqrt{\frac{2 \times 0.0269}{0.223(1-0.0269)}} = 0.498.$$

Example 16.3 For a parallel crimped blade damper using typical empiracle values $a = -1.5$, $b = 0.08$, determine the value the damper sizing constant, G, with a start angle of $10°$ ($\varphi = 10°$).

Solution

The loss coefficient $K_{d\varphi}$ is given by Equation 16.5:

$$\log_e K_{d\varphi} = a + b\varphi$$
$$= -1.5 + 0.08 \times 10$$
$$\therefore K_{d\varphi} = 0.497.$$

At $\theta = 50°$, the value at $\gamma' = 0.5$; using Equation 16.6:

$$\gamma = e^{\varphi - \theta)b/2}$$
$$= e^{(10-50)0.08/2}$$
$$= 0.202.$$

From Equation 16.8:

$$N = \frac{3\gamma^2}{1-\gamma^2}$$

$$= \frac{3 \times 0.2022^2}{1-0.2022^2} = 0.1275.$$

Using Equation 16.10:

$$G = \sqrt{\frac{2N}{K_{d\varphi}(1-N)}}$$

$$G = \sqrt{\frac{2 \times 0.1275}{0.497(1-0.1275)}} = 0.767.$$

Typical values of G are summarized in Table 16.1.

Equation 16.5 applies to only a limited range of blade angles, but this range is for the most critical part of the characteristic. If Equations 16.4 and 16.6 (with $\gamma'=0.5$) are used for blade angles greater than approximately 60°, there will be little loss in accuracy for opposed blade dampers, eg, at a blade angle of 90°, $\gamma=0.01$ and $\gamma'=0.04$ which are typical of the *leakage* of commercial dampers. On the other hand, for the parallel blade damper, the corresponding values are 0.03 and 0.10 and these figures are unlikely to be acceptable, in which case it will be necessary to take account of a more appropriate leakage rate.

Use of damper sizing equation

An example of sizing dampers for the free-cooling, variable, recirculation system is now given.

Example 16.4   For the system shown in Figure 16.4, determine the sizes of the three dampers if the design flow rates are to be 2.0 m³/s for the supply and 1.8 m³/s for exhaust and recirculation. The pressure drops have been calculated to give the following internal duct air pressures relative to atmospheric pressure:

at nodal point: X -30 Pa
              Y $+40$ Pa.

Solution

For a damper start angle of zero degrees, an air density of $1.2\,\text{kg/m}^3$ and using Equation 16.9, the calculation is set out in the following table:

parameter	damper					
	opposed			parallel		
	D_1	D_2	D_3	D_1	D_2	D_3
sizing constant, G		0.498			0.908	
flow rate, $\dot{V}$ (m³/s)	2	1.8	1.8	2	1.8	1.8
system pressure drop (Pa)	30	70	40	30	70	40
damper cross section area, A_d (m²)	0.803	0.473	0.626	0.440	0.259	0.343

Choice of parallel or opposed blade dampers

Example 16.4 shows that, for a given design system flow rate and pressure drop, an opposed blade damper will be approximately twice the size of a parallel blade damper. Having considered this, the choice of damper will then depend on a number of other factors, including the following:

– convenience of making the duct connections: the ductwork sized for other plant items may more easily accommodate one damper rather than another;
– the costs of the alternative arrangements;
– the noise characteristics of the damper/system;
– the importance of the length of duct required for full static pressure regain, eg pressure controllers might be influenced adversely in the low pressure region immediately downstream of the damper. It has been shown that full pressure recovery for an opposed blade damper occurs between three and six equivalent duct diameters downstream of the damper, whereas ten diameters are required for a parallel blade damper[10];
– the ability of the dampers to mix two air streams at two temperature levels, to prevent stratification in the total air volume flow before the next section of the system. This is particularly the case where the dampers are arranged to discharge directly into a mixing chamber.

With regard to this last point the alternative arrangements to achieve mixing are shown in Figure 16.6.

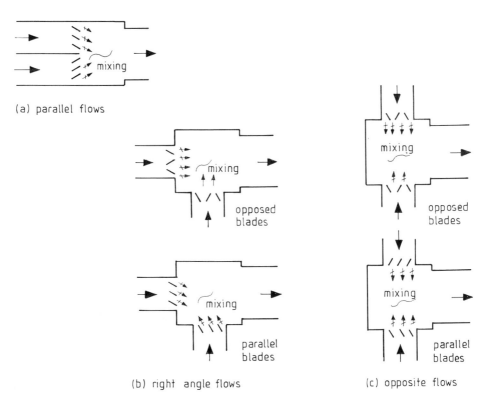

(a) parallel flows

(b) right angle flows

(c) opposite flows

16.6 Mixing arrangements for two dampered inlet ducts

For throttling applications, provided they are sized at an appropriate authority, opposed and parallel blade dampers will work equally well. Note that for plenum discharge, the empiracal values of **a** and **b** are likely to be rather higher than those given in Table 16.1.

Leakage

When considering liquid control valves, the ratio of maximum flow to minimum controllable flow at constant pressure drop across the valve is known as *rangeability*. When the valve is installed in the system, its authority affects this minimum flow and *turndown* is the term used to define the ratio of maximum flow to minimum controllable flow with a constant pressure drop across the circuit. These terms not yet been accepted as part of damper performance. It is more usual to refer to the flow through the damper in the fully closed position as *air leakage* for both the inherent and installed characteristics. For efficient plant operation leakage should be minimized wherever complete shut off is required, eg, this will normally be the case for the recirculation damper, D_2, in Figure 16.4. A leakage rate of 2% on the inherent characteristic is normal for commercial dampers so that for an authority of 0.05 to 0.1 a leakage rate of 6 to 8% can be expected which in turn may increase the annual refrigeration energy consumption by about 5%.

CONTROL VALVES

Regulation of cooling and heating systems in building services takes place mainly at the heat exchanger, through the adjustment of fluid flow rates and temperatures. The modulation of a fluid supply temperature can be accomplished by on/off control of boilers and chillers resulting in a cyclical or sawtooth temperature pattern. The smooth variation in temperature usually required is achieved more effectively by mixing streams of fluids at different temperatures. A valve of the three-port mixing type is needed for this purpose just as a two-port control valve is, on first sight, required for flow rate modulation. In practice, the main method used for the control of thermal output is, in both cases, the variation of a valve opening.

Where the output of a heat exchanger is controlled by varying the flow rate, the primary medium inlet temperature to the heat exchanger remains constant. Therefore, the output of heat only falls off because of the reduction in heat transfer coefficients which in turn depend on the reducing flow rate. This is offset, somewhat, by the increase in temperature drop consequent upon the longer primary medium residence time in the heat exchanger. As a result, flow rate modulation is of a very non-linear form, with typically 80% of flow reduction needed for a 40 to 50% heat output drop. In this case controllers can find difficulty in producing accurate and stable outputs.

It follows that there are several conflicting requirements in engineering the correct application of control valves. On the one hand, close, linear, stable regulation of output is wanted but on the other, liquid circulation needs to be maintained for system balance, freedom from sediment, minimizing pump and ancillaries sizes and power use. Valve types, characteristics, sizing and circuit design, all contribute to obtaining an optimum solution.

Types

The two main types of valve are:

– flow restrictors;
– flow mixers.

Flow restrictors

Two examples of this type are shown in cross-section in Figure 16.7. The two-port valve allows the reduction of flow as the valve stem is lowered and the *plug* pushed onto the *seat*. The disadvantage of this type of valve is that flow may be entirely obstructed. This in turn leads to pumps having to operate against closed valve conditions and to boilers and water chillers operating with insufficient flows. Scaling and deposition of solids tend to occur, in the worst possible place, ie on the exchanger heat transfer surfaces, leading to permanent loss of output and eventual blockage of the small tubes. Maximum pressure drops against which the valve may close are also limited, particularly where system pressure may rise steeply as the pump operates at a lower flow rate. These objections to two-port valves do not apply in the case of steam flow control. In this case the valve may be used directly to govern a heat exchanger output required for air or domestic hot water heating. The relatively high velocity and modest pressure of process steam obviate the difficulties otherwise experienced.

Flow mixers

An example of a typical flow mixer is shown in Figure 16.8. As the stem position is altered, the opening of one inlet port increases while the flow area of the other decreases. The blend of fluid varies progressively from one inlet condition towards the other, hence this is known as a *three-port changeover valve*. The temperature, t_m of the mixed outlet fluid is given by:

$$t_m = \frac{\dot{m}_1 t_1 + \dot{m}_2 t_2}{\dot{m}_1 + \dot{m}_2}$$ (16.11)

where $\dot{m}_1$ and $\dot{m}_2$ are the inlet mass flow rates and t_1 and t_2 are the inlet temperatures respectively.

In general, mixing valves should be designed to deliver a constant outlet mass flow rate irrespective of stem position. This allows the outlet temperature to be the principal variable, independent of mass flow effects, a particular requirement for good heat exchanger control.

The near-constant mass flow rate from a mixing valve is an advantage in the part of the circuit following the valve. However, both inlet flows will vary and provision may have to be made in a pumped primary circuit to allow an alternative fluid return path when all valves are closed to flow from the main. An example of this is the pressure relief valve which allows a primary main to discharge direct to the return main if flow rate becomes insufficient and the pressure difference between flow and return mains becomes excessive.

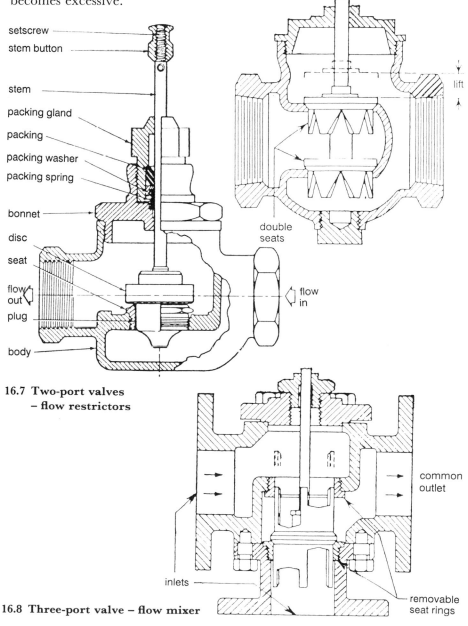

16.7 Two-port valves
– flow restrictors

16.8 Three-port valve – flow mixer

PIPE CIRCUITS

Two main types of circuit exist to which three-port mixing valves may be applied. Each produces a distinctly different effect which must be understood if the piping/valves are to be designed and sized correctly.

Mixing

If the control valve is situated in the heat exchanger supply branch so as to mix primary fluid with return fluid, then the *circuit* is said to be of the mixing type as shown in Figure 16.9. In order for the return fluid to mix, pressurisation is required by means of sub-circuit pumping. A balancing double regulating valve (DRV) is also required in the return branch to equalise pipe pressure losses around the alternative circuits. Frequently, a balance pipe is installed to avoid interaction between primary and sub-circuit pumps.

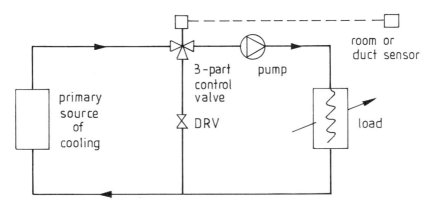

16.9 Chilled water mixing circuit

The blending of the two flows maintains a constant mass flow rate to the heat emitters, whilst allowing linear variation in temperature. With heat transfer equipment, there is a near-linear relationship between output and flow water temperature – a satisfactory characteristic. Conveniently linear valve and actuator characteristics may therefore be employed, leading to good control though more expensive than the diverting circuits described below. Acceptable authorities are held to lie in the range $0.5 < N < 0.7$.

Diverting

To avoid the need and expense of sub-circuit pumping, the alternative position for a mixing valve may be chosen, ie in the return, as shown in Figure 16.10. Here, return fluid from the heat emitters is mixed with by-pass fluid to give a blended return to the primary system. By-pass balance must be achieved by a DRV to maintain constant total flow at all control valve positions. This is a simpler and more economical arrangement than the mixing circuit.

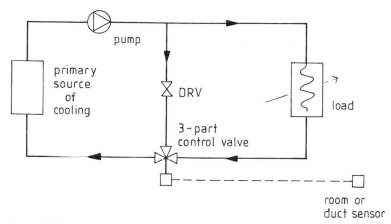

16.10 Chilled water diverting circuit

Because of the mixed return flow, the change in the sub-circuit is now made to the flow rate rather than the temperature flow. In this case, the output of the heat exchange equipment is not linear to flow rate. As the flow rate reduces, the fluid's time inside the exchanger increases and so the contact factor (or effectiveness) and hence temperature change increase. This non-linearity has to be taken into account when selecting the appropriate valve/actuator characteristics. Acceptable authorities for these applications are in the range $0.3 < N < 0.5$. The diverting circuit is more demanding in terms of technical design but is cheaper in capital and running costs.

VALVE CHARACTERISTICS

A valve test consists of measuring the pressure drop through a valve for a range of flow rates at different positions of the valve stem (or percentage opening), for authorities of unity. When these points are plotted, the valve inherent characteristic is obtained. Three broad types of characteristic are found, each of which has a particular application. The types are:

Linear In practice the installed characteristic of a linear valve may only be a approximately linear at normal authorities. As the authority falls below unity, every valve becomes more *quick opening* and the inherent (or test) characteristic should only be moderately parabolic in shape. Not surprisingly, the shape of the valve plug/seat combination mirrors that of the flow characteristic. Valve plugs of this type are shown in cross-section in Figure 16.11 and the installed characteristics for various authorities are shown in Figure 16.12.

The applications of this type of valve plug are in three-port valves used in mixing circuits, where linear changes in blending ratio produce linear variations in, say, heat exchanger output. Two-port steam control valves should also have a generally linear behaviour, as flow rate is directly proportional to output.

(a) double port

(b) single port

16.11 Plugs for linear valves

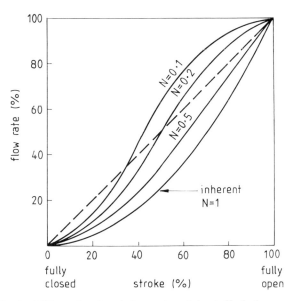

16.12 Typical 'linear' valve inherent and installed characteristics

Quick opening A quick opening valve, also known as a *bevel disc* type of valve, has the characteristic of giving near maximum flow at relatively low stem lift. The gate valve is usually of this type and, combined with its poor authority, is difficult to use for manual flow regulation. Examples of the type of plug are shown in Figure 16.13 and typical characteristics are shown in Figure 16.14. As an illustration of the reason why the valve opens up so rapidly, the balance between flow areas through the seat and above the plug may be examined in Equation 16.12, using the dimensions from Figure 16.13.

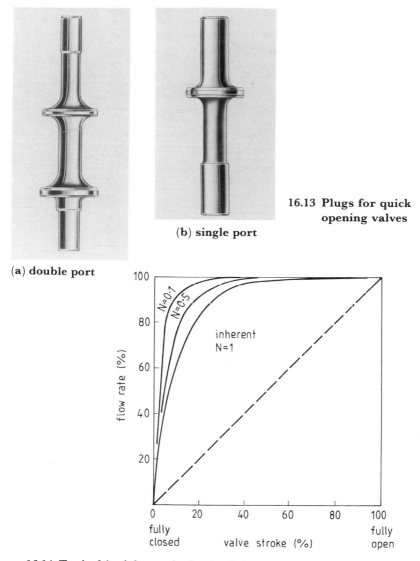

16.13 Plugs for quick opening valves

(b) single port

(a) double port

16.14 Typical 'quick-opening' valve inherent and installed characteristics

At full flow rate:
    area of cylinder above plug = cross section area of plug

$$\pi DL = \pi D^2/4$$

$$\therefore L = D/4 \tag{16.12}$$

Thus the lift need only be perhaps 25% of the maximum, usually of the same order as the diameter, to allow full flow. This type of characteristic is ideally suited to the solenoid operated on/off type of valve. The sharpness of operation may bring some risk of water hammer where there are long pipe runs and high static pressures.

Equal percentage As the contact factor or effectiveness of a heat exchanger rises as the flow rate falls, a valve that mirrors and therefore reverses this exponential effect is needed. Because heat output falls only slowly as feed is reduced, a valve characteristic which is opposite to this is called for, ie one which increases flow rate slowly as it opens, with perhaps 15 to 20% of full flow at the mid-stem position. An *equal percentage* valve is designed to give this effect. It is illustrated in Figure 16.15 and has characteristics shown for various authorities in Figure 16.16. The application most desirable for this type is in the mixing valve used in a diverting circuit. In order to retain system balance, the combined or blended total flow through such a valve must stay reasonably constant for any stem position. For this reason, only the valve plug facing the heat exchanger should be of the equal percentage type.

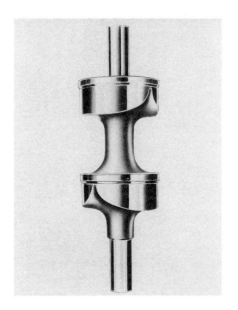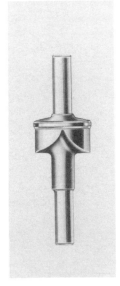

16.15 Plugs for equal percentage valves (a) double port (b) single port

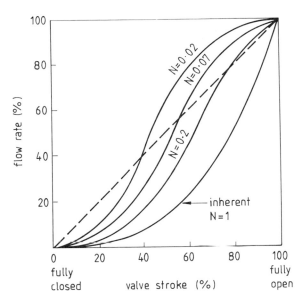

16.16 Typical 'equal percentage' valve inherent and installed characteristics

SYMBOLS

A', B'	constants of fan characteristic'
A	representative area of FCD
A_d	area of damper cross section
a	intercept of Equation 16.5
b	slope of Equation 16.5
D	plug diameter
G	damper sizing constant (defined by Equation 16.9)
K	pressure loss coefficient (defined by Equation 12.12)
L	stem lift
$\dot{m}$	mass flow rate
N	authority
r	resistance
t	temperature
$\dot{V}$	volume flow rate
$\bar{v}$	mean velocity
γ	inherent characteristic
γ'	installed characteristic
Δp	pressure drop
ρ	air density
θ	valve (stem) position or damper blade angle
φ	valve (stem) fully open or damper start angle

Subscripts

d	damper
s	system
θ	position of valve stroke or blade angle relative to axis of duct
φ	initial position of valve stroke or start angle of damper blades

Appendix

DERIVATION OF EQUATIONS FOR INHERENT AND INSTALLED CHARACTERISTICS OF A FLOW CONTROL DEVICE

Inherent characteristic

The inherent characteristic, γ, is defined by the relationship between the stroke of the FCD (or angular position of the damper blades) and the flow rate relative to the maximum flow in the fully open position, whilst maintaining constant pressure drop across the device. It is given by the equation:

$$\gamma = \frac{\dot{V}_\theta}{\dot{V}_\varphi}.$$

If the pressure drop across the control device is defined in terms of pressure loss coefficients, then:

$$\Delta p = K \ 0.5 \ \rho \bar{v}^2.$$

From the definition of the installed characteristic:

$$\Delta p_\varphi = \Delta p_\theta$$

$$\therefore \ K_\varphi \ 0.5 \ \rho \bar{v}_\varphi^2 = K_\theta \ 0.5 \ \rho \bar{v}_\theta^2. \tag{16.13}$$

If A is a representative area of the device, then:

$$\bar{v} = \dot{V}/A.$$

The inherent characteristic is therefore given by:

$$\gamma = \frac{\dot{V}_\theta}{\dot{V}_\varphi} = \frac{\bar{v}_\theta}{\bar{v}_\varphi}$$

$\therefore$ from Equation 16.13

$$\gamma = \sqrt{\frac{K_\varphi}{K_\theta}} \qquad \text{(Equation 16.2)}$$

In a similar way, the inherent characteristic can be expressed in terms of resistances, ie:

$$\gamma = \sqrt{\frac{r_\varphi}{r_\theta}} \tag{16.14}$$

354

Installed characteristic

The installed characteristic, γ', is defined by the relationship between the stroke of the FCD and the flow rate through the system relative to the flow rate with the FCD in the fully open position, ie:

$$\gamma = \frac{\dot{V}_\theta}{\dot{V}_\varphi}.$$

Consider a general case where the total pressure available in the system varies with the flow rate. A typical example of this is where the control device is used to *throttle* the total flow rate of a pump or fan. If the prime mover characteristic is defined by Equation 14.7:

$$p_F = A' - B'\dot{V}^2. \tag{16.15}$$

At maximum flow rate, with the control device fully open $(\theta = \varphi)$ the total system pressure loss, Δp_{st}, can be expressed as:

$$\Delta p_{st} = \Delta p_s + \Delta p_\varphi \tag{16.16}$$

where Δp_s = pressure drop in the system excluding the pressure drop across the FCD

Δp_φ = pressure drop across the fully open FCD.

Using resistances in series, given by Equation 12:

$$\Delta p_{st} = r\dot{V}_s^2 + r_\varphi \dot{V}_\varphi \tag{16.17}$$

At the operating point $p_F = \Delta p_{st}$. Therefore equating 16.15 and 16.17: at maximum flow rate, with the FCD fully open at stroke φ:

$$A' - B'\dot{V}_\varphi^2 = r_{st}\dot{V}_\varphi^2 + r_\varphi \dot{V}_\varphi^2 \tag{16.18}$$

at the reduced rate with the stroke at θ:

$$A' - B'\dot{V}_\theta{}^2 = r_{st}\dot{V}_\theta{}^2 + r_\theta \dot{V}_\theta{}^2. \tag{16.19}$$

From Equations 16.18 and 16.19 the installed characteristic is obtained as:

$$\gamma' = \frac{\dot{V}_\theta}{\dot{V}_\varphi} = \sqrt{\frac{r_{st} + r_\varphi + B'}{r_{st} + r_\theta + B'}}$$

With the inherent characteristic expressed in terms of resistances, from Equation 16.14:

$$r_\theta = \frac{r_\varphi}{\gamma^2}$$

The FCD authority, N, is defined by Equation 16.1. Therefore,

$$N = \frac{\Delta p_\varphi}{\Delta p_s + \Delta p_\varphi} = \frac{r_\varphi}{r_s + r_\varphi} \tag{16.19}$$

$$\therefore \; \gamma' = \sqrt{\frac{r_\varphi(1/N-1)+r_\varphi+B'}{r_\varphi(1/N-1)+r_\varphi/\gamma^2+B'}}$$

$$\gamma' = \sqrt{\frac{r_\varphi/N+B'}{r_\varphi(1/N-1+1/\gamma^2)+B'}}. \qquad\qquad \text{(Equation 16.3)}$$

For a system with a constant pressure drop, $B'=0$. Therefore,

$$\gamma' = \frac{\gamma}{\sqrt{(N+\gamma^2(1-N))}}. \qquad\qquad \text{(Equation 16.4)}$$

17 Energy Consumption

The ability to calculate the predicted annual energy consumption and energy costs of air conditioning systems is required for the following reasons:

- to allow a comparison of systems on a total cost basis;
- to optimize the design and control of the selected system to achieve minimum energy consumption;
- to inform the client of the expected costs for budgeting purposes;
- to provide a basis for the energy performance of the installed system during its working life.

The method of calculating the average annual energy consumption for heating systems using the concept of degree-days is well established.[1] There is no such commonly accepted method for calculating the energy used by air conditioning systems. It is possible to use computer programs with detailed mathematical models of a building, the weather, the systems and hour-by-hour operation to assess the consumption, but this method produces custom-built models which involve a large amount of time and money for the model's construction. Two other possible approaches are given in this chapter. The first makes use of *equivalent* operating hours of the various plant items based on historical data from surveys of existing systems. The second method analyses the variations of plant loads in conjunction with the frequency of outdoor air conditions.

EQUIVALENT HOURS OF FULL LOAD OPERATION

Equivalent hours of full load operation, H_{eq}, based on surveys of existing plants, can be used to estimate energy consumption. To obtain the annual consumption, the installed maximum operating load of the energy using equipment is multiplied by equivalent full load hours.

For fuel consumption, the hours of equivalent full load operation of boiler plant ranges from about 750 to 1500 annually, with a mean figure of 1000 hours. For electricity consumption of air conditioning plant items for office buildings in the United Kingdom, the equivalent full load operating hours are given in Table 17.1.[2]

357

plant	offices with air conditioning		offices with heating only
	prestige	standard	
refrigeration plant and cooling towers	1040	1040	-
fans	3800	3000	-
pumps	4000	3000	2460

Table 17.1 Estimation of electrical energy consumption in office buildings: annual hours of equivalent full load operation

Example 17.1 Estimate the annual electrical consumption of an air conditioning system for a *prestige* office block with the following installed loads.

(a) refrigeration plant   − 50 kW compressor;
(b) cooling tower fan   − 3 kW;
(c) supply and extract fans − 20 kW motors (sum of individual
                     motors);
(d) pumps   − 4 kW.

Solution

The annual hours of equivalent full load are obtained from Table 17.1. The calculation is tabulated as follows:

plant item	installed capacity (kW)	equivalent full load operation, hours (H_{eq})	electrical units per annum (kWh)
(a) refrigeration compressor	50	1040	52000
(b) cooling tower fan	3	1040	3120
(c) supply and extract fans	20	3800	76000
(d) pumps	4	4000	16000
		total	147120

This method of using equivalent hours based on historical data is useful for making preliminary estimates of total annual energy consumption. Because the operating hours given in Table 17.1 are global average figures based on a variety of systems, the method is not particularly suitable for the *comparison* of system energy costs. Present published data is also limited to office buildings.

ANNUAL ENERGY CONSUMPTION USING FREQUENCY OF OCCURRENCE OF OUTDOOR AIR CONDITIONS

The approach described in the remainder of this chapter is a technique (originally described by ROBERTSON[3]) to estimate annual energy consumption from which cost comparisons between alternative designs can be made. The technique is similar to the BIN method given in the *ASHRAE Guide*,[4] but whereas the BIN method relates to a base of dry-bulb temperature, that given here offers greater flexibility in the choice of outdoor conditions provided the relevant data are available.

The annual energy cost for the complete air conditioning system(s) is obtained through the summation of the energy consumption/cost of each individual energy using plant item within the system. These may include the following:

- heaters and boiler plant;
  coolers and refrigerating plant;
- steam humidifiers;
- fans for central plant supply and extract;
- fans for room air conditioning units;
- fans for cooling towers;
- pumps for heating, cooling and humidification water systems;
- compressors for unitary systems.

Some of these plant items will run at constant load in which case the annual consumption is simply the annual hours of operation multiplied by the installed load. Where plant load varies because of variations in outdoor conditions and internal heat gains then it will be necessary to analyse the load variations according to their annual frequency of occurrence.

The method is as follows:

(a) establish the relationship between energy demand and an appropriate climate factor;
(b) integrate the variations in energy demand with the annual frequency distribution of the climate factor. (This may take the form of the frequency data described in Chapter 4.)

An example of the technique is illustrated in Figure 17.1. The load on the plant item being considered is plotted against the outdoor climate factor for which there is a known frequency of occurrence. Since the load is increasing with outdoor conditions, this diagram represents the load handled by a cooler battery in an air conditioning system.

Some of the features of this diagram are:

- The load profile is shown on the upper chart as **ABCDE**;

– The plant is switched on at **A**;

– The load between points **A** and **B** is constant and is termed a *base load*. A base load may be due to miscellaneous heat gains and losses to pipework. The frequency of such occasions is obtained by summing each of the frequency of occurences between points **A** and **B** – this is identified on the lower chart between points **a** and **b**.

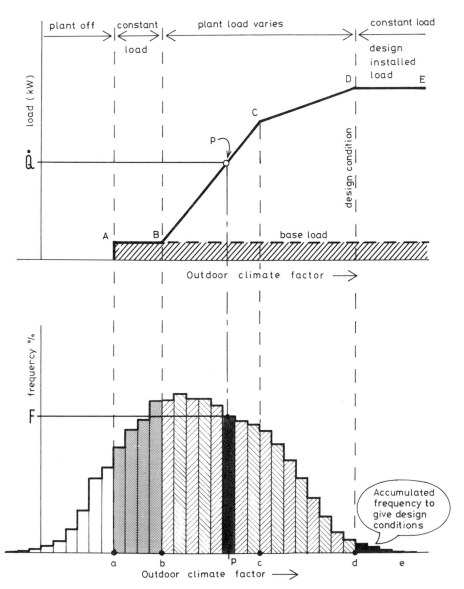

17.1 Plant load versus outdoor climate factor with known frequency of occurrence: BIN method

- The load varies between **BC** in the frequency range **bc**. The mode of operation of the plant changes at **C** and the load varies between **CD** in the frequency range **cd**.
- For any load the expected frequency may be obtained, eg at point **P** load $\dot{Q}$ occurs with frequency **f**.
- The load **DE** is constant at the design or installed load (ie the plant is unable to increase the load above this). This occurs for a total accumulated frequency in the range **de**.
- The choice of the climate factor will depend on the preferred operating characteristics of the system.
- A *bin*, used in the BIN method for calculating energy consumption, is any one of the frequency columns of the lower chart; any plant load, part-load efficiency, heat gain/loss associated with that particular range of the climate factor is *placed in the bin!*

Where both the load and frequency of occurrence of the load are varying it is necessary to accumulate the sum of their product. The average annual energy consumption is then given by:

for cooling loads:

$$E_c = \frac{H}{100} \Sigma \left(\frac{fQ}{COP} \right) \qquad (17.1)$$

for heating loads:

$$E_h = \frac{H}{100} \Sigma \left(\frac{fQ}{\eta} \right) \qquad (17.2)$$

where

   COP = cooling system part-load coefficient of performance
   η = heating system part-load efficiency
   H = operating time of plant in hours.

From this calculation the annual energy consumption is in units of kWh; the annual energy costs may be obtained from the unit costs of the fuel.

The theoretical approach is particularly useful for the analysis of the cost-benefit of additional energy saving equipment in a system. In this case only the analysis of the *difference* in energy costs between the alternative schemes would be required.

For manual calculations a range (or interval) of the climate factor is chosen to suit the calculation. Too small a range leads to a lengthy calculation; too large a range leads to reduced accuracy and flexibility.

A number of examples now follow which illustrate the general approach and method of calculation.

Example 17.2 The air conditioning system (System 1) shown in Figure 17.2 provides a constant ratio of outdoor to recirculated air quantities and the off-coil condition, **B**, is controlled by the dew-point sensor T_{dp}. Calculate the energy consumption for the cooler for the following criteria:

mass air flow rate (assumed constant)   2.5 kg/s
air enthalpies:
   recirculation, h_R 42 kJ/kg$_{da}$
   off-coil (T_i), h_B 30 kJ/kg$_{da}$
   outdoor air, design condition   54 kJ/kg$_{da}$
outdoor air mass flow rate fraction (x) 0.25
coil contact factor   1.0
climate data for Heathrow (Table 3.4)
hours of plant operation per annum   8760
COP (assumed constant)   3.8

The operation of this system is described in Chapter 6.
Assume no miscellaneous heat gains or losses from fans, ductwork or pipes.

Solution

Plot typical psychrometric processes, as in Figure 17.3. It is seen from these that in order to maintain the space conditions, the cooler will be required to operate throughout the year down to low outdoor air enthalpies.

The year-round cooling load is given by:

$$Q = \dot{m}_a(h_M - h_B).$$

Since the mass flow rate, $\dot{m}$, remains constant this can be omitted temporarily from the calculations until the numerical integration has been completed. Therefore the load can be expressed in terms of an enthalpy difference:

$$\Delta h_1 = h_M - h_B$$
$$\Delta h_1 = (1-x)h_R + xh_O - h_B$$
$$= (1 - 0.25)42 + 0.25h_O - 30$$
$$\Delta h_1 = 0.25h_O + 1.5 \qquad (17.3)$$

Using Equation 17.3 the design load is given by:

$$\Delta h_{1d} = 0.25 \times 54 + 1.5 = 15 \text{ kJ/kg}_{da}$$

The load profile for the operation of this system is given in Figure 17.4.

The numerical integration is completed in Table 17.2, with the frequency of occurrence of hourly values of outdoor air climate factor, grouped in convenient ranges of specific enthalpy from Table 4.4 in Chapter 4. The notes column allow the engineer to keep track of the operating conditions of the plant.

The accumulated sum of the product $(f_3\Delta h_1) = 812.5$.

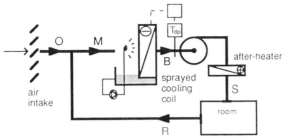

17.2 Air conditioning system (System 1): Example 17.2

17.3 Psychrometric processes: Example 17.2

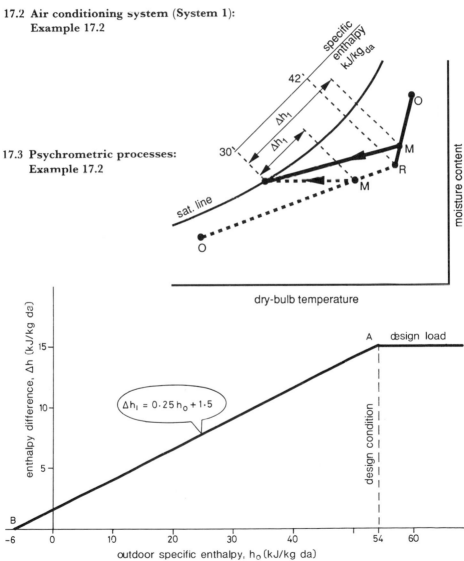

17.4 Load profile: Example 17.2

Using Equation 17.1, the annual energy consumption is given by:

$$E_c = \frac{H}{100} \Sigma \left(\frac{fQ}{COP} \right)$$

since $\dot{m}$ and COP are constant,

$$E_c = \frac{\dot{m}_a H}{COP \ 100} \Sigma (f\Delta h_l)$$

$$= \frac{2.5 \times 8760}{3.8 \times 100} \times 812.5 = 46,826 \ \text{kWh/annum} \ (169 \ \text{GJ})$$

range of outdoor air specific enthalpy (kJ/kg$_{da}$)	mid-point of range h_O (kJ/kg$_{da}$)	frequency f_3 (%)	Δh_1	$f_3 \ h_1$	notes
< 2.0	-3	0.32	0.75	0.2	use Equation 17.3
2 - 5.9	2	1.38	2.5	3.5	
6 - 9.9	8	4.65	3.5	16.3	$\Delta h_1 = 0.25 h_O + 1.5$
10 - 13.9	12	7.78	4.5	35.0	
14 - 15.9	16	1.41	5.5	57.3	
18 - 21.9	20	12.39	6.5	80.5	
22 - 25.9	24	12.68	7.5	95.1	
26 - 29.9	28	12.19	8.5	103.6	
30 - 33.9	32	11.12	9.5	105.6	
34 - 37.9	36	10.42	10.5	109.4	
38 - 41.9	40	8.12	11.5	93.4	
42 - 45.9	44	5.15	12.5	64.4	
46 - 49.9	48	2.27	13.5	30.6	
50 - 53.9	52	0.84	14.5	12.2	
> 54.0	-	0.36	15.0	5.4	design load

$$\Sigma(f_3 \Delta h_1) = 812.5$$

Table 17.2 Calculation of average hourly energy demand of cooler operating in System 1 (Example 17.2)

Example 17.3 In the recirculation air conditioning system (System 2) shown in Figure 17.5 the cooler and the control dampers D_1, D_2 and D_3 are operated sequentially by the sensor T_{dp} to maintain the off-coil condition, **B**. The damper change-over is controlled by C_1 with a set point at the room condition of 42 kJ/kg. Calculate the annual energy consumption of the cooler for the data given in Example 17.2.

The operation of this system is described in Chapter 6.

Solution

Plot typical psychrometric processes, as in Figure 17.6.

summer operation – as system 1
mid-year operation – 100% outdoor air
winter operation – dampers modulate, no cooler.

With the outdoor air condition above the set point of C_1 the cooling load is the same as for System 1 in Example 17.2, ie use Equation 17.3.

When the outdoor air enthalpy is between the set points of T_1 and C_1 the system operates on 100% outdoor air (x = 1.0) and the cooling load becomes:

$$Q = \dot{m}(h_O - h_B)$$

or specifically, since the mass flow rate is constant:

$$\Delta h_2 = h_O - 30 \qquad (17.4)$$

The load profile for the operation of this system is given in Figure 17.7. The numerical integration is completed in Table 17.3.
The accumulated sum of the product $(f_3 \Delta h) = 278.5$.
Using Equation 17.1, the annual energy consumption is obtained:

$$E_c = \frac{H}{100} \Sigma \left(\frac{fQ}{COP} \right)$$

since $\dot{m}_a$ and c_{pas} and η are constant,

$$E_c = \frac{\dot{m}_a H}{COP\ 100} \Sigma (f\Delta h)$$

$$= \frac{2.5 \times 8760}{3.8 \times 100} 278.5 = 16{,}776 \text{ kWh/annum } (60.4 \text{ GJ}).$$

The two systems in Examples 17.1 and 17.2 are alternative methods of treating the recirculated air, the addition of the modulating dampers making System 2 more energy efficient than System 1. The difference between the energy consumption between the two systems is then given by:

$$46{,}826 - 16{,}050 = 30{,}776 \text{ kWh/annum } (111 \text{ GJ}).$$

The theoretical percentage energy saving, expressed as a percentage, is:

$$\frac{30{,}776}{46{,}826} \times 100 = 66\%.$$

The difference in annual energy consumption can be costed as part of a cost benefit calculation for the additional dampers, ductwork and controls.

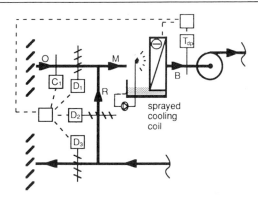

17.5 Air conditioning system: Example 17.3

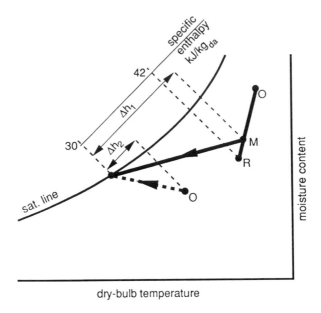

17.6 Psychrometric processes: Example 17.3

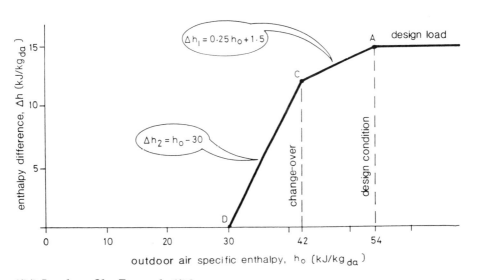

17.7 Load profile: Example 17.3

range of outdoor air specific enthalpy (kJ/kg_{da})	mid-point of range h_O (kJ/kg_{da})	frequency f_3 (%)	Δh	$f_3\Delta h$	notes
< 30	-	-	-	-	cooler off
30 - 33.9 34 - 35.9 38 - 41.9	32 36 40	11.12 10.42 8.12	2 6 10	22.2 62.5 81.2	cooler operates on 100% outdoor air; use Equation 17.4 $\Delta h_2 = h_O - 30$
42 - 45.9 46 - 49.9 50 - 53.9	44 48 52	5.15 2.27 0.84	12.5 13.5 14.5	64.4 30.6 12.2	cooler operates on min. outdoor air; use Equation 17.3 $\Delta h_1 = 0.25h_O + 1.5$
> 54.0	-	0.36	15.0	5.4	design load

$$\Sigma(f_3\Delta h) = 278.5$$

Table 17.3 **Calculation of average hourly energy demand of cooler operating in System 2 (Example 17.3)**

Short route to determine energy consumption difference

If only the *difference* between the systems is required this can be obtained by a shorter route as shown in the following example, which compares the two systems in the previous examples.

Example 17.4 Determine the difference in energy consumption between Systems 1 and 2 of Examples 17.2 and 17.3.

Solution

The load profiles for Systems 1 and 2 are shown in Figures 17.4 and 17.7 respectively. If these two diagrams are superimposed, Figure 17.8 is obtained. The difference in energy consumption is then represented by the area BDC. The intervals of enthalpy difference of this area are then integrated with the frequency of specific enthalpy in convenient ranges.
  For $h_O < 30$ Equation 4 applies.
  In the enthalpy range $30 - 42$ kJ/kg_{da} the difference given by:

$$\Delta h_3 = \Delta h_1 - \Delta h_2$$
$$= (0.25 h_O + 1.5) - (h_O - 30)$$
$$\Delta h_3 = 31.5 - 0.75 h_O. \qquad (17.5)$$

The numerical integration is completed in Table 17.4.
The accumulated sum of the product $(f_3\Delta h) = 534.0$.
Using Equation 17.1, the annual energy consumption is obtained:

$$E_c = \frac{H}{100} \Sigma \left(\frac{fQ}{COP} \right)$$

$$E_c = \frac{\dot{m}_a H}{100\ COP} \sum (f\Delta h)$$

$$= \frac{2.5 \times 8760}{3.8 \times 100} \times 534 = 30{,}776 \text{ kWh/annum}$$

which is the same as the energy saving calculated previously.

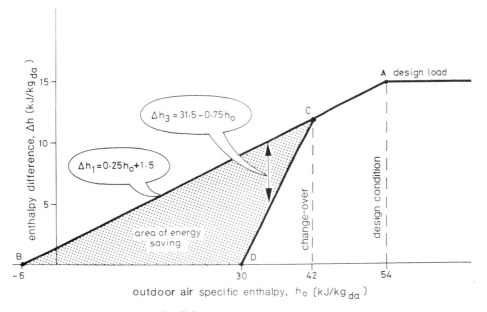

17.8 Load profiles: Example 17.4

range of outdoor air specific enthalpy (kJ/kg$_{da}$)	mid-point of range h_O (kJ/kg$_{da}$)	frequency f_3 (%)	Δh	$f_3\Delta h$	notes
< 2.0	−3	0.32	0.75	0.2	use Equation 17.3
2 − 5.9	2	1.38	2.5	3.5	$\Delta h_1 = 0.25h_O + 1.5$
6 − 9.9	8	4.65	3.5	16.3	
10 − 13.9	12	7.78	4.5	35.0	
14 − 17.9	16	10.41	5.5	57.3	
18 − 21.9	20	12.39	6.5	80.5	
22 − 25.9	24	12.68	7.5	95.1	
26 − 29.9	28	12.19	8.5	103.6	
30 − 33.9	32	11.12	7.5	83.4	use Equation 17.5
34 − 37.9	36	10.42	4.5	46.9	$\Delta h_3 = 31.5 - 0.75h_O$
38 − 41.9	40	8.12	1.5	12.2	

$$\sum (f_3\ \Delta h) = 534.0$$

Table 17.4 Calculation of the difference in average hourly energy demand of cooler operation, System 1 compared to System 2 (Example 17.4)

The next example considers a system heating requirement including an air-to-air heat recovery unit. It is complicated slightly by assuming that the output of the unit is insufficient to meet all the preheat load.

Example 17.5 The air conditioning system shown in Figure 17.9 is required to maintain two animal rooms at 21°C for 24 hours per day and for 7 days per week. The heat recovery unit and preheater are controlled by thermostat T_1 to maintain a constant temperature of 14°C after the coil. Using the design information determine the annual heating energy saved by using a heat recovery unit for part of the preheat load.

Design data

air mass flow rate (assumed constant)	2.5 kg/s
outdoor air design condition	-4°C
climate data for Heathrow (Table 4.2)	
boiler firing efficiency (average)	75%

The temperature rise across the HRU related to the outdoor dry-bulb temperature, t_O, by the equation:

$$\Delta t_{hru} = 10 - 0.4 t_O. \tag{17.6}$$

Solution

The load on the preheater is given by Equation 2.4.

$$Q_p = \dot{m}_a c_{pas}(t_B - t_A).$$

Since the mass flow rate, $\dot{m}_a$, and the humid specific heat, c_{pas}, are considered constant they can be omitted temporarily from the calculations until the numerical integration has been completed. The load variations can therefore be expressed in terms of temperature differences. The preheater load is then given by:

$$\begin{aligned}
\Delta t_p &= t_B - t_A \\
&= 14 - t_A \\
&= 14 - (t_O + \Delta t_{hru}) \\
&= 14 - (t_O + 10 - 0.4 t_O) \\
\therefore \Delta t_p &= 4 - 0.6 t_O. \tag{17.7}
\end{aligned}$$

At the outdoor design condition of -4°C the design temperature rise across preheater from Equation 17.7:

$$\Delta t_{pd} = 4 - (0.6 \times -4) = 6.4 \text{ K}.$$

The load profile for the operation of the HRU and the preheater are given in Figure 17.10. The line **ABD** would be the load profile for a preheater with no HRU and the line **EFG** is the load profile with the HRU. Therefore the energy saving is represented by the shaded area **ABDGFE**. For the temperature range -4 to 6.7 K the HRU operates at maximum output and the energy saving is obtained from Equation 17.6. For the temperature range 6.7

---

to 14 K the preheater is off and the HRU output is modulated, the energy saving is obtained from:

$$\Delta t_{hru} = 14 - t_O. \qquad (17.8)$$

The numerical integration is completed according to Table 17.5.

The accumulated sum of the product $(f_1 \Delta t_{hru}) = 404$.

Using Equation 17.2, the annual energy consumption is obtained:

$$E_h = \frac{H}{100} \Sigma \left(\frac{fQ}{\eta} \right)$$

since $\dot{m}$ and c_{pas} and η are constant,

$$E_h = \frac{\dot{m} c_{pas} H}{\eta 100} \Sigma (f\dot{Q})$$

$$= \frac{2.5 \times 1.02 \times 8760}{0.75 \times 100} \times 404 = 120,327 \text{ kWh/annum } (433 \text{ GJ}).$$

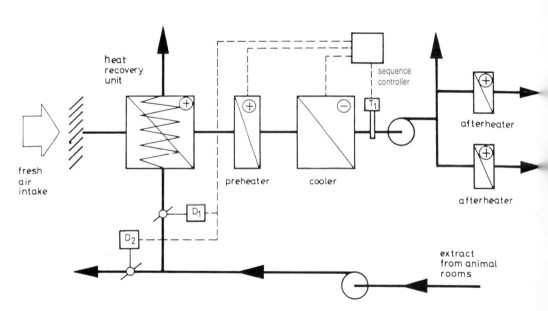

17.9 Air conditioning system (System 3): Example 17.5

range of outdoor air dry-bulb temperature (°C)	mid-point of range t_1 (°C)	frequency f_1 (%)	Δt_{hru}	$f_1\Delta t_{hru}$	notes
< -4.1	-	0.33	11.6	3.9	design load
-4.0 to -1.1	-2.5	1.64	11.0	18.0	HRU at full output;
-1.0 to 1.9	0.5	6.35	9.8	62.2	use Equation 17.6
2.0 to 4.9	3.5	11.31	8.6	97.3	$\Delta t_{hru} = 10 - 0.4t_o$
5.0 to 6.6	5.8	8.78	7.7	67.6	
6.7 to 7.9	7.3	7.62	6.7	51.1	HRU at reduced load;
8.0 to 10.9	9.5	18.30	4.5	82.4	use Equation 17.8
11.0 to 13.9	12.5	16.86	1.5	25.3	$\Delta t_{hru} = 14 - t_o$

$$\Sigma (f_1\Delta t_{hru}) = 403.9$$

Table 17.5 Calculation of average hourly energy saving with heat recovery unit in System 3 (Example 17.5)

Note: design load is that of a conventional heating coil

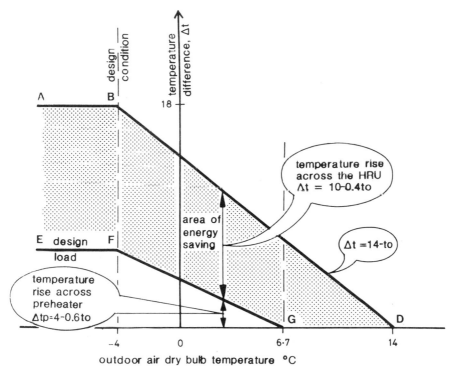

17.10 Load profiles: Example 17.5

This annual energy saving can be costed and compared with the capital cost of the heat recovery unit and associated ductwork, filters and controls. However, account must also be taken of increased fan and miscellaneous equipment (eg motor for thermal wheel, pump for run-around coils) energy consumption costs.

Variations in system efficiency

In the examples given in this chapter, COP and boiler efficiency have been considered constant. If the variation of these two parameters are known, they can easily be associated with each bin and included in the calculations. The technique has been used to investigate the economic application of variable compressor speed for refrigeration system capacity control[5].

LOAD DIAGRAMS

The variation of heat gains and losses to a building, zone of a building or an individual room, can be plotted against outdoor air temperature in a similar fashion to Figure 17.10 and the resulting graph is known as a *load diagram* or *load chart*. These diagrams were developed by HIGHAM[6] and APPLEBY[7], and now accepted in the latest edition of the *CIBSE Guide*[8]. Load profiles may be obtained from design calculations or from BIN data such as those published by LETHERMAN and DEWSBURY.[9] The diagrams are used to assist with system design, planning the economic operation of the plant and for estimating annual energy demands.

Refer to Figure 17.11. The line **AB** is the transmission heat loss (or gain) $\Sigma(UA/\Delta t)$; this line may also include heat loss due to air infiltration. Note that **AB** passes through point **R** where the outdoor air temperature is equal to the room temperature t_R and where the transmission loss is zero. It is usual for the design total internal sensible heat gain from occupants, lights and electrical equipment to be considered constant throughout the year. When the total internal heat gain is added to the transmission line it produces line **CD** parallel to **AB**. To complete the diagram, the maximum solar heat gains calculated at the summer and winter design conditions are added to line **CD** to produce **EF** the line of maximum heat gain (or minimum heat loss). The area **ABFE** represents the heat gain/loss load variations for the space or zone, between the limits of the zone being occupied in summer and the zone being unoccupied at the winter design conditions.

Load diagrams depend on the orientation. Figure 17.11 is a load diagram suitable for a zone on a west face of a building. For most orientations **EF** will be a *dog-leg* EFF' as in Figure 17.12 where the maximum solar gain does not coincide with the maximum outdoor temperature.

For constant air flow rate systems, supply air temperature lines may be drawn in conjunction with the load diagram as shown in Figure 17.13. The

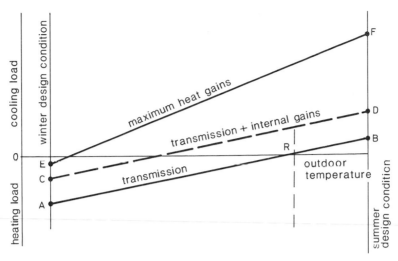

17.11 Load diagram for west orientation of a building

minimum supply temperature, t_S, corresponds to the maximum heat gain at **F**, the temperature difference $(t_R - t_S)$, Δt_c, being the design cooling temperature differential used to determine the air flow rate. In a constant air flow rate system, as the room sensible heat gain decreases the supply air temperature increases to give the line **S'T**. Similarly the maximum supply air temperature will correspond to the maximum heat loss at **A** to give the point **U**, while the supply temperature line **UV** corresponds to the transmission line at a particular value of t_O. The supply air temperature will therefore be at any point located within the area **STUV**, depending on the actual sensible heat gain or loss to the air conditioned space.

The energy consumption of most systems will depend on the frequency of occurrence of the loads within the load diagram. The BIN method would allow this to be done relatively easily by incorporating within the diagram the predicted frequency of occurrence of the heat gains within each *bin*. The corresponding frequency of occurrence of the air supply temperatures for each of these loads can then be obtained. However, the method given in the *ASHRAE Guide*[10] is to produce a *mean load line*. Here, the heat gains are multiplied by load diversity factors based on occupancy levels, use of heat generating equipment and solar heat gains. Thus data on the average levels of bright sunshine hours published by the Meteorological Office (Table 4.1) may be used to obtain summer and winter load factors, k_{s1} and k_{s2}, to be applied to the solar gains. Typical figures for the United Kingdom are 40% for summer months and 17% for winter design months but since these figures would underestimate the effect of indirect radiation, figures of $k_{s1} = 0.5$ and $k_{s2} = 0.2$ might be used.

This is illustrated in Figure 17.14 where average load lines have been

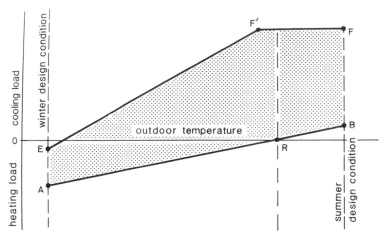

17.12 Load diagram for a zone of a building whose peak solar gain does not coincide with the outdoor design temperature

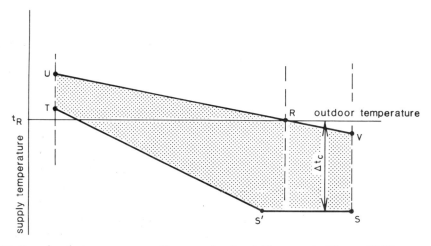

17.13 Supply air temperature diagram for load diagram in Figure 17.12

added to the load diagram of Figure 17.11. Line GH is obtained by adding the average internal gains to the transmission line:

$$q_{AG} = q_{BH} = k_i q_{BD}$$

where k_i = internal heat gain load diversity factor.

The average heat gain at the summer design condition is obtained:

$$q_K = q_H + k_{s1} q_{DF}.$$

The average heat loss at the winter design condition is obtained:

$$q_J = q_G + k_{s2} q_{CE}.$$

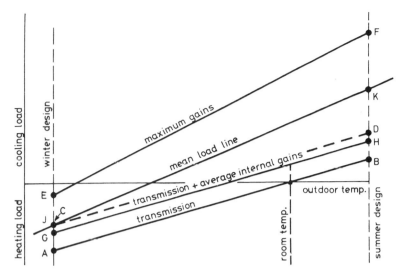

17.14 Determination of mean load line

The mean load line is arrived at by joining points **J** and **K**. Note that **JK** extends beyond the design conditions when the outdoor temperatures are more severe than the design conditions.

SCHEDULING

To improve the operating efficiency of some systems, temperatures within the plant can be scheduled to outdoor air dry-bulb temperatures. The schedules are determined from an analysis of the load and temperature diagram. The analysis can also include the calculation of the energy consumption and this is illustrated by the following example of the use of an air-to-air heat recovery unit in an 100% outdoor air system.

Example 17.6 The air conditioning system shown in Figure 17.9 is required to maintain two laboratories at 20°C for 24 hours per day and for 7 days per week. The output of the heat recovery unit is controlled by sensor T_1, reset by sensor T_2, to maintain a scheduled temperature at point B. Using the design data, determine the annual heating energy saved by using a heat recovery unit in place of a conventional preheater. (*Note*: the preheater shown in Figure 17.9 is not required for this example.)

Design data

mass air flow rate (assumed constant)   2.5 kg/s
outdoor air design condition   −4°C
climate data for Heathrow
boiler firing efficiency (average)   75%

The temperature rise across the HRU related to the outdoor dry-bulb temperature, t_O, by the equation:

$$\Delta t_{hru} = 14 - 0.7 t_O. \tag{17.9}$$

Solution

The temperature analysis is prepared on Figure 17.15. Line **STU** is the variation of minimum supply temperatures required to meet maximum heat gains of the load diagram and is the *ideal* schedule line. Line **XY** is the dry-bulb temperature of the outdoor air entering the system. Line **LNR** is the temperature after the heat recovery unit, based on the temperature rise expressed by Equation 17.6. To meet the requirements of the schedule line **STU**, the output of the heat recovery unit has to be reduced from point **N** until zero load at point **P**. The heating requirements of the HRU are then represented by the shaded area **XLNP**; (note that the plant is operating at maximum load at outdoor temperatures lower than the winter design condition). This will give the energy saving compared with a conventional preheater.

Since the mass flow rate, $\dot{m}_a$, and the humid specific heat, c_{pas}, remain constant, they can be omitted temporarily from the calculations until the numerical integration has been completed. The load variations can therefore be expressed in terms of temperature differences.

At the outdoor air conditions between $-4°C$ and $10°C$ the temperature rise would be that for a conventional heater and Equation 17.9 applies. Between $10°C$ and $14°C$ the temperature rise across the heat recovery unit will be the difference between line **UNPT** and the outdoor temperature.

Equation of line **UNPT**:

$$t_s = 19.6 - 0.35 \, t_o$$
$$\Delta t_{hru} = 19.6 - 0.35 \, t_o - t_o$$
$$\therefore \Delta t_{hru} = 19.6 - 1.35 \, t_o. \tag{17.10}$$

The numerical integration is completed according to Table 17.7. The accumulated sum of the product $(f_1 \Delta t_{hru}) = 578$.
Using Equation 17.2, the annual energy consumption is obtained:

$$E_h = \frac{H}{100} \sum \left(\frac{fQ}{\eta_b} \right)$$

$\therefore$ since $\dot{m}$ and c_{pas} and η_b are constant,

$$E_h = \frac{\dot{m} c_{pas} H}{\eta_b 100} \sum (f_1 \Delta t_{hru})$$

$$= \frac{2.5 \times 1.02 \times 8760}{0.75 \times 100} \times 578 = 172,152 \text{ kWh/annum (618 GJ)}.$$

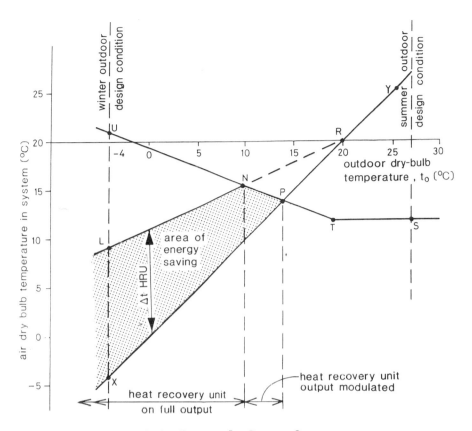

17.15 Temperature analysis diagram for System 3

range of outdoor air dry-bulb temperature (°C)	mid-point of range t_1 (°C)	frequency f_1 (%)	Δt_{hru}	$f_1 \Delta t_{hru}$	notes
< -4.1	-	0.33	16.8	5.5	design load
-4.0 to -1.1	-2.5	1.64	15.8	25.8	HRU at full output;
-1.0 to 1.9	0.5	6.35	13.7	89.1	use Equation 17.9
2.0 to 4.9	3.5	11.31	11.6	130.6	$\Delta t_{hru} = 14 - 0.7t_0$
5.0 to 7.9	6.5	16.40	9.4	155.0	
8.0 to 9.9	9.0	12.22	7.7	94.1	
10.0 to 11.9	11.0	11.84	4.6	55.1	HRU at reduced load;
12.0 to 13.9	13.0	13.10	1.8	22.9	use Equation 17.10 $\Delta t_{hru} = 19.6 - 1.35t_0$

$$\Sigma (f_1 \Delta t_{hru}) = \underline{578.1}$$

Table 17.6 Calculation of average hourly energy saving of heat recovery unit in Sstem 3 (Example 17.6)
Note: **design load is that of a conventional heating coil**

FAN AND PUMP ENERGY CONSUMPTION

The annual cost of electrical energy for a fan or a pump driven by an electric motor at constant speed and supplying a constant volume flow rate is calculated from the equation:

$$E_f = \frac{\dot{V} P_{tF} H}{\eta_F 1000} \qquad (17.11)$$

where suffix F denotes fan or pump.

The SI units used to specify fan and pump duties will not always be identical. Those applying to Equation 17.11 would be:

volume flow rate, $\dot{V} - m^3/s$
pressure, $P_{tF1} - Pa$.

The BIN method described for heating and cooling plant can be used to calculate the energy consumption of variable duty fans and pumps. An example of the analysis of a VAV system is given in the *ASHRAE Handbook*.[3]

SYMBOLS

COP	coefficient of performance	q	room heat gain/loss
c_{pas}	specific heat of humid air	Q	heat flow rate
E_c	annual energy consumption, cooler	$\dot{m}_a$	air mass flow rate
E_F	annual energy comsumptions of fan or pump	P_{tF}	fan total pressure
		t	dry-bulb temperature
		$\dot{V}$	volume flow rate
E_h	annual energy consumption, heater	x	fraction of outdoor air
		Δh	enthalpy difference
f_1	frequency of occurrence of outdoor air dry-bulb temperature	Δt	temperature difference
		η_b	boiler firing efficiency
		η_F	fan efficiency
f_3	frequency of occurrence of outdoor air enthalpy	**Abbreviations**	
H	annual hours of plant operation	HRU	heat recovery unit
		p	preheater
H_{eq}	equivalent annual hours of plant operation	pd	preheater design
h	specific enthalpy	**Subscripts**	
k_i	internal heat gain load diversity factor	B	cooler off-coil condition
		O	outdoor air condition
k_{s1}	solar heat gain load factor, summer	M	mixed air condition
		R	room air condition
		S	supply air condition
k_{s2}	solar heat gain load factor, winter	d	design load
		p	preheater

18 Commissioning, Operation and Maintenance

COMMISSIONING

As air conditioning systems and other building engineering services have increased in size and complexity, it is important that not only do they work but that they operate efficiently. To bring the installed systems into full working order they have to be *commissioned*. Commissioning in the context of building engineering services is a term used in the United Kingdom for the activities that are involved in bringing the systems into use and ensuring that the systems comply with the specification before they are handed over to the client. The *ASHRAE Handbook*[1] uses the term *Testing, Adjusting and Balancing* for these activities.

The CIBSE Codes[2] give the following definitions for commissioning:

The advancement of an installation from the stage of static completion to full working order to the specified requirements.

Commissioning includes the setting-to-work and regulation of an installation:

(a) *Setting-to-work* The process of adjusting the system to the specified tolerances.
(b) *Regulation* The process of adjusting the rates of fluid flow in a distribution system within specified tolerances.

Commissioning is deemed complete at the *conclusion of regulation*.

PROCEDURES

Commissioning should be viewed as a continuous process during the project cycle. The interdependence of the various stages in the project cycle and the requirements of the building and other services should be recognized and the work planned accordingly. In many ways commissioning can be considered as a management process in that it requires good organization, documentation and definition of responsibilities for each part of the work; much of this is related to Quality Assurance.[3]

Notes appropriate to each stage of the project cycle are given below, in order that some of the technical aspects can be seen in perspective, the emphasis having been placed on the procedures for the air handling side of air conditioning. The approach for other services associated with air conditioning, eg boiler plant, automatic controls, refrigeration systems and water distribution, are given in the appropriate Commissioning Codes.

The system(s) has to be set-to-work, balanced and tested before the installation is accepted by the client as complete and satisfactory, but it should be noted that testing could be carried out during the guarantee period, after the plant has been handed over.

Design requirements

A system which operates to the design requirements depends on sound design. The commissioning of an installation will not make a poor design operate efficiently, but will emphasize any shortcoming and failures. More specifically, the system must be designed for the procedures to be adopted. This means that the facilities for measuring and regulating the variables have to be incorporated on the drawings and in the specifications, to ensure that the work entailed is carried out efficiently and effectively. The ideal will be a self-balancing system, so that when the system is started up, each outlet or unit will deliver its design output. This approach can be more expensive than a system which requires manual adjustment, but the additional cost could well be offset by the reduced time required for setting the systems to work, the benefit of long term stability and reduced maintenance problems. Self-balancing systems are not always a practical solution, but there is scope in the arrangement of duct and piping layouts to ensure easier regulation of the flow rates to achieve the system balance required by the design.

Decisions will often have to be made at the conceptual stage of a project so that the detailing of the design can proceed. The following is a list of the more important provisions required in the design; some of these facilities will be those required for maintenance:

- measurement stations for system variables, eg flow rates, pressures and temperatures;
- instrumentation for balancing, testing performance and efficient operation;
- regulating devices, eg dampers and valves;
- access and inspection openings;
- filling, drainage, air venting and flushing of water circuits;
- electrical provisions for convenience, safety and isolation of sub-systems.

The CIBSE Codes include appendices which provide a guide to the design implications of each service. An understanding of the proportional balancing methods described in Chapter 15 is essential to the correct provision of the design requirements.

Construction and installation

Having then completed the design drawings and specifications, tenders are invited and a contract will be placed for the execution of the work. No matter how complete the design and specification may appear to be, there will always be on-site problems. Adequate supervision is essential to ensure correct installation and good liaison between the designer's office, the contractor's office and site. The main contractor and specialist sub-contractors should act together as a team so that the work can be adequately planned and executed. Some members of the design team can be members of the construction team to ensure the correct interpretation of the design.

Supervisors representing the design engineer and contractor should be appointed and they must be supplied with the appropriate design information. To ensure that all plants are adequately constructed and installed, check lists are used to ensure systematic inspections. A typical list for a plant item is shown in Figure 18.1.

Any faults which are found can be summarized (a *snagging list*) for action by the contractor and a later check made to verify that they have been rectified. It is essential that these inspections are carried out early enough, since, for example, it would be impossible to inspect adequately ducts covered with builder's work or hidden by other services. It may also be necessary to witness tests on certain plant items at the factory, before delivery to site. There should be adequate physical protection of plant which has to be installed well before being made operational. In these ways a smooth transfer to the stage of setting the plant to work can be achieved with a minimum of backtracking.

Setting the plant to work

Pre start-up checks will be required to ensure that the plant is complete, clean and safe to operate. These checks are listed in the various codes and are usually in addition to those inspections carried out during installation. Once the plant has been found to be physically correct and to have the necessary supply services available it should be systematically started up and balanced. The procedure would be tailored to suit individual projects but a typical approach for an air conditioning system may be as follows:

- water circuits made operational, including initial setting-to-work of the heating and cooling plants;
- inspection of plant under running conditions;
- air circuit balancing;
- water circuit balancing;
- final checks on boiler and refrigeration plants;
- controls calibration.

It will not be possible to complete each stage before moving on to the next. There will naturally be a good deal of overlapping between the various phases and it will take experience and good planning to ensure a satisfactory sequence of events.

CHECKS ON THE BUILDING	Doors and windows fitted	☐
	Suspended ceilings fitted	☐
	Recirculation and air transfer openings — correct size and position	☐
	Structure not interfering with access to dampers and terminals	☐
	Leakage test completed on builder's work shafts and other structural air channels and plenums	☐
	Keys available for all locked rooms — including the plant room	☐
CHECKS ON THE SYSTEM	Power supply to fan motors and other electrical equipment	☐
	All main plant components fitted	☐
	Drain seals fitted	☐
	Clean filter medium in place	☐
	Fan rotation correct	☐
	Plant and ducting clear of debris	☐
	Air intake and mixing chamber clear	☐
	Control systems, particularly authomatic dampers, operable	☐
	Ducts adequately sealed to builder's work shafts and terminals	☐
	Access doors, inspection covers and test hole plugs fitted	☐
	No obvious gaps in ductwork, eg flexible connections left out or duct-mounted components missing	☐
	Regulating dampers working correctly	☐
	Duct leakage tests completed	☐
	All terminals in correct positions	☐
SETTING UP FOR REGULATION	All regulating dampers open	☐
	All fire dampers open	☐
	Automatic control dampers set for full fresh air or full recirculation	☐
	Grille louvres set square to the face	☐
	Ceiling diffuser cones set for full downwards discharge	☐
	Supply and extract plants running at the same time	☐
	Windows and doors in the building closed	☐

18.1 Basic pre-regulation check list[4]

Water circuits operational

The commissioning of boiler plant and heating circuits is a priority if warm air is required during the cold months of the year. To have chilled water available while the air circuits are being balanced will not normally be necessary, though the commissioning of the refrigeration plant can also commence at this stage.

Flushing the piped water systems through to ensure cleanliness is of utmost importance to the eventual satisfactory performance of the system.

Inspection of plant under running conditions

Check lists should be available similar to those required for the static inspections. A typical list covering several components in a system is shown in Figure 18.2. It may be necessary to bring in the manufacturer for specialist plant items, but the *system* approach is desirable in order that the *performance as a whole* can be considered.

Details of fan and air handling unit

PERFORMANCE		units	design	measured	comments
volume flow rate					
fan speed					
motor current					
static pressure	upstream				
	downstream				
pressure drop across filter					condition of filter when measured

SUPPLEMENTARY DATA	units	design	measured	comments
motor speed				
air temperature				
pressure drop across other plant items				
1				
2				
3				

18.2 Plant performance summary check list[5]

Air and water circuit balancing

The balancing procedures for air and water systems are described in detail in Chapter 15. The regulation of water distribution systems by thermal methods is excluded from the CIBSE Code as it is considered to be unreliable and wasteful of commissioning time (as in the case of systems with a close approach of air and water temperatures, such as chilled water systems). But, if the water circuits are associated with heaters in an air conditioning system, then the measurements can sometimes be simplified by using the temperature rise across the heaters to regulate the water flow rates. When this course is adopted it implies that design air flow rates should first be obtained by balancing the air flow circuits. To obtain full output from the heater control, valves must be fully open and this can be achieved by a manual adjustment of the thermostat setting.

Automatic controls

Automatic controls are covered by the CIBSE Commissioning Code (series C). This code is of a general nature and reference to the specialist manuals prepared by the controls equipment manufacturer is recommended.

Though some pre start-up checks can be made prior to setting-to-work the remainder of the system, final calibration of the control systems can only be completed after the regulation of the air, water, boiler and refrigeration systems.

It is probable that final tuning to obtain maximum efficiency is possible only after the building has been occupied and the building and system have had teething troubles rectified. The client and/or building occupier must be aware of this when contracts are placed and responsibilities defined.

ORGANIZATION OF COMMISSIONING

Commissioning as defined by the CIBSE includes the setting-to-work and regulation of the flows and the calibration of the controls. Setting-to-work is the process of bringing a static system into motion, and regulation is the process of adjusting the rates of fluid flow in a distribution system within specified tolerances required by the design. Commissioning does not replace the inspections required during the installation of the plant, nor is it an optional extra. Adequate documentation and design information must be available to the personnel responsible for this section of the work. Records of the measurements and adjustments to the system must be kept at the time they are made, as commissioning is often an intermittent operation. These records will show the state of progress at any stage. Records are also necessary to provide a base for comparing the state of the system during maintenance.

Sufficient time must be included in the contract programme to allow for satisfactory completion. If the construction has been delayed it is not advisable to try to make up for the lost time by reducing the time allowed; even partial occupation of a building can cause problems.

For successful commissioning it is important that the responsibility of each engineer involved in the project is clearly defined and incorporated in the appropriate contracts and terms of appointment. The responsibility for deciding by whom each section of the work should be done, generally rests with the design engineer. Several alternatives are available, the most appropriate of which should be selected for each project. These include the following:

Design and site engineers

The designer will decide how variables are to be measured and controlled so that the necessary facilities are built into the installation. The site engineer will be responsible for inspections, setting the plant to work, and regulating the flow circuits. The designer, or the client's representative, will then verify that the installation is correct and satisfies the design intent, either by examining records or making spot checks. The advantages of this arrangement are that the designer and site engineer, being intimately connected with the installation, will know the location of all equipment and controls. The designer will know precisely what is expected of the plant and will be able to recognize and quickly correct any faults which occur.

Manufacturers

It is common practice, at the present time, for suppliers of specialized equipment such as controls to commission their own apparatus. This idea could be extended to cover practically all equipment but to be effective, co-operation, goodwill and co-ordination between all parties is essential.

Commissioning teams

Some organizations maintain their own commissioning teams who carry out all the commissioning and performance testing on plant items, even if it was designed and installed by an outside contractor. Independent, commercial commissioning teams are gaining acceptance in many countries. The advantages with such an organization are that the testing and commissioning can be done quickly and efficiently, with all the necessary instruments to hand, and that design-in-use surveys can be carried out. The design and site engineers will be freed earlier to concentrate on their own specialities. At the design stage the expert commissioning engineers will be able to give prompt and up-to-date advice on all provisions to be made for the measurement and regulation of variables. There is also a move today to employ commissioning consultants to advise the designers and oversee site work to achieve best management of the activities involved in this area of work.

PERFORMANCE TESTING

The commissioning process should ensure that the plant is set to work in a methodical manner and balanced to the design engineer's requirements. That is, water, heating, refrigeration and electrical circuits are made operational, and equipment is started and inspected under running conditions. Air volume flow rates are regulated to ensure specified output for heaters, coolers and humidifiers. The controls are finally calibrated to give design conditions in the building.

This balancing process in itself will not show that the plant is capable of operating satisfactorily throughout the year. To do this a certain number of performance tests should be applied to the plant. It is the maintenance of the space conditions that counts finally – balancing is only a step towards this end.

The main problem encountered in checking the plant will satisfy the building requirements at the summer and winter design conditions is that, when the plant is set-to-work it will rarely be operating at these conditions. In fact outdoor design conditions may only occur, say, once in ten years. Any tests must be carried out at off-peak conditions and a prediction made from the test results of the ability of the plant to maintain the indoor conditions when the outdoor design conditions are reached. Other problems are, that immediately after the commissioning process, the building has not had time to *settle down* and to dry out; systems may not be subject to normal occupancy loads and the client will want to occupy the building with a minimum of delay.

It is important to distinguish between two types of tests associated with plant installed in a building. One of these is to ascertain whether the plant delivers its rated output. The second type of test is to ascertain whether the installed plant will satisfy the building's requirements. The first of these is a check on the manufacture's data, with the plant item integrated with other components; the second is to check on the designer's own calculations.

At present there is no agreement within the building services industry as to what tests, if any, should be adopted. The CIBSE Commissioning Code defines *testing* as:

The evaluation of the performance of a commissioned installation.

Testing is not covered by the Codes and is a separate consideration. Though some testing is inherent in the commissioning process, no attempt has yet been made to produce documentation for system testing as defined above. BS 5720: 1979[6] recommends that performance testing should be carried out, listing some appropriate tests. These tests are similar to those carried out on the engineering systems for an experimental hospital ward[7] and are the basis of the approach described below.

Test method

In general, to devise a suitable test, it is necessary to identify the variables upon which the output of the piece of equipment depends to obtain a simple mathematical relationship. If possible a linear relationship should be used and a statistical linear regression line calculated from the results. This will be of the form:

$$y = a + bx \qquad (18.1)$$

where y = the dependent variable
x = the independent variable
b = the slope of the line
a = the intercept on the y axis

Calculation of the regression line will usually be necessary, due to the scatter of the test results. The standard error (SE) of the estimate of y can also be calculated to enable confidence limits to be placed on the results.

In analyzing the overall performance it is necessary to break down the system into component parts. Consider the system shown in Figure 18.3.

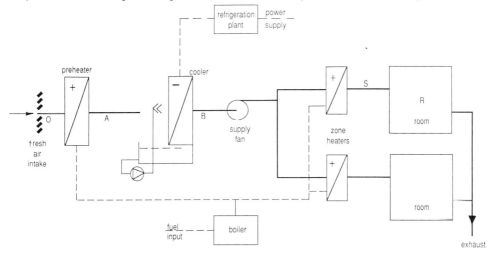

18.3 System to illustrate test requirements

The preheater and zone heaters are supplied from a hot water boiler, the cooler by a refrigeration plant and to investigate the complete system the following sub-systems would be considered:

– boiler capacity to satisfy total humidifying and heating loads;
– zone heater capacity to meet room heating loads;
– plant cooling capacity to meet room cooling loads;
– preheat and cooling section humidifying efficiency and cooling capacity to meet relative humidity requirements;
– refrigeration capacity to satisfy the total air conditioning cooling load.

Before testing commences checks should be made that plant items function correctly; a faulty control valve or ghost circulation,* for

*See Appendix, page 396.

instance, could invalidate a test. Instruments must, likewise, be accurate and calibrated where necessary. The technique is illustrated by considering the boiler and reheater loads.

Boiler capacity and total heating load

Reference to a typical psychrometric diagram (Figure 18.4) shows that the main independent variable, x, on which the heating load depends, is the difference between the supply and the outdoor air enthalpies. As the supply condition is relatively close to the room condition, an approximation for x could be achieved by using the difference between the room air and the outdoor air enthalpies. If the room condition is considered to be constant then the independent variable can be taken as the outdoor air enthalpy. The use of this approximation makes data collection relatively simple.

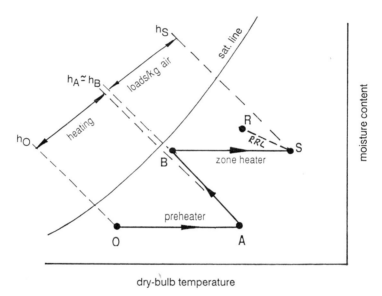

18.4 System psychrometric process in winter

The dependant variable, y, is the fuel consumption per unit time. To obtain valid data it is necessary to obtain time-averaged values of the outdoor air enthalpy over daily periods for two or three weeks during winter and spring/autumn; these results are plotted as in Figure 18.5. The regression line is calculated and extrapolated to the outdoor air design enthalpy condition at **A**. If this intersection is below the maximum capacity (obtained from a full load plant test) at **B** then the boiler is shown to have adequate capacity.

The confidence limits represent the band within which the boiler load will lie on 95% of days of a given outdoor air enthalpy. The band is due to effects of miscellaneous heat gains and to experimental error.

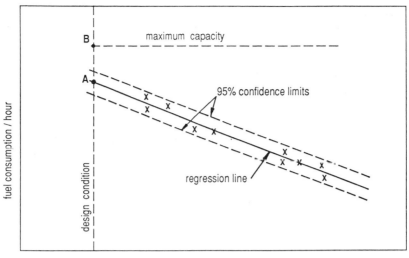

18.5 **Boiler capacity and heating load**

Zone heating and cooling load

For a heating system, the conductive heat loss, q_s', from a room is evaluated from the equation:

$$q_s' = \Sigma(UA)(t_R - t_O) \tag{18.2}$$

and the heat input, Q, to an air heated room is approximately given by:

$$Q = \rho \dot{V} c_{pas}(t_S - t_R). \tag{18.3}$$

For a heat balance $Q = q_s'$. With the supply of air volume maintained at a constant level and A, U, ρ and c_{pas} constant, equating Equations 18.2 and 18.3:

$$(t_S - t_R) = C(t_R - t_O)$$
$$t_S = (C + 1)t_R - Ct_O \tag{18.4}$$

where C is a constant $= \Sigma(UA)/\rho(\dot{V}c_{pas})$.

If the controlled room temperature t_R is also considered to be constant then t_O can be used for the independent variable and t_S the dependent variable. Measurements are made of the supply and outdoor air temperatures over a number of daily periods, each of which is time-averaged to minimize the effect of the building's thermal lag. A typical result is plotted in Figure 18.6. The intersection of the regression line AG with the winter design temperature (point A) is below the full-load output of the heater (point C). This result indicates that the heating capacity of the zone heater in relation to the building is satisfactory. It is also within the 95 per cent confidence limit line (point B), indicating that the heater will maintain the design temperature on 95% of the days of external design conditions.

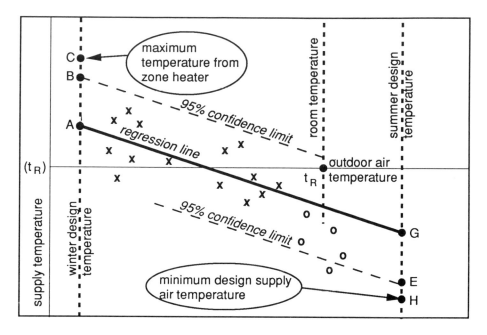

x test points, time averaged over daily period
o test points, time averaged over 1 hour during periods of continuous sunshine

18.6 Zone heating and cooling requirements

To investigate maximum zone cooling requirements, lines are extended to cut the summer design condition line. Individual conditions are plotted for days with continuous sunshine, these test results based on, say, one-hour periods with an appropriate allowance made for the thermal lag of the building. The results lie within the 95% confidence limit line which intercepts the summer design condition above the design supply temperature (point H). It can be inferred that the air cooling capacity is satisfactory.

It will be observed that the plot of supply temperatures is similar to those obtained from theoretical load diagrams described on page 372f. Since load and temperature diagrams are used for design analysis then it should be possible to verify them from test measurements.

Discussion
If plants are to perform as required by the designer and client, and if future designs are to be improved, it is desirable that tests are carried out

under actual operating conditions, usually with the building occupied. This will undoubtedly raise a number of contractual problems which need to be resolved within the guarantee or faults liability period.

Tests should be kept relatively simple, since any testing which is carried out as part of the designer's (or installer's) contract must also be simple and inexpensive. With some care, it is possible that the maintenance staff could take the actual readings using prepared data sheets.

Other methods of test have been suggested and used from time to time, eg heaters can be installed in the spaces to load-up the cooling system; a preheater can produce a load for the cooling coil. Control settings can be altered to obtain design temperature differences, when the external conditions are not at the design values. Though there might be some advantages in these methods it still has to be demonstrated that the plant will operate satisfactorily under actual running conditions, and artificial loading of the plant/building may not be able to do this.

The ability of the plant to meet design conditions during its life will depend on adequate maintenance. If regular tests are carried out as part of a planned maintenance scheme, failures and deteriorations can be detected by comparing the results with the original tests. Appropriate remedial action can then be taken.

The tests described here are indicative only of the problem of verifying the performance of air-conditioning systems. Other system types, and buildings which are intermittently heated, would have to be treated differently. It is envisaged that with more sophisticated instrumentation, data logging and computer techniques using mathematical models of the system performance, testing will attain far greater accuracy in the future. These will go hand-in-hand with optimization of the control settings. Testing of other aspects of system performance (conforming to relevant Standards and Codes of Practice), will also be required. These include internal space conditions, air diffusion patterns, noise and sound control.

OPERATION AND MAINTENANCE

Hand-over and documentation

Designers and contractors can help to ensure adequate maintenance by stressing its importance to the client. The client should be advised about the personnel required and provided with the requisite information for maintenance.

For plant to continue working properly throughout its life it is necessary for the maintenance and operating staff to be familiar with the principles and methods of running the plant. In addition to a full set of record drawings, two manuals should be available, one for the operating staff (eg office manager, caretaker) and another for the maintenance staff. The former should describe such things as the design principles, the method of operation, details of alarms and safety precautions. The latter, larger service manual, should contain the information listed by BS 5720:1979[8] and reproduced below.

List of documentation required for system maintenance engineer

---

- the designer's description of the installation, including simplified line flow and balance diagrams for the complete installation;
- *as fitted* installation drawings and the designer's operational instructions;
- operation and maintenance instructions for equipment, manufacturer's spare parts lists and spares ordering instructions;
- schedules of electrical equipment;
- schedules of mechanical equipment;
- test results and test certificates as called for under the contract, including any insurance or statutory inspection authority certificate;
- copies of guarantee certificates for plant and equipment;
- list of keys, tools and spare parts that are handed over.

British Standard 5720:1979 provides further advice regarding the organization and content of maintenance manuals. These should be available in draft form for checking at the commissioning stage, in addition to *as fitted* drawings. This will assist those concerned in setting the plant to work efficiently and, at the same time, the manual can be revised to suit any operational changes that may have been necessary, before they are issued to the client.

Before the plant is handed over, it is necessary that the staff responsible for operating the plant are given verbal instruction and demonstrations on the principles and operation of the systems. Any such verbal instructions should be in addition to the documentation.

Design for maintenance

As with commissioning, decisions will often have to be made at the conceptual stage of the project, in order that the installed system can be operated and maintained efficiently and safely. The following is a list of some of the more important aspects to be considered.

Plant rooms Of prime importance is the provision of adequate building space for the installation of equipment and for its regular servicing as well as for the possible removal, replacement and repair of major plant items. BURBERRY[10] gives guidance on the size of plant rooms. Plant rooms should be adequately heated and ventilated.

Lifting and strong points These should be provided, where necessary, to allow dismantling and assembling and possible removal and replacement of major plant items.

Access   Reasonable access is required to all major plant items and most distribution ducts. This provision is required both in the building work chambers, ducts and false ceilings, as well as in the plant and ductwork system themselves. *Access drawings*, akin to builders work drawings, are useful for highlighting this important aspect of maintenance. Such drawings also identify access *routes* through the building; these routes are for delivery of large plant items at the installation stage and for the subsequent removal and replacement of equipment.

Instrumentation to test the systems for efficient operation certain instruments will be needed. Such instruments must be certified and tested for accuracy.

Standby plant   Care is needed in providing the correct amount of standby plant to ensure that the system as a whole continues to operate, when required, in the event of a component failure. An over-generous provision means incurring unnecessary capital cost, with plant standing idle and depreciating in value. Partial standby can often be achieved by selecting plant items having more than one stage to meet the total capacity.

Spares   Spare parts for essential repairs, together with replacement items such as clean filter cells, are required.

Safety features and isolation of sub-systems   Systems should be easy to operate, understandable and safe. Adequate provision is required to minimize fire risk and to isolate electrical circuits. Various interlocks and alarms will be required with boilers, pressure vessels, refrigeration plant and fans. The requirements to protect personnel and equipment will follow normal design practice, eg vents, isolating valves, pressure switches and safety guards. These requirements are contained in the appropriate national and international standards.

Materials   The materials of each item of plant, duct, pipe, insulation and so on, must be compatible with, and suitable for, the conditions under which the system will be operating.

Maintenance organization

Once the client has accepted an installation from the contractor, the maintenance of a plant of any complexity should be organized to ensure continued efficient operation of the plant, aiming to protect the capital investment at a minimum economic cost.

The basis of any planned maintenance scheme is a filing system whereby the checks and services to be carried out on any piece of plant come to light at the appropriate time. This can be done by a card index system or by using a computer. When staff complete a piece of maintenance they ought to note down anything which they see to be in need of attention or likely to be in need of attention in the near future, such as a bearing running hot which would eventually fail. This enables the repair to be carried out at a convenient time, rather than in a period when everything seems to fail at once.

Contract maintenance by specialist firms is often used as an alternative to directly employed labour, either for a part or the whole of the service.

Frequency of servicing

Routine maintenance includes inspections, cleaning, water treatment, adjustment and overhaul. The frequency at which these should be made is normally given in the manufacturers' manuals but these are average values which are best modified by the actual site conditions and in the light of operating experience.

The frequency at which plant is serviced depends on the following:

- plant and system efficiency and hence efficient energy consumption;
- effect on reliability of service;
- routine maintenance costs;
- fault repair costs;
- safety inspection;
- hours of system operation.

As part of the routine maintenance inspections, standby and emergency plant must be checked but not necessarily brought on-line for long operating periods.

Following shut-downs for repairs and plant modifications it may be necessary to re-commission the system or part thereof, in which case the appropriate procedures should be followed.

It is particularly important to include insurance inspections of pressure vessels and the testing of fire alarms in routine maintenance.

Fault-finding

Fault-finding procedures may be included in the service manual. Though these procedures for individual plant items will often be available from manufacturers, they should also relate to the system in which the plant item is placed. Some guidance to these procedures have been produced by HAWES[11] and by RUDYARD.[12]

Maintenance support

In support of maintenance, it is recommended that consideration be given to the following:

- engineer's office;
- workshop with appropriate tools;
- equipment spare parts;
- maintenance materials;
- instruments;
- site tools.

Operating efficiency

An important objective of maintenance is to operate the systems as efficiently as possible after successful commissioning.

In this respect there is an overlap between commissioning and maintenance. It is not always possible to achieve full system efficiency prior to the occupation of the building and it is probable that maximum efficiency can be achieved only with the building in normal use. The building use may change after first occupancy and this may affect the optimum plant operation.

If the system is not operating as efficiently as it should, or if economies need to be made, an energy audit of the plant may be undertaken, which would indicate where to place most emphasis to achieve energy savings. For normal performance monitoring it is necessary that maintenance staff have as good an understanding as possible of the overall and detailed system design.

SYMBOLS

A	area of building fabric	U	thermal transmittance coefficient
a	slope of regression line		
b	intercept of regression line	$\dot{V}$	air volume flow rate
C	global constant (Equation 18.4)	x	independent variable
c_{pas}	specific heat of humid air	y	dependent variable
h	enthalpy	ρ	air density
$\dot{m}_a$	mass flow rate of dry air	**Subscripts (for temperature)**	
Q	heater battery duty	A, B plant conditions	
q_s'	sensible heat loss from air conditioned space	O	outdoor air condition
		R	room condition
t	dry-bulb temperature	S	supply air condition

Appendix

GHOST CIRCULATION IN PIPEWORK SYSTEMS

Incorrect piping connections in a pumped circuit can cause unnecessary heating and cooling loads, resulting in reduced efficiency. Consider the example of a preheater followed by a cooler which are controlled in sequence by a thermostat, as shown in Figure 18.7. When the cooler is in operation, the heater should not provide heat; if it does operate due to a fault in some part of the system then additional cooling must be supplied to maintain condition B. Hence both heating and cooling circuits are working inefficiently, even though the design conditions are being maintained, ie by the controller B.

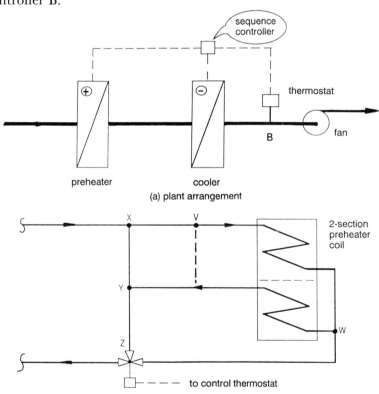

(a) plant arrangement

(b) piping arrangement of preheater

18.7 Ghost circulation through a two section air heater battery

The ghost circulation is caused in the two section heater in the following way. The three-way control valve is closed to heater and open to bypass. Flow is therefore along **XYZ** causing a pressure difference between **X** and **Y** and hence circulation along the path **XVWY**. The correct piping connection should be made at point **V** instead of **Y**.

These unwanted circulations may be recognized through appropriate checks during inspection of plant under running conditions.

References

Chapter 1

[1] EDEN, P, Independent Radio News; personal communication, 1989. (Pressures adjusted to equivalent sea level pressures)
[2] National Engineering Laboratory, *Steam Tables*, HMSO, 1964
[3] JONES, WP, *Air Conditioning Engineering*, 3rd edition, Edward Arnold, 1985, Chapter 2
[4] CIBSE, *Guide Book C*, 1986, Chapter 1

Chapter 3

[1] BURBERRY, R, *Environment and Services*, 6th edition, Batsford, 1988, p 83
[2] BEDFORD, T, *The warmth factor in comfort at work*, HMSO, 1936
[3] HICKISH, DE, Thermal sensations of workers in light industry in summer: a field study in Southern England, *Journal of Hygiene*, 1955, Vol 53, p 11
[4] FANGER, PO, *Thermal Comfort*, McGraw-Hill, 1973
[5] CROOME-GALE, DJ and BM ROBERTS, *Air Conditioning and Ventilation of Buildings*, Pergamon, 1981
[6] BEDFORD, T, 'Requirements for satisfactory heating and ventilating', *Annual Occupational Hygiene*, 1960, Vol 1, p 172
[7] BEDFORD, T, *Basic Principles of Ventilation and Heating*, Lewis, 1984, p 132
[8] SEELEY, LJ, 'Change in temperature of air in nasal cavity', *American Society of Heating and Ventilating Engineers Transactions*, 1940, Vol 46, p 285
[9] ASHRAE, *Applications Handbook*, 1982, Chapter 12
[10] LEGG, RC, 'Analysis of energy demands of air conditioning systems', *Heating and Ventilating Engineer*, 1977, Vol 51, p 6
[11] CIBSE, *Guide Book A*, 1986, p A1 6
[12] BEDFORD, T, 'Requirements for satisfactory heating and ventilating', *Annual Occupational Hygiene*, 1960, Vol 2, p 167
[13] ASHRAE, *Fundamentals Handbook*, (SI edition), 1985, p 8.19.

Chapter 4

[1] Meteorological Office, *Tables of Temperature, Relative Humidity, Precipitation and Sunshine for the World*, HMSO, 1973
[2] CIBSE, *Guide Book A*, Section A2, (Weather and solar data), 1986, pp A2–18
[3] FOWLER, KJV, *Practical Building Services Design*, Vol 1 *Inside/outside design conditions*, Godwin, 1983
[4] METEOROLOGICAL OFFICE (UK), *Climatological Memoranda*, Stations with percentage frequencies of hourly values of dry-bulb temperature and associated wet-bulb temperature. Various dates
[5] COLLINGBOURNE, RH and RC LEGG, 'The frequency of occurrence of hourly values of outside air conditions in the United Kingdom', IoEE Technical Memorandum No 58, South Bank Polytechnic, 1979
[6] CIBSE, op.cit., p A2–19
[7] ASHRAE, *Fundamentals Handbook*, (SI edition), 1985, Chapter 24
[8] JONES, WP, *Air Conditioning Engineering*, 3rd edition, Arnold, 1985, p 123.

Chapter 5

1. BURBERRY, P, *Environment and Services*, 6th edition, Batsford, Chapter 5, (Heating), 1991
2. CIBSE, *Guide Book A*, Section A9, (Estimation of plant capacity), Chartered Institute of Building Services Engineers, 1986
3. ASHRAE, *Fundamentals Handbook* (SI edition), Chapter 26, (Air-conditioning cooling load), 1985
4. Carrier Air Conditioning Co, *Handbook of Air Conditioning Design*, McGraw-Hill, 1965, Chapter 2
5. CIBSE, *Guide Book B*, op.cit., Section A5
6. HARDY, AC and MITCHELL, HG, 'Building a climate: the Wallsend project', *JIHVE*, 1970, Vol 38, p 71
7. CIBSE, *Guide Book B*, Section B3, Chartered Institute of Building Services Engineers, 1970, p B3–4
8. ASHRAE, *Environmental Control Principles*, 1972
9. Department of Health and Social Security (UK), *Ventilation of Operating Departments – A Design Guide*, 1983.

Chapter 8

1. OUGHTON, RJ, 'Legionnaires' disease in refrigeration and associated equipment', *Proceedings, Institute of Refrigeration*, 1986/7, Vol 83, p 84
2. ANON, 'Polished water fights Legionnaires' disease', *H and V Engineer*, Vol 62, No 697, p 5.

Chapter 11

1. BS 6540: 1985, *Air filters used in air conditioning and ventilation, Part 1: Methods of test for atmospheric dust spot efficiency and synthetic dust arrestance*, British Standards Institution
2. ASHRAE 52-76, *Methods of testing air-cleaning devices used in general ventilation for removing particulate matter*
3. EUROVENT 4/5, *Method of testing air filters used in general ventilation*, 2nd edition, HEVAC Association, 1980
4. BS 3928: 1969, *Method for sodium flame test for air filters*, British Standards Institution.
5. Eurovent 4/4, *Sodium chloride aerosol test for filters using flame photometry technique*, HEVAC Association, 1980.

Chapter 12

1. CIBSE, *Guide Book C*, Section C2 (Properties of water and steam), 1986
2. MILLER, DS, *Internal Flow Systems*, BHRA Fluid Engineering, 1990
3. BS 848: PART 1: 1980, *Fans for general purposes. Part 1: Methods of testing performance*
4. BS 1042: PART 2A: 1973, *Methods of Measurement of Fluid Flow in Pipes, Part 2 Pitot Tubes: 2A class A accuracy*, British Standards Institution.

Chapter 13

1. CIBSE, *Guide Book C*, Section 4, (Flow of fluids), 1986
2. MILLER, DS, *Internal Flow Systems*, BHRA Fluid Engineering, 1990
3. ASHRAE, *Fundamentals Handbook* (SI edition), 1985, Chapter 33
4. HEATING AND VENTILATING CONTRACTORS ASSOCIATION, *Specification for Sheet Metal Duct Work: low, medium and high pressure/velocity air systems*, 1982.

Chapter 14

1. Fan application Guide, HEVAC Association (UK), 1981
2. BS 848: Part 1: 1980, *Fans for general purposes. Part 1: Methods of testing performance*, British Standards Institution.

Chapter 15

[1] HARRISON, E and NC GIBBARD, 'Balancing air flow in ventilating duct systems', *Journal of Institute of Heating and Ventilating Engineers*, Vol 33, 1965, p 201

[2] CIBSE Commissioning Code: Series A, *Air Distribution*, Chartered Institution of Building Services Engineers, 1971

[3] CIBSE Commissioning Code: Series W, *Water Distribution*, Chartered Institution of Building Services Engineers, 1989

[4] MA, MYL, 'The averaging pressure tubes flowmeter for the measurement of air flow in ventilating ducts and the balancing of air flow circuits in ventilating systems'. *Journal of Institute of Heating and Ventilating Engineers*, Vol 34, 1967, p 327

[5] HOLMES, MJ, 'Air systems balancing regulating variable flowrate systems'. *CIBSE symposium: Testing and Commissioning of Building Services Installation*, 1977

[6] LEGG, RC, 'The measurement of air flow at the face of a grille', *Building Services Engineer*, Vol 47, 1976, p 273

[7] CAMPBELL, J and DJ NEWSON, 'Flow Balancing in Water Systems'. CIBSE Symposium, *Testing and Commissioning of Building Services Installations*, 1977

[8] BS 1042 Section 1.1: 1981, *Methods of Measurement of Fluid Flow in Closed Conduits, Part 1 Pressure Differential Devices*, British Standards Institution

[9] BSRIA, *Manual for Regulating Water Systems*, Application Guide 1/79, 1979.

Chapter 16

[1] HITTLE, DC and DL JOHNSON, 'Performance measurement and system simulation applied to the design of standard air-conditioning control systems', Symposium: *Performance of HVAC systems and controls in buildings*; Building Research Establishment, UK, June 1984

[2] INT-HOUT, D, 'Measurement of airflows in air-handling systems', *ASHRAE Transactions*, 1985, Vol 91, p 1116

[3] KAO, JL and ET PIERCE, 'Sensor errors – effect on energy consumption', *ASHRAE Journal*, Dec, 1983, Vol 25, p 42

[4] ALYEA, HW and NJ JANISSE, 'Matching damper to system by damper characteristic', *Air Conditioning, Heating and Ventilating*, Feb, 1964, Vol 44

[5] STOLA, L, 'Energy conservation at Courtaulds', IChemE Symposium series No 48: *Energy in the 80s*, 1977, p 19.1

[6] LEGG, RC, 'Characteristics of single and multi-blade dampers for ducted air system', *BSER and T*, 1987, Vol 7, p 129

[7] DICKEY, PS and HL COPLEN, 'A study of damper characteristics', *Transactions of ASME*, 1942, Vol 64, p 137

[8] BROWN, EJ and JR FELLOWS, 'Pressure losses and flow characteristics of multi leaf dampers', *ASHRAE Transactions* No. 1637, 1958, Vol 64, p 299

[9] KARADY, P. 'La egolazione dell'aria mediante serrande (air flow control by dampers)', *Condisionamento Dell'aria*, 1967, Vol 11, p 23

[10] LEGG, RC, 'Multi-blade dampers for ducted air systems', IMechE Conference: *Installation Effects in Ducted Fan Systems*, 1984, p 67.

Chapter 17

[1] Department of Energy, *Degree Days Fuel Efficiency*, Booklet No. 7, 1977

[2] CIBSE, *Guide Book B*, 1970, p B18.18

[3] ASHRAE, *Fundamentals Handbook*, Chapter 28, (Energy Estimating Methods), 1985

[4] ROBERTSON, P, 'Evaluation of air conditioning costs'. *The Building Services Engineer*, Vol 42, 1974, p 195

[5] WONG, AKH and RC LEGG, 'The economic evaluation of variable compressor speed control', *IoEE Research Memorandum* 106, South Bank Polytechnic, 1987

[6] HIGHAM, DJ, 'Design notes on load diagrams', *Design notes*, IoEE, South Bank Polytechnic, 1974

[7] APPLEBY, PH, 'Using load and plant operation charts', *The Building Services Engineer*, 1986, May p 65 and July p 55

[8] CIBSE, *Guide Book B*, Section B3, Chartered Institution of Building Services Engineers, 1986

[9] LETHERMAN, KM and J DEWSBURY, 'The *bin* method – A procedure for predicting seasonal energy requirements in buildings', *BSER and T*, Vol 7(2), 1986, pp 55–64

Chapter 18

[1] ASHRAE, *Systems Handbook*, Chapter 37 (Testing, Adjusting and Balancing), 1984

[2] CIBSE COMMISSIONING CODES:
SERIES A – AIR DISTRIBUTION, 1971
SERIES B – BOILER PLANT, 1975
SERIES C – AUTOMATIC CONTROL, 1973
SERIES R – REFRIGERATION SYSTEMS, 1972
SERIES W – WATER DISTRIBUTION, 1989

[3] SCURRY, P, *Quality Assurance in Building Services*, BSRIA, Technical Memorandum TM1/83, 1983

[4] *Documents for air system regulation*, BSRIA Application Guide 1/77, 1977

[5] *Manual for regulating air conditioning installations*, BSRIA Application Guide 1/75, 1975

[6] BS 5720: 1979, *Mechanical ventilation and air conditioning in buildings*, Code of Practice, British Standards Institution

[7] LEGG, RC, 'Performance testing of air conditioning systems', *Heating and Ventilating Engineer*, Part 1, Vol 48, 1974, p 569, Part 2, Vol 49, 1975, p 12

[8] BS 5720: 1979, op.cit. p 60

[9] Ibid., p 80

[10] BURBERRY, P, *Environment and Services*, 6th edition, Batsford, 1988, p 170ff

[11] HAWES, DS, 'The choice, installation and commissioning of air handling units'. *Heating and Ventilating Engineer*, Vol 46, 1972, p 61

[12] RUDYARD, HA, 'Organisation and training for building services maintenance'. *BSE*, Vol 43, 1975, p 61.

Bibliography

There is an extensive literature on air conditioning and related topics. The following selective list includes publications from which the author has obtained information, not otherwise given as a reference, together with those from which the reader may obtain additional information on the subject areas in this book.

Air conditioning and refrigeration
DOSSAT, RJ, *Principles of Refrigeration*, 2nd edition, John Wiley, 1981
GOSNEY, WB, *Principles of Refrigeration*, Cambridge University Press, 1982
STOECKER, WF and JW JONES *Refrigeration and Air Conditioning*, McGraw Hill, 1987
'Refrigeration – its role in Environmental Control', Proceedings of CIBSE Conference, June, 1988

Equipment
GARLAND, GC, 'Humidifiers past and present', *BSE*, Vol 44, 1976, p A14
The Fan Manufacturers' Association, *Guide to Fan Noise and Vibration*, HEVAC, 1984
OSBORNE, WC, 'The selection and use of fans', *Engineering Design Guide* No 33, Oxford University Press, 1979
OWENS, PGT, 'The spinning disc humidifier', *JIHVE*, Vol 33, 1965, p 85
REAY, DA, *Heat recovery systems: A directory of equipment and techniques*, 1979

Instrumentation, testing and commissioning
Meteorological Office (UK), *Handbook of Meteorological Instruments*, Vol 1 *Measurement of Atmospheric Pressure*, 1980; Volume 2, *Measurement of Temperature*, 1980; Vol 3, *Measurement of Humidity*, HMSO, 1981
OWER, E and RC, PANKHURST, *The measurement of air flow*, Pergamon Press, 1977
Commissioning by Design, CIBSE Symposium, April, 1986
Manual for regulating water systems, BSRIA Application Guide 1/79, 1979
Site Testing of Fans and Systems, Conference sponsored by the Fluid Machinery Group, IMechE, 1978.

SI Units

All quantities in this volume are given in SI units. This system of units has been used extensively for many years in Europe and more recently in North America. The style of the book is to use numerical examples to illustrate theory and it is hoped that this will prove useful to those readers who are unfamiliar with SI units.

The SI system is based on six units of measurement, ie:
length – metre (m)
mass – kilogram (kg)
time – second (s)
electrical current – ampere (A)
temperature – degree Kelvin (K)
luminous intensity – candela (cd)

From these are derived the remainder of the units necessary for measurement, eg, area from length (m^2), velocity from length and time (m/s). Special units are given to some of these derived units, as follows:

quantity	unit	symbol	basic units involved
frequency	hertz	Hz	1 Hz = 1/sec (1 cycle per sec)
energy	newton	N	1 N = 1 kg m/s^2
work, quantity of heat	joule	J	1 J = 1 N m
power	watt	W	1 W = 1 J/s
pressure	pascal	Pa	1 Pa = 1 N/m^2

Multiples of SI units are increased or decreased by the use of named prefixes, each of which has an agreed symbol. Those most relevant to this book are given in the table below. (Note that *kilogram*, which is a basic unit, departs from the general rule.)

multiplying factor	prefix name	prefix symbol
10^9	giga	G
10^6	mega	M
10^3	kilo	k
10^{-3}	milli	m
10^{-6}	micro	μ

Conversion factors

The conversion factors given in the table are for those physical quantities most commonly used in air conditioning, mechanical ventilation and refrigeration. To convert a quantity in *British units* to *SI units*, multiply by the conversion factor. To convert a quantity in *SI units* to *British units*, divide by the conversion factor.

physical quantity	SI unit description	SI unit symbol	British unit	conversion factor
space				
length	metre	m	foot	0.3048
			inch	25.4
area	square metre	m^2	square foot	0.09
volume	cubic metre	m^3	gallon(USA)	0.0378
			gallon(UK)	0.0455
mass				
mass	kilogram	kg	pound	0.454
			ton	1016
moisture content	kilogram per kg	kg/kg	grain per pound	1.43×10^{-4}
density	kilogram per cubic metre	kg/m^3	pound per cubic foot	16.02
motion				
velocity	metre per second	m/s	foot per minute	5.08×10^{-3}
flow rate				
volume flow[+]	cubic metre per second	m^3/s	cubic foot per minute (cfm)	0.472
mass flow	kilogram per second	kg/s	pound per hour	1.26×10^{-4}
pressure[*]	pascal	Pa	inch water gauge	249.1
temperature				
scale, zero $0\,^\circ C$	degree Celsius	$^\circ C$	degree Fahrenheit, $^\circ F$	$\dfrac{5(^\circ F - 32)}{9}$
interval	degree Kelvin	K	degree Fahrenheit	0.56

Continued

*Note: It is customary to take the unit of atmospheric pressure as the millibar (mbar); 1 bar $= 10^5$ Pa. Occasionally other pressures also use this unit.

Continued

physical quantity	SI unit		British unit	conversion factor
	description	symbol		
viscosity (kinematic)	square metre per second	m^2/s	square foot per minute	$1.55 \ 10^{-3}$
energy quantity of heat	kilojoule	kJ	British thermal unit (Btu)	1.055
consumption	gigajoule	GJ	therm	0.1055
consumption	megajoule	MJ	kilowatt hour (kWh)	3.6
power heat flow rate	watt	W	Btu per hour	0.293
motor power	kilowatt	kW	horsepower	0.746
refrigeration	kilowatt	kW	ton	3.52
heat specific heat capacity	kilojoule per kilogram Kelvin	kJ/kgK	Btu per pound deg. Fahrenheit	4.19
specific enthalpy	kilojoule per kilogram	kJ/kgK	Btu per pound	2.33
latent heat	kilojoule per kilogram	kJ/kgK	Btu per pound	2.33

Flow rates are sometimes quoted in litres per second (l/s); 1 l/s = 0.001 m^3/s

Index

Index